FOOD IN SOCIETY

Food Diary : p261, 266-7, 269 (Table 20:8), 286-7, 298.

FOOD IN SOCIETY

Economy, Culture, Geography

PETER ATKINS

Department of Geography, University of Durham, UK

and

IAN BOWLER

Department of Geography, University of Leicester, UK

A member of the Hodder Headline Group
LONDON
Distributed in the United States of America by
Oxford University Press Inc., New York

First published in Great Britain in 2001
This impression printed in 2005 by
Arnold, a member of the Hodder Headline Group,
338 Euston Road, London NW1 3BH

http://www.hoddereducation.com

Distributed in the United States of America by
Oxford University Press Inc.,
198 Madison Avenue, New York, NY10016

British Library Cataloguing in Publication Data
A catalogue record for this book is available from the British Library

Library of Congress Cataloging-in-Publication Data
A catalog record for this book is available from the Library of Congress

ISBN-10: 0 340 72004 2
ISBN-13: 978 0 340 72004 2

5 6 7 8 9 10

Typeset in 10 on 12 pt Sabon by Cambrian Typesetters, Frimley, Surrey
Printed and bound by Replika Press Pvt. Ltd., India

Contents

Preface

The authors of this book have known each other for 40 years. We lived in neighbouring streets in Great Crosby, a suburb of Liverpool, and attended the same schools. At Merchant Taylors' School we acquired a taste for geography under the influence of Granville Jones, Ron Roby and others, and we are grateful for the formative nature of that experience. After school our careers diverged, but we have always continued to share an interest in agriculture and food. The present book arose at the invitation of Laura McKelvie at Arnold and it is based on our research and teaching at the Universities of Durham and Leicester.

We have attempted to write a volume with varied subject material. The linking theme is food, an issue on which all of the social sciences, and for that matter all of the general public, have a view. There is therefore great scope in terms of approach. The study of food is rather like a 'barium meal' for X-raying social, political, economic and cultural issues, a kind of marker dye for broad structures and processes. In this sense food is the bearer of significance, as well as a material object of consumption.

The book is divided into six parts. Following an introduction, attention is devoted in turn to the political economy of the supply chain, global and geopolitical issues, the political ecology of food, and food consumption. Each part is sub-divided into thematic chapters, each of which is supported by suggestions for further reading and by lists of references. The book is principally aimed at students attending courses on various aspects of food studies, but we also intend the material to have a broader appeal.

A list of relevant websites is posted at http://www.arnoldpublishers.com/support/food/. Surfing the web has become commonplace, but perhaps we should remind ourselves of two points. First, websites are not subjected to the same academic refereeing process in terms of their contents as are most journal articles, and therefore one cannot always be confident of their accuracy and objectivity. Anyone can set up a website and this led to a vast outpouring of the electronic publication of text and images related to food in the 1990s which was both stimulating and of varied quality. The second point is that no listing of the Worldwide Web can be wholly accurate or comprehensive. Websites come and go and their addresses (URLs) change. The inevitable delays in writing, publishing and marketing this book mean that the listings will be out of date by the time you read this. Readers are invited to use one of the many search engines to update links and find additional food sites.

We wish to thank Steven Allen and Chris Orton at Durham for drawing most of the art work. As ever, Nicholas Cox has been a source of much information, and we wish to thank Bruce Scholten and Arouna Ouédraogo for discussing issues with us. Every effort has been made to trace the copyright holders of quoted material. If, however, there are inadvertent omissions, these can be rectified in any future editions.

The book is dedicated with love to Liz and Val.

Setting the scene

Who can deny the significance of food? It has a central role in our sustenance and pleasure, and it touches the deepest of nerves in our economy, politics and culture (Rozin 1999). Yet Mennell, Murcott and van Otterloo (1992) have observed that until recently food has received little academic attention. They give two possible reasons for this neglect. First, our regular consumption of food makes it one of the most taken-for-granted aspects of life and, as a topic for research, it may therefore have appeared to lack the novelty, scarcity or exotic qualities that tend to propel new themes on to the agenda of social science. Second, food preparation and presentation have always been women's work and, in a male-dominated academy, they have therefore acquired a lowly status. We may add the further observation that, in the past, social scientists have tended to reflect the preferences of their own cultural context. Thus, there has been a relative indifference to culinary culture shown in the Anglo-American realm, by comparison with a greater commitment shown by scholars in France, Germany and Italy.

We are pleased to say that the recent outpouring of food-related publications, particularly in sociology and to some extent in geography, has broken with all three of these constraints and scholars from many countries are now actively researching food issues. Food studies can claim to have a growing constituency and we hope that this book may make a contribution to popularizing the social science of food in its varied aspects.

Some themes

The literature on food is now so vast that a selection of themes is inevitable in a book of this kind. We have decided to blend some familiar economic and political issues with environmental and cultural perspectives that are less well known, at least with the slant that we intend to give them.

In Chapter 1 we will give a brief background to the lineage of food studies by discussing the approaches adopted in the last few decades. The inertia of intellectual capital invested in these empirical and theoretical traditions means that entirely new departures are rare, but the recent cultural turn in ethnographically inspired food studies looks set to become one such revolutionary change.

Part II adopts the stance of political economy in dissecting the food chain. The argument here is that an understanding of the role of capital in the restructuring of agriculture, food processing, manufacturing, and distribution is fundamental. This provides a foundation for the rest of the book since the structures, processes and manifestations of capitalism bear heavily upon issues raised in the remaining chapters.

Part III is closely related because it builds upon the globalized food networks introduced in Chapter 4. The drama of famine is familiar to all consumers of the media and most readers will already have been introduced at some level to the food security debates rehearsed in Chapters 9–12. The novelty lies in our attempt to find where the balance lies between the pessimistic and optimistic viewpoints which are often presented as being mutually incompatible. After that, Chapters 12 and 13 suggest that global food trade and international food policies have a major economic and political significance that cannot be accommodated in a descriptive framework of commodity systems.

Part IV gives an environmental dimension to the notion of political economy by shifting the focus to the 'political ecology' of food. This is not an inventory of the environmental damage caused by over-intensive farming, but rather an investigation of the place of food studies in the current rethinking of the boundaries between nature and culture. Chapters 15 and 16 see food as central to the human environment through a discussion of food quality and health hazards. Chapter 17 contributes to this by looking at genetically modified organisms in agriculture and food manufacture, and Chapter 18 touches on the vital area of food regulation and food policy.

There is a change of gear in Part V, which deals with food consumption. There are biological considerations here of course but Chapter 20 points to other factors affecting the quantity and quality of food intake, such as income, age and ethnicity. Chapters 21 and 22 look at cultural aspects, including historical and geographical influences, and Chapter 23 focuses on the vital issue of food and gender.

Peter Atkins
Ian Bowler

Abbreviations

ACP Africa–Caribbean–Pacific
AJV Alliances and Joint Ventures
ANT Actor Network Theory
BMR Basal Metabolic Rate
BSE Bovine Spongiform Encephalopathy (Mad Cow Disease)
BST Bovine Somatotropin
CAC Codex Alimentarius Commission
CAP Common Agricultural Policy
CIA Central Intelligence Agency
DES Dietary Energy Supply
EAGGF European Agricultural Guidance and Guarantee Fund
EC European Commission
ECU European Currency Unit
EEC European Economic Community
ESRC Economic and Social Research Council
ET Embryo Transfer
EU European Union
EWS Early Warning Systems
FAD Food Availability Decline
FAO Food and Agriculture Organization
FEWS Famine Early Warning Systems
FFW Food For Work
FIVIMS Food Insecurity and Vulnerability Information and Mapping Systems
FSA Food Standards Agency
GATT General Agreement on Tariffs and Trade
GIEWS Global Information and Early Warning System
GIS Geographical Information Systems
GMO Genetically Modified Organism
GNP Gross National Product
HIC High Income Country

HYVs	High Yielding Varieties
ICM	Integrated Crop Management
IFS	Integrated Farming Systems
ILM	Integrated Livestock Management
IMF	International Monetary Fund
IPC	Integrated Pest Control
LDC	Less Developed Country
LIC	Low Income Country
M&A	Mergers and Acquisitions
MAFF	Ministry of Agriculture, Fisheries and Food
MCH	Mother and Child Health
MNC	Multi National Corporation
MSR	Mode of Social Regulation
NAC	New Agricultural Country
NGO	Non-Governmental Organization
NIC	Newly Industrialising Country
OGA	Other Gainful Activity
PGI	Protected Geographical Indication
PDO	Protected Designation of Origin
PYO	Pick Your Own
QAS	Quality Assurance Scheme
RDA	Recommended Daily Allowance
SCP	Simple Commodity Production
SMP	Skimmed Milk Powder
SoP	Systems of Provision
TNC	Trans National Corporation
TV	Traditional Variety
UA	Unit of Account
UK	United Kingdom
UN	United Nations
USA	United States of America
USAID	United States Agency for International Development
USSR	Union of Soviet Socialist Republics
WFP	World Food Programme
WFS	World Food Summit
WTO	World Trade Organization

PART I

HORS D'OEUVRE

1

A background to food studies

Cooking is a moral process, transferring raw matter from 'nature' to the state of 'culture', and thereby taming and domesticating it . . . Food is therefore 'civilised' by cooking, not simply at the level of practice, but at the level of the imagination (Lupton 1996, 2).

Introduction

From the points of view of epistemology (theories of knowledge and method) and general zeitgeist (spirit of a particular time), there have been many influences over the years on the study of food. The following list is merely a sample of recent trends:

- The French *Annales* school of history, especially in the 1960s and 1970s through the work of Fernand Braudel, who was interested in all aspects of 'material life' (Forster and Ranum 1979).
- The political economy of food systems and the associated concepts of food filières, food régimes, food networks, and systems of provision. Marxist structuralist writers in particular focused on this in the 1980s and 1990s.
- The 'cultural turn' in social science in the 1990s, which unleashed a new series of writings on food, some using the qualitative methodologies of ethnography, and many having a 'post-modern' flavour.
- The surge of interest in food amongst the general public in the last 20 years. This has been driven to some extent by the recent increase in exotic foodstuffs available in supermarkets and the number of cooking programmes on the television and cookery books in the shops.
- Recent worries about food safety, which have created a climate of fear and distrust surrounding the activities of commercial food companies and government food regulation. This has been balanced to some extent

by the discovery of a number of 'functional foods' that are attributed with a health-giving property.
- The connexions between food and the body touch on more than just health. A number of eating problems, such as anorexia nervosa and bulimia, are associated with the role of food in crises of bodily identity.
- Awareness of hunger and malnutrition in poor countries, and the explanation of famine given by Nobel Prize-winning author Amartya Sen.

This chapter seeks to demonstrate the rich variety of disciplinary and theoretical contexts in which food studies have flourished as a result of these and other intellectual currents.

Historical approaches

Although some historical studies of food are either antiquarian or are intended to illuminate a time-specific setting, many scholars have used food as an evolutionary marker of change over long periods, with the aim of making generalizations about socio-economic behaviour. Burnett (1979) in particular has shown the central role of food in the study of social history, for instance as a major contributor to the changing cost of living. Glennie (1995) prefers to investigate the changing nature of consumption by identifying the various stages in the evolution of the mass market. This must include material considerations of wealth and the technology of production, but studies of the culture of consumption are also important, as are supply-side factors such as the emergence of new retail forms (Lancaster 1995).

Mennell, Murcott and van Otterloo (1992) identify a 'developmentalist' food literature. In their view this includes some of the writings of Marvin Harris (1986), Stephen Mennell himself, and others such as Sidney Mintz and Jack Goody. The orientation here is towards the explanation of socially and geographically varied patterns of food consumption in terms of their historical evolution in particular contexts of economy and the exercise of power. Thus food avoidances and preferences are not random and beyond rational explanation, but can be elicited from a series of historical events that have left their trace in present-day diets. Mennell's influential book (1985) is discussed critically by Warde (1997), who sees it as an extension of Norbert Elias's work on the civilizing process. Mennell's implicit underpinning is the supply-side commodification of food within a structure of manufacturing and service-sector capitalism. Mintz (1985), on the other hand, works within the framework of World Systems theory, a materialist approach to the study of change initiated by Immanuel Wallerstein. Grew (1999) has edited a collection of papers with a similar, global outlook.

Other historians have looked at the changing role of particular commodities over long periods of time. Salaman's (1949) study of the potato is justly famous (see also McNeill 1999), along with Mintz (1985, 1999) on sugar. Others have concentrated on the evolution of national diets (Drummond and Wilbraham 1957; Levenstein 1988, 1993), the long-running Historians' and Nutritionists' Seminar being a particularly wide-ranging project in this regard (Yudkin and McKenzie 1964; Barker, McKenzie and Yudkin 1966; Barker, Oddy and Yudkin 1970; Barker and Yudkin 1971; Oddy and Miller 1976, 1985; Geissler and Oddy 1993; Burnett and Oddy 1993). A particularly encouraging recent development has been the emergence of international societies whose aim is the study of food in a comparative context. The International Commission for Research into European Food History (founded 1989) has been particularly active (Burnett and Oddy 1993; Hartog 1995; Schärer and Fenton 1998; Teuteberg 1992) and new the Association for the Study of Food and Society seeks to achieve the same on a global scale.

Popular enthusiasm for the history of cooking has encouraged extensive publication in this area, along with public events such as the Oxford Symposium on Food and Cookery, which has been organized every year since 1981.

Cultural and sociological approaches

During the twentieth century, many sociologists and anthropologists took an interest in food, from the functionalists to the structuralists. Among the functionalists, in their first flowering, were empiricists who described food habits in terms of the kind of customary and ritualized behaviour that underpins the reproduction of a stable society (Lupton 1996). They identified certain values and norms in eating patterns that are symbolic of broader structures in society as a whole, and argued that what to outsiders may appear to be strange food customs may in fact have a function that helps to bind society together (Goody 1982).

Functionalism emphasizes the utilitarian nature of food and gives primacy to its physical qualities. This whole approach has been criticized for analysing patterns and processes within a static framework, and allowing little room for the explanation or even recognition of the importance of origins, change and conflict. It has also been attacked for the claim that we can identify the functional needs of a social system from its customs and institutional structures without entering into a circular argument (Beardsworth and Keil 1997). Much of the early food-related functionalist work was undertaken by social anthropologists, amongst whom the most prominent were writers such as Audrey Richards and Margaret Mead.

By comparison, structuralism seeks broader and deeper causes and meanings of food habits, especially how 'taste is culturally shaped and socially controlled' (Mennell, Murcott and van Otterloo 1992). Flavour, texture, nutritional qualities and other biological properties are underplayed in favour of social context. One approach has involved scholars in applying Saussure's linguistic analysis to an understanding of food culture. In particular, Claude Lévi-Strauss (b. 1908) analysed the universality of oppositional meanings of food such as raw, cooked and rotten (his 'culinary triangle'). He thought that certain attitudes to food are 'hard wired' into the human mind and therefore generate universal structures of thought and of action. But Lévi-Strauss has been criticized for making 'fanciful' assumptions and generalizations from the myths of tribal peoples, and for failing satisfactorily to elucidate the foodways of advanced societies (Bearsdworth and Keil 1997).

Roland Barthes (1915–80) was one of the most entertaining and insightful of the structuralists. He brought his semiotic eye to bear in interpreting popular food preferences and food in media such as advertising. He saw foods as a system of signs (Barthes 1967, 28). Like any language, diet has rules of exclusion, signifying opposites (such as savoury/sweet), rules of association for how individual dishes and menus should be assembled, and rituals of use. Again, Saussure's linguistic structuralism was an influence but, unlike Lévi-Strauss, Barthes discusses concepts such as capitalism and imperialism, and his analysis therefore has greater immediacy for western readers. One of Barthes' lasting contributions was his identification and interpretation of certain 'mythologies' that he drew from everyday life in France itself. He analysed soap powder, the Eiffel Tower, the world of wrestling, and many others. A central theme was food and drink, with commentaries on ornamental cookery, steak and chips, and margarine. He wrote of wine as his country's totem-drink, corresponding to milk for the Dutch or tea for the British, and therefore standing as a national symbol (Barthes 1972). For Barthes, food was central to various aspects of life touching the body and the mind, all of which are susceptible to a unified method of enquiry, a psychosociology (Barthes 1975).

Like Lévi-Strauss and Barthes, Douglas (1975, 1982, 1984, 1998) deciphers the 'grammar' of meals, as if they were coded texts to be dismantled into their significant components, but she prefers a 'thick' description based upon participant observation. She has been called a structural functionalist because she draws upon elements of both approaches. Her 'meaning of the meal' is derived from its role as a structured social event.

The linguistic flavour of much of the early structuralism was later leavened by the political economy project of the Marxist structuralists, who, following influential thinkers such as Louis Althusser, were especially important in the philosophical climate of the 1970s and 1980s. They tended to privilege theoretical over empirical insights and were particularly absorbed with the need to uncover the complex structures and processes of capital accumulation. As one example of this approach, Chapter 3 investigates the concept of food régimes.

Bourdieu (1984) has proved to be one of the most significant theorists of relevance to food studies. Like many other writers, he recognized the need to move away from a reliance upon the production-orientated explanations of society, which had for so long dominated materialism, towards a framework that can accommodate considerations of consumption and lifestyle. However, Bourdieu does still see class as important, and interprets taste and the nature of consumption behaviour as both expressions of class identity and as means of reproducing the class distinctions in society. For him food habits represent a naturalization of ideology (Miller 1995).

Bourdieu has one foot in structural marxism and the other in cultural studies. There are scholars who would prefer to move further, to privilege the latter at the expense of the former. They do so because they believe that individuals in many societies increasingly have freedom of choice over their own consumption decisions, in ways that are not fundamentally constrained either by the production and marketing decisions of food corporations or by loyalty to a class norm. In this view, consumption is becoming more individualized and informal, and less disciplined (Baumann 1988; but *see* Warde 1994).

Since the 1970s, feminism has added a dimension to food studies that previously was sorely lacking. Feminist writers have analysed the role of women within the household, and especially the fundamental part played by their food preparation tasks in the reproduction of the family (Chapter 23). But they have also addressed the relationship between food and body shape in the construction of female identity within a framework of patriarchal expectations. One wing of feminism even sees dietary items themselves as significant, arguing that killing animals and eating meat are patriarchally inspired activities (Adams 1990).

The cultural turn in social science has affected aspects of food studies in the 1990s. Bell and Valentine (1997) illustrate the various themes well in their book, concentrating mainly on the relationship between food geographies and consumption. They adopt a scale-focused approach and use sites of analysis starting with the body, then moving to the home, the community, the city, the region, the nation and, ultimately, the globe. Their subtitle is 'we are where we eat', a geographical modification of the German dictum 'man ist was man ißt – you are what you eat'. For most social scientists with an interest in food, this cultural shift has meant the adoption of ethnographic methodologies of data collection.

Post-modern and post-structuralist approaches to food

The cultural turn, post-modernism, and post-structuralism are all terms that have been used to summarize recent methodological and theoretical developments in the social sciences. The variety of publications has been exciting and stimulating, with a number of key themes emerging:

- Understanding the nature of food-related knowledge through its social construction rather than relying upon 'objective' scientific or descriptive historical accounts alone.
- A critical approach to food practices which acknowledges their socially, culturally, economically and politically embedded nature, and which seeks out the competing interests that drive change.
- Discourses: studying patterns of language and practices in the production of meaning about food.
- An interest in diverse sites of activity: popular culture, texts, individuals' accounts.
- The privileging of the body as a crucial site of significance in understanding health and identity issues.
- The analysis of identity and subjectivity through the medium of food studies.
- Stress upon the fragmentation of food consumption patterns rather than their coherence in economic or cultural terms. Niche markets have been established for slimmers, vegetarians, consumers of organic foods, and so on, but food choice has also become a means of resistance against broader trends in society.

Along these lines, Warde (1997, 14–18 and 30–41) has produced a succinct and helpful summary of the competing explanations of recent food consumption trends. He identifies four themes, the first of which deals with a tendency towards arbitrary individual diversity. The argument here is that in recent times the formerly strong social bonds on lifestyles have been loosened and the diversity of consumption behaviour has therefore greatly increased (Beck 1992). The agency of consumers is emphasized over both the social and economic structures in which they find themselves, and this is a crucial means of establishing an individual's identity (Giddens 1991). The disintegration of constraints (dearth, seasonality, ritual, taboo, cultural foodways) has led to what Fischler (1980) calls 'gastroanomy' (*see* Chapter 22). This may cause anxiety, for instance about the paradox of plentiful food supplies set against worries about what to choose because of health considerations and body self-consciousness.

The second theme addresses what Warde calls post-Fordist food. Here the emphasis is upon the formation of new groups in society, who share types of lifestyle rather than social class. Generational differences and also fashion play their part and, because people may now choose their own package of consumption habits, it is likely that variety within groups of the like-minded will decrease. There is an argument here as to whether niche consumption is voluntarist or the result of trends in capitalism towards more flexible production (Harvey 1989), which would imply that the initiative still lies with the industrialist rather than the consumer.

Third, there is the much-discussed phenomenon of mass consumption in a mass society. Ritzer (1993), in particular, identifies a process of

'McDonaldization' in which corporations cater to the 'lowest common denominator' of mass consumer culture and therefore sell bland, unchallenging products that transcend class and taste boundaries by their broad acceptability. Considerations of mass consumption also include government health campaigns and the construction of public taste through the media.

Finally, we must not ignore the persistence of social differentiation. There are increasing income divides between the 'West' and the less developed countries, and even in Britain the disparities widened in the 1980s and 1990s. Consumption patterns for the poor remain largely unchanged: they are excluded from the possibilities of quality and variety. According to this perspective, class is still an important aspect of diet (Bourdieu 1984).

Food systems

The idea of a 'food system' is a convenient means of conceptualizing the relationship between the different forces acting upon the commodity flows from producer to consumer. The idea is not new. There is a sizeable literature stretching back 150 years, in which writers have described the features of particular food chains. George Dodd's (1856) book, for instance, was a pioneering attempt to reconstruct the food supply of a particular city across a range of products. Raison's (1933) two-volume work is an excellent example of the single industry genre, in which his contributors covered every aspect of the British dairy industry from milk production to delivery. More recently, it has been possible to reconstruct food system structures through the complex analysis of industrial input–output matrices (Figure 1.1), and scholars have begun to theorize food system dynamics (Malassis 1973, 1975, 1986). The most recent popular account of 'the food system' is by Tansey and Worsley (1995).

Fine *et al.* (1996) have commented that food studies hitherto have been highly fragmented, according to the approaches traditionally adopted by individual disciplines, and that they have also been lacking in theoretical coherence. They argue that the time has come for greater cross-fertilization between geography, sociology, anthropology, economics, psychology and the other social sciences that have found some common interest in food. The commodity chain in its various manifestations provides one convenient locus for such cross-fertilization, both conceptually and empirically. It encompasses both production and consumption, and it provides clear links with the spatial conceptions of society.

Leslie and Reimer (1999) provide helpful critiques of the commodity chain literature (*see also* Jackson and Thrift 1995, 212–17; Hartwick 1998; Doel 1999). They identify, first of all, a tradition that derives its inspiration from the world systems theory of writers such as Wallerstein. The agenda here is the tracing of commodity flows at a global scale, in order to uncover the usually biased and exploitative relationship between the raw material

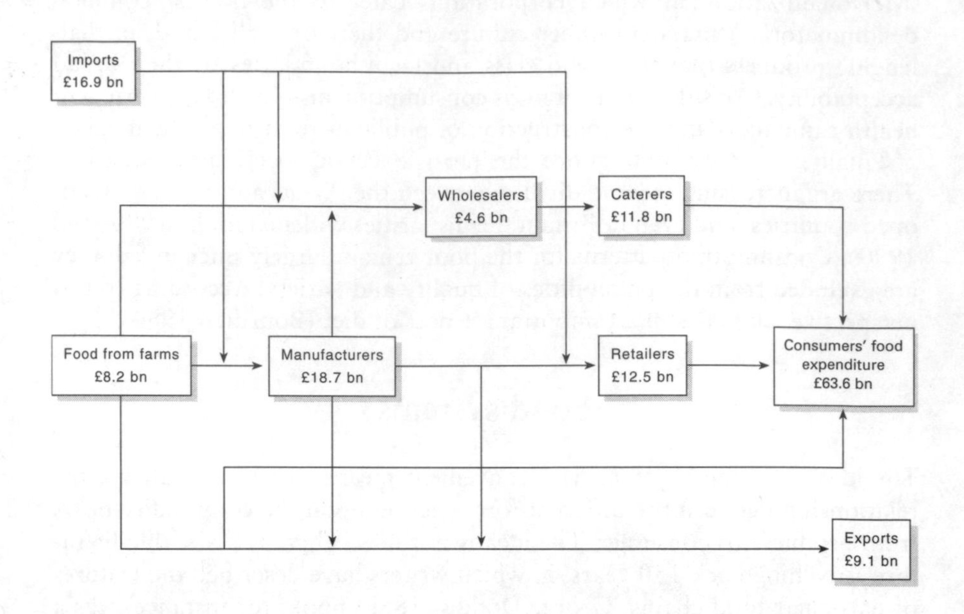

Fig. 1.1 Estimates of the main components of the UK food system, 1997. *Source*: Ministry of Agriculture.

provider and the site of consumption. Leslie and Reimer argue that the scale of this analysis leads to the neglect of particular nodes in the chain. Such superficial analyses at the level of the system are reductionist and ignore the role of human agency.

Second, Leslie and Reimer unpack the notions of commodity circuits and arenas of consumption. Here the physical commodity is not followed through from plough to plate as it were, but is seen as symbolic of the interactions in society, which are variable in time and space according to contingent articulations. This approach emanates from cultural geographers (see Bell and Valentine 1997, 199–200), but there are possibilities of theoretical hybridities with political economy (Crang 1996; Cook and Crang 1996). It is argued that 'biographies' of foods need to take account of their social constructions as commodities, which will inevitably mean some mutually reflexive interactions between these 'objects' and the hands through which they are passing (Cook *et al.* 1998). As an example, Cook *et al.* (1999) discuss the identity politics of ethnic foods and restaurants and their role in the gradual establishment of a multicultural Britishness.

Third, we must highlight the 'food networks' literature (Chapter 4). There is overlap here, particularly with the political economy of food, but there is an important input from the methodologically innovative Actor Network Theory (ANT) of Bruno Latour and fellow travellers. Latour

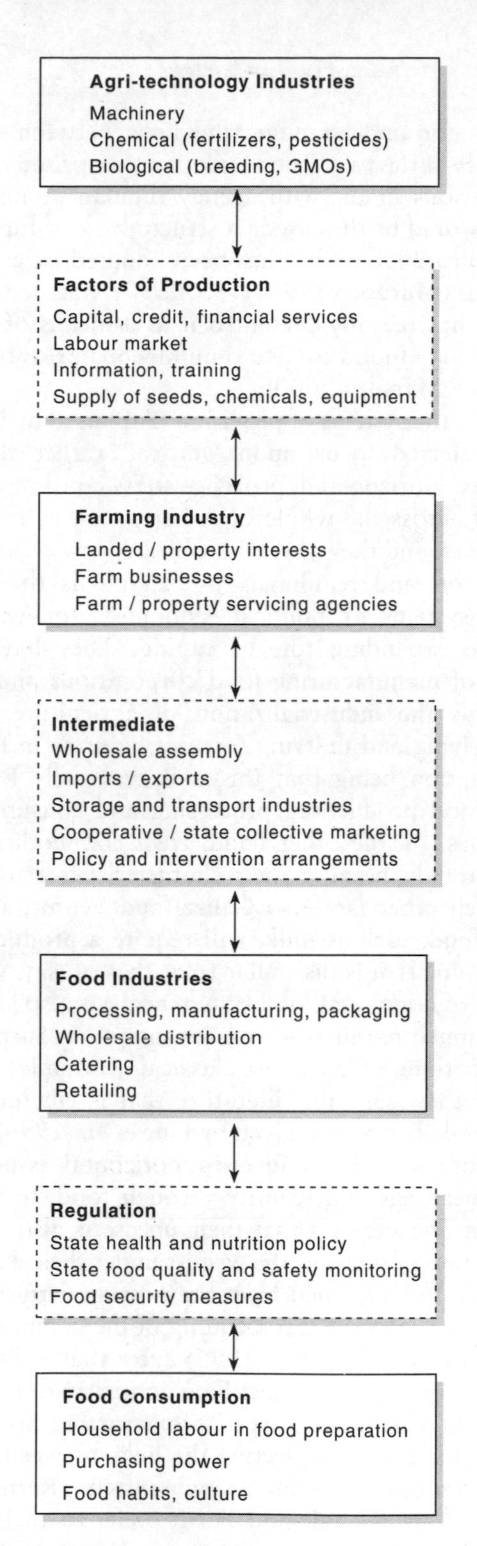

Fig. 1.2 The agri-food system. *Source*: Modified after Whatmore 1995.

provides a means for understanding the links between actors: humans, objects and hybrids of the two. These links are comprised of influences and interactions of various kinds, with agency (human or non-human) more transparent than would be the case in a structuralist or functionalist analysis (Lockie and Kitto 2000). ANT has been adopted by a number of writers on food systems (Murdoch 1994, 1995, 1997; Marsden and Arce 1995). Goodman (1999) has recently presented it as a means of overcoming the inability of agro-food studies to give simultaneous priority to both nature and society (but see Marsden 2000).

Fourth, there are the systems of provision (SoP) used by Fine *et al.* (1996) to describe a preferred focus upon 'vertical' rather than 'horizontal' commodity themes. Horizontal approaches stress consumption factors that may be identified across the whole of society, such as the social class or gender of consumers, and they also take a broad view of food system activities like production and retailing. An example is the attention given recently to food systems by political economists, many of whom have focused upon the expanding role of capital. They have described the economic power of manufacturing food corporations and have identified tendencies towards the industrialization of agriculture. This approach emphasizes the driving and unifying force of industry in food systems, an underlying assumption being that this will eventually lead to a greater homogeneity in food production, processing and consumption practices. Vertical approaches, on the other hand, treat commodities or groups of commodities separately because their characteristics remain significantly different from each other (*see also* Galizzi and Venturini 1999). Thus, a perishable dairy food, such as milk, will require a production, processing and marketing régime that is dissimilar from that of, say, wheat.

Fine's rejection of horizontal food system analysis also leads him to question the value of much recent work on consumption. Surprisingly, in view of his relentless critique of positivist classical economics, he seems to be wary of much of the qualitative literature that is emerging with various perspectives on food. The reason given by Fine *et al.* (1996) is that the post-modernist literature slavishly addresses horizontal issues, such as the construction of meanings and identity through food, to the exclusion of other matters. But one suspects that their unease is also derived from the non-materialist stance of many of the writers under their microscope. In our view, there is much to be learned from the post-modernist and post-structuralist literature, not least the transcending of the commodity focus.

Glennie and Thrift (1992, 1993, 1996) agree that SoPs have a place in studies of consumption but they reject Fine's criticism of horizontal studies. The two are not mutually exclusive. Reconstructing SoPs for individual commodities runs the risk of neglecting the links between systems and the 'new effects that cannot be traced to a single system'. Rather than adopting Fine's commodity focus, Glennie and Thrift prefer to anchor their work in the repetitive and intuitive aspects of consumption. Consumers are an important object of study, especially the practical and embodied aspects of their knowledge rather than an intellectualized understanding of them as individualistic subjects. For the pre-1900 period at least, Glennie and Thrift avoid analyses of consumption at the abstract level, such as the semiotics of advertising for instance, and instead devote their energies to recovering the chronologies of consumption history. Glennie (1998) opposes reductionist consumption theory and engages with a detailed historical study of retailing that includes comments on food shopping.

Fine is a political economist, but he does not subscribe to the consensus of his colleagues on food systems that has emerged over the last 20 years or so. The latter has been driven by the observation that food is increasingly commoditized and that the agrarian economy in general is being drawn into and subsumed within the broader structures of capitalism, to such an extent that industrial conditions of production now prevail in some agricultural sectors. The role of finance capital, the deployment of labour and technology, and modes of capital accumulation are other aspects of these agri-industrial relations that invite styles of analyses also familiar in studies of the restructuring of Fordist and post-Fordist industry. At the global scale, this group of political economists identify what they call 'food régimes' (Chapter 3). These are shifts in the patterns of international trade that have evolved from mercantile and colonial structures to an increasing domination by the purchasing and organizational power of Transnational Corporations and by the regulatory systems of the World Trade Organization.

Fine and his co-workers recognize the importance of this work by political economists, many of them economic geographers, but they disagree with some of the assumptions. Their most important departure is in arguing that food as a commodity is so different from other products that it deserves separate treatment. They assert that food has 'organic' properties that set it apart from the manufacture of cars or textiles (Fine 1993). The food system is squeezed into the fault line between environment and society, and as such is conditioned by natural phenomena. Human control is increasing but food systems are still largely dependent, first, upon production conditions that are influenced by soil, sun and water, and, second, upon the perishability of foods, which means that they must be handled with great care at every stage from the field to the table. These considerations impose organizational and technological constraints that are of a different order from those of other industrial sectors. Fine *et al.* (1998) also stress the biological needs of the human consumers at the end of the food chain. This is highly significant because of the frequency of food consumption, coupled with food being literally a matter of life or death for us all. A second essential consideration is that there is an upper physical limit to food consumption of a kind that applies to no other goods. There is only so much we can eat or drink at a sitting, or for that matter in a lifetime, and this is different from our insatiable demand for other items.

The present authors agree with Fine and his co-writers that a holistic analysis of food systems (SoPs in their terms) is productive. Tracing the chain from production through to consumption helps us to grasp the contingent combinations of factors that bear in: economic, social, cultural, political, technological, and so on. We would go further and argue that patterns of evolution are absolutely fundamental to these contingencies, with the result that historical geographies of food systems must not be neglected. However, this argument is not developed further here as we wish to develop other approaches to the study of food.

Food studies as an inter/multidisciplinary project

There is no doubting that in developed countries the general public's interest in food is growing. In Britain the television schedules are sprinkled with cookery programmes and the best-seller lists are regularly topped by cookery books. The regular links now made between food and health, coupled with the appearance of food issues on the political agenda in the 1990s, have made us all care and worry about our diet.

This popular upsurge of food awareness has no doubt encouraged the parallel expansion of academic research. Particularly important was the Economic and Social Research Council's research programme, 'The Nation's Diet', which ran from 1992 to 1998. Anne Murcott (1998a) has given a full account of its objectives and results, but in sum it funded a multidisciplinary agenda of work that sought to answer the question 'why do we eat what we do?' (Table 1.1). There was involvement by practitioners of economics, geography, psychology, social administration, anthropology, sociology, education, marketing and media studies. According to Murcott (1998a, 9–13), this plurality of disciplines demonstrated the strength of diversity in the social sciences, and she warned against any raised expectations that this programme might represent a statement of integration.

We do not share Murcott's pessimism and prefer to argue that intellectual barriers have been exaggerated and that, if anything, there has been a recent convergence of the social scientific and cultural theories of relevance

Table 1.1 The main themes of the Nation's Diet research programme

- The formation and impact on food choice of individual attitudes, beliefs and knowledge
- Cultural definitions and the symbolic use of food
- Social processes and food choice
- Micro-economic influences on food choice

Source: Murcott (1998a, 10).

to food studies. Indeed, our own research experience suggests tremendous opportunities for cross-fertilization between disciplines. Practical issues, mainly constraints of time and funding, continue to exist but they are not insurmountable. The fostering of inter-disciplinary 'food studies' centres would be a start, along with the academic superstructure of learned journals and societies. Several already exist and more are likely to follow.

Further reading and references

A number of recent textbooks explore the general issues of food studies in greater detail than we have time for here. In particular the reader is referred to Beardsworth and Keil (1997); Bell and Valentine (1997); Caplan (1997); Fine, Heasman and Wright (1996); Tansey and Worsley (1995); and Warde (1997).

Adams, C. 1990: *The sexual politics of meat: a feminist–vegetarian critical theory.* New York: Continuum Publishing.

Atkins, P.J. and Raw, M. 1995: *Agriculture and food.* London: Collins Educational.

Barker, T.C., McKenzie, J.C. and Yudkin, J. (eds) 1966: *Our changing fare.* London: MacGibbon & Kee.

Barthes, R. 1967: *Elements of semiology.* London: Cape.

Barthes, R. 1972: Wine and milk. In Idem., *Mythologies.* London: Cape, 58–61.

Barthes, R. 1975: Toward a psychosociology of contemporary food consumption. In Forster, E. and Forster, R. (eds): *European diet from pre-industrial to modern times.* New York: Harper & Row, 47–59.

Bauman, Z. 1988: *Freedom.* Milton Keynes: Open University Press.

Beardsworth, A. and Keil, T. 1997: *Sociology on the menu: an invitation to the study of food and society.* London: Routledge.

Beck, U. 1992: *Risk society.* London: Sage.

Bell, D. and Valentine, G. 1997: *Consuming geographies: we are where we eat.* London: Routledge.

Bourdieu, P. 1984: *Distinction: a social critique of the judgement of taste.* London: Routledge & Kegan Paul.

Burnett, J. 1979: *Plenty and want: a social history of diet in England from 1815 to the present day.* Revised edn. London: Scolar Press.

Caplan, P. (ed.) 1997: *Food, health and identity.* London: Routledge.

Cook, I. and Crang, P. 1996: The world on a plate: culinary culture, displacement and geographical knowledges. *Journal of Material Culture* **1**, 131–54.

Cook, I., Crang, P. and Thorpe, M. 1998: Biographies and geographies: consumer understandings of the origins of foods. *British Food Journal* **100**, 162–67.

Cook, I., Crang, P. and Thorpe, M. 1999: Eating into Britishness: multicultural imaginaries and the identity politics of food. In Roseneil, S. and Seymour, J. (eds) *Practising identities: power and resistance.* Basingstoke: Macmillan, 223–48.

Crang, P. 1996: Displacement, consumption and identity. *Environment and Planning A* **28**, 47–67.

Dodd, G. 1856: *The food of London.* London: Longman, Brown, Green and Longman.

Doel, C. 1999: Towards a supply-chain community? Insights from governance processes in the food industry. *Environment and Planning A* **31**, 69–85.

Douglas, M. 1975: Deciphering a meal. *Daedalus* **101**, 61–81.

Douglas, M. 1982: Food as a system of communication. In Idem: *In the active voice.* London: Routledge, 81–118.

Douglas, M. (ed.) 1984: *Food in the social order: studies of food and festivities in three American communities*. New York: Russell Sage Foundation.
Douglas, M. 1998: Coded messages. In Griffiths, S. and Wallace, J. (eds): *Consuming passions: food in the age of anxiety*. Manchester: Mandolin, 103–10.
Drummond, J.C. and Wilbraham, A. 1957: *The Englishman's food*. Revised edn. London: Cape.
Fine, B. 1993: Resolving the diet paradox. *Social Science Information* **32**, 669–87.
Fine, B., Heasman, M. and Wright, J. 1996: *Consumption in the age of affluence: the world of food*. London: Routledge.
Fine, B., Heasman, M. and Wright, J. 1998: What we eat and why: social norms and systems of provision. In Murcott, A. (ed.) *'The Nation's Diet': the social science of food choice*. London: Longman, 95–111.
Fischler, C. 1980: Food habits, social change and the nature/culture dilemma. *Social Science Information* **19**, 937–53.
Forster, R. and Ranum, O. (eds) 1979: *Food and drink in history: selections from the Annales E-S-C*. Baltimore: Johns Hopkins University Press.
Galizzi, G. and Venturini, L. (eds) 1999: *Vertical relationships and coordination in the food system*. New York: Physica Verlag.
Giddens, A. 1991: *Modernity and self-identity*. Cambridge: Polity Press.
Glennie, P. 1995: Consumption within historical studies. In Miller, D. (ed.) *Acknowledging consumption*. London: Routledge, 164–203.
Glennie, P. 1998: Consumption, consumerism and urban form: historical perspectives. *Urban Studies* **35**, 927–52.
Glennie, P.D. and Thrift, N.J. 1992: Modernity, urbanism, and modern consumption. *Society and Space* **10**, 423–43.
Glennie, P.D. and Thrift, N.J. 1993: Modern consumption: theorising commodities and consumers. *Society and Space* **11**, 603–06.
Glennie, P.D. and Thrift, N.J. 1996: Consumers, identities, and consumption: spaces in early modern England. *Environment and Planning A* **28**, 25–45.
Goodman, D. 1999: Agro-food studies in the 'age of ecology': nature, corporeality, bio-politics. *Sociologia Ruralis* **39**, 17–38.
Goody, J. 1982: *Cooking, cuisine and class: a study in comparative sociology*. Cambridge: Cambridge University Press.
Grew, R. (ed.) 1999: *Food in global history*. Boulder, CO: Westview.
Harris, M.B. 1986: *Good to eat: riddles of food and culture*. London: Allen and Unwin.
Harriss-White, B. and Hoffenberg, R. (eds) 1994: *Food: multidisciplinary perspectives*. Oxford: Blackwell.
Hartwick, E. 1998: Geographies of consumption: a commodity-chain approach. *Society and Space* **16**, 423–37.
Harvey, D.W. 1989: *The condition of postmodernity*. Oxford: Blackwell.
Jackson, P. and Thrift, N. 1995: Geographies of consumption. In Miller, D. (ed.) *Acknowledging consumption*. London: Routledge, 204–37.
Lancaster, W. 1995: *The department store: a social history*. London: Leicester University Press.
Leslie, D. and Reimer, S. 1999: Spatializing commodity chains. *Progress in Human Geography* **23**, 401–20.
Levenstein, H. 1988: *Revolution at the table: the transformation of the American diet*. New York: Oxford University Press.
Levenstein, H. 1993: *Paradox of plenty: a social history of eating in modern America*. New York: Oxford University Press.
Lévi-Strauss, C. 1978: *The raw and the cooked*. London: Cape.
Lockie, S. and Kitto, S. 2000: Beyond the farm gate: production-consumption networks and agri-food research. *Sociologia Ruralis* **40**, 3–19.

Lupton, D. 1996: *Food, the body and the self*. London: Sage.
McNeill, W.H. 1999: How the potato changed the world's history. *Social Research* **66**, 67–83.
Malassis, L. 1973: *Économie agro-alimentaire, tome I: économie de la consommation et de la production agro-alimentaire*. Paris: Éditions Cujas.
Malassis, L. 1975: *Groupes, complexes et combinats agro-industriels*. Paris: ISMEA.
Malassis, L. 1986: *Économie agro-alimentaire, tome III: l'économie mondiale*. Paris: Éditions Cujas.
Marsden, T.K. 2000: Food matters and the matter of food: towards a new food governance? *Sociologia Ruralis* 40, 20–29.
Marsden, T.K. and Arce, A. 1995: Constructing quality: emerging food networks in the rural transition. *Environment and Planning A* **27**, 1261–79.
Mennell, S. 1985: *All manners of food: eating and taste in England and France from the middle ages to the present*. Oxford: Blackwell.
Mennell, S., Murcott, A. and van Otterloo, A.H. 1992: *The sociology of food: eating, diet and culture*. London: Sage.
Miller, D. 1995: Consumption and commodities. *Annual Review of Anthropology* **24**, 141–61.
Mintz, S.W. 1985: *Sweetness and power: the place of sugar in modern history*. London: Viking.
Mintz, S. 1999: Sweet polychrest. *Social Research* **66**, 85–101.
Murcott, A. (ed.) 1998a: *'The Nation's Diet': the social science of food choice*. London: Longman.
Murcott, A. 1998b: The nation's diet. In Griffiths, S. and Wallace, J. (eds): *Consuming passions: food in the age of anxiety*. Manchester: Mandolin, 110–18.
Murdoch, J. 1994: Some comments on 'nature' and 'society' in the political economy of food. *Review of International Political Economy* **1**, 571–77.
Murdoch, J. 1995: Actor-networks and the evolution of economic forms: combining description and explanation in theories of regulation, flexible specialization, and networks. *Environment and Planning A* **27**, 731–58.
Murdoch, J. 1997: Inhuman/nonhuman/human: actor-network theory and the prospects for a nondualistic and symmetrical perspective on nature and society. *Society and Space* **15**, 731–56.
Raison, C. (ed.) 1933: *The milk trade: a comprehensive guide to the development of the dairy industry*. 2 vols. London: Virtue & Co.
Ritzer, G. 1993: *The Macdonaldization of society*. Thousand Oaks, CA: Pine Forge Press.
Rozin, P. 1999: Food is fundamental, fun, frightening, and far-reaching. *Social Research* **66**, 9–30.
Salaman, R.N. 1949: *The history and social influence of the potato*. Cambridge: Cambridge University Press.
Schärer, M.R. and Fenton, A. (eds) (1998) *Food and material culture: Proceedings of the Fourth Symposium of the International Commission for Research into European Food History*. East Linton: Tuckwell Press.
Tansey, G. and Worsley, T. 1995: *The food system: a guide*. London: Earthscan.
Warde, A. 1994: Consumers, identity and belonging: reflections on some theses of Zygmunt Bauman. In Abercrombie, N., Keat, R. and Whiteley, N. (eds) *The authority of the consumer*. London: Routledge, 58–74.
Warde, A. 1997: *Consumption, food and taste: culinary antimonies and commodity culture*. London: Sage.
Whatmore, S. 1995: From farming to agribusiness: the global agri-food system. In Johnston, R.J., Taylor, P.J. and Watts, M.J. (eds) *Geographies of global change: remapping the world in the late twentieth century*. Oxford: Blackwell.

PART II

THE POLITICAL ECONOMY OF FOOD

2

Introduction

For most of human history, food has been consumed at or near its location of production. Early hunters and gatherers, for example, were dependent on food sources within their tribal territories, while even today most of the food produced globally is consumed within the country of origin. Nevertheless, as trade relations between societies developed, so the bartering and then commercial trading of food increased in significance, as exemplified historically by the fifteenth to seventeenth century spice trade between Asia and Europe and the nineteenth century supply of agricultural raw materials to European states by their overseas colonies. In recent times the internationally traded component of national and regional food supplies has again been increasing under a process termed 'the globalization of food' (*see* Chapter 4).

These general trends mask a complex heterogeneity as regards food production, distribution and consumption within the global economy. On the one hand, developed capitalist, socialist and less developed market economies can be differentiated in their historical experience as regards food. For example, in broad terms the industrial model of capitalist agricultural development has been more successful in delivering food to consumers than either socialist or subsistence-based agricultural systems. On the other hand, a wide range of farming regions, food distribution systems and patterns of consumption can be found, even within individual economies. Taking farming regions as an example, within France they range from large-scale, specialized cereal production in the north, through small-scale livestock farming in the Massif Central, to intensive wine, fruit and vegetable production in the Mediterranean south.

Political economy offers one approach to understanding the development of nationally and globally orientated food systems that replaced, first, subsistence agriculture and then commercial agriculture geared to regional markets. Political economy provides a theoretically informed, structuralist perspective on global tendencies, with 'economy' understood as 'social economy', that is to say a way of life founded in production (Peet and Thrift 1989). The central concern of political economy with the capitalist world economy 'locates economic analysis within specific social formations and explains the development process in terms of the benefits and costs they carry for different social classes' (Redclift 1984, 5). In particular, political economy focuses attention on the roles of capital and the state in the restructuring of economy and society, together with the consequences for different social groupings. When used in a study of the agro-food system, a political economy approach takes the analysis beyond the farm gate to consider financial institutions, food processors and food retailers, along with food trade and transport, and intervention by the state (Fine 1994).

Political economy itself comprises a number of perspectives (Peet and Thrift 1989) but in this book attention will be confined to theorizations of regulation, globalization and agrarian transformation. In the next chapter, for example, we turn to regulation theory and its expression in the organizing concept of 'food régimes'. This concept provides a structured interpretation of global food tendencies since the mid-nineteenth century, especially as regards changing relations *between* nation states. Chapter 4 examines the contemporary 'globalization' of food in more detail, including the role of transnational corporations in the restructuring of the global agro-food system and leading to the concept of 'food networks'. Chapter 5 investigates the 'agrarian question' *within* individual states, focusing on the relations between capital and agriculture as mediated by state intervention. The role of processors/manufacturers in the production of food is examined next, taking a food systems approach to uncover the varied power relations within the vertical structure of different commodity chains. The concluding chapter completes the analysis of the agro-food system by considering food wholesaling, retailing and catering.

Further reading and references

Fine, B. 1994: Towards a political economy of food. *Review of International Political Economy* 1, 519–45.

Peet, R. and Thrift, N.J. (eds) 1989: *New models in geography*. London: Unwin Hyman.

Redclift, M. 1984: *Development and the environmental crisis: red or green alternatives?* London: Methuen.

3

Food régimes as an organizing concept

Introduction

The concept of 'food régimes' developed in the 1980s out of the French school of 'regulation theory', which itself had been applied mainly to the analysis of the non-agricultural economy (Aglietta 1979, Lipietz 1987). Following Noel (1987), regulationists interpret capitalist development broadly as a sequence of time periods, each period having a specific institutional framework with corresponding norms as regards the organization of production, income distribution, exchange of products and consumption. These distinct time periods within capitalism are termed 'régimes of accumulation'. The first ('extensive') régime is interpreted as lasting from the mid-nineteenth century until World War I; a second ('intensive') régime was established from the end of World War II until the early 1970s; and a third régime has been in the making since the 1980s. Régimes of accumulation are separated by crises of capitalism, as illustrated by the 1930s depression between the first and second régimes and the global recession of the 1970s, following the oil crisis of 1973–74, between the second and third régimes. The terms 'Fordist' and 'post-Fordist' have been used as convenient labels for the second and third régimes, erroneously in the view of Goodman and Watts (1994), although a number of alternatives have been developed to describe the third régime, including 'flexible accumulation' and 'flexible specialization'.

Each régime of accumulation has its stability based on a 'mode of social regulation' (MSR), that is the institutional forms and procedures through which a society organizes and conducts production and reproduction and how social relations are maintained given the class and other antagonisms which they produce (Kenney *et al.* 1989). In this interpretation, national regulatory frameworks and state rules are the product of class forces, while international regulatory structures are created from and are sustained by nations and other transnational entities. For example, as Le Heron (1993,

64–8) explains, the Fordist régime was characterized by mass production (semi-automatic assembly line production) and mass consumption of standardized products, in which significant gains were made by worker productivity with a resulting class struggle over the allocation of the surplus between capital and labour. There were some parallels between agriculture and industry in these trends, for instance in the emergence of 'industrialized' agriculture in sectors such as horticulture, pigs, poultry, dairy and beef feedlot production systems. These systems contained such features as labour specialization as well as high input–output and assembly line-like production techniques. On the marketing side, large supermarket chains developed to distribute the relatively low cost, standardized, uniform quality of foods to consumers. Once a régime is terminated, the formation of a new MSR develops to establish the stable terms under which a new round of capitalist accumulation can take place. Old social forms and institutions are transformed or destroyed and new ones built.

The concept of 'food régimes'

The concept of food régimes draws on regulation theory by recognizing three similar historical periods in international agricultural development (i.e. pre-World War I; from the 1940s to the 1970s; and from the 1980s to the present), each régime being characterized by particular farm products, food trade structures linking production with consumption, and regulations governing capitalist accumulation. Summarizing the seminal work of Friedmann (1982, 1987) and McMichael (1992, 1994), the main features of the food régimes concept include:

- A reading of the recent history of agriculture that privileges relations between a system of independent, liberal nation states in explaining the trajectory of economic change.
- The development of the industrialization of agriculture to permit the sustained international expansion of capitalist agricultural accumulation.
- The changing relationship between capital formations and regulatory modes, both national and international, to determine the speed and characteristics of the globalizing tendencies of agro-food capital.
- The integration of the farm sector into external economic processes of food production (i.e. circuits of capital based on new agri-technologies in chemicals, machinery, animal feed and food processing).
- The role of the state in mediating the circulation of capital, cultural norms and environmental values so as to influence the development of agriculture.

Thus, for Aglietta (1979, cited in Friedmann and McMichael 1989), the concept of food régimes 'links international relations of food production and consumption to forms of accumulation in broadly distinguishing periods of capitalist transformation since 1870'; while, for Friedmann (1993, 300), it is a 'rule-governed structure of production and consumption of food on a world scale'. Le Heron (1993, 73) has elaborated on these rather abstract notions using the more concrete constructs of principal tendencies, governing premises, main historical features, main international policy features and main national policy features. These constructs, summarized in Figure 3.1, are woven together in the following account drawn from the interpretations of Le Heron (1993) and Robinson (1997).

The first food régime (pre-1914)

The first food régime was based on an extensive form of capitalist production relations under which agricultural exports from white 'settler' countries, in Africa, South America and Australasia, supplied unprocessed and semi-processed foods and materials to metropolitan states in North America and Western Europe. The introduction of refrigerated ships in the 1880s increased both the range of produce that could be supplied by distant colonies and the distance over which perishables such as butter, meat and tropical products could be transported to the metropolitan economies (Peet 1969). However, the régime was geared to industrial capitalism so that European imports of wheat and meat ('wage-foods') were exchanged for exported European manufactured goods, labour and capital. Trade became increasingly multilateral, thereby breaking down the prior trading monopolies of the European colonial system and leading to a reconstitution of the world economy as an international economy. In effect, international trade facilitated the relocation of commercial farm production from Europe to extensive settler frontiers and produced three new relations for agriculture (Friedmann and McMichael 1989, 102):

- Complementary product exchange gave way to competitive product trade according to comparative economic advantage.
- The settler family farm represented a new form of specialized commercial agriculture but industrial capital began to 'appropriate' parts of the agricultural labour process (chemicals and machinery).
- Home markets began to be organized within nationally organized economies into agri-industrial complexes.

Friedmann and McMichael (1989, 96) go so far as to argue that the first food régime, with its international division of labour based on more cost-efficient agricultural production in the settler countries, is central to an understanding of the creation of the system of national economies governed by independent nation states.

The first food régime was undermined by the global economic recession of the late 1920s and early 1930s, but aspects of the régime survive, for instance in the food trade in dairy produce, meat and cereals originating in

FOOD REGIME	FIRST (PRE WORLD WAR II)	SECOND (1950s - 1970s)	TRANSITION TO THIRD (1980s - 1990s)
Principal tendencies	Culmination of colonialism Rise of nation-state system	Extension of state system to former colonies Transnational restructuring of agricultural sectors by agro-food capitals	Contradictions between productive forces and consumption trends Disintegration of national agro-food capitals
Governing premises	Acceptability of alien rule Propriety of accumulation domain Importance of balancing power Legitimacy of neo-mercantilism Non-interference in others' colonial administration	Respect for free international markets and free enterprise National absorption of adjustment imposed by international markets Qualified acceptance of extra market channels of food distribution Avoidance of starvation Free flow of scientific and crop information Low priority for national self-reliance National sovereignty Low concern about chronic hunger	Multi-polarity of power (eg US, EC, Japan) Global transmission of adjustments Rise of new protectionism Retreat from distributional issues Restricted flow of technological information Renewed interest in national self-reliance
Main historical features	Centred on European imports of wheat and meat from settler states between 1910 and 1914 and imports by settler states of European manufactured goods, labour and capital	Based on strong state protection and organization of world food economy under US hegemony after 1945	Crisis in world agricultural trade, featuring price instability, breakdown in multilateral agreements, increased competition in export markets and limited imposition of structural adjustment policies
Main international policy features	Imperial preference, with vertical hierarchical relations	Bretton Woods Agreement, GATT, post-war reconstruction programmes, multilateralism Nondiscrimination and legal approach to regulation Commodity agreements and conventions US management of international agricultural trade system via agenda setting in international negotiations	Attempts to resolve world agricultural trade issues through GATT framework
Main national policy features	Assistance for land settlement and infrastructure	Cheap food policies Credit expansion Production control mechanisms Market creation via concessionary export sales and food aid	Opposing trends of further protection and deregulation of agricultural sector

Fig. 3.1 The characteristics of the three food régimes.
Source: based on Le Heron 1993.

the extensively farmed rangelands of the Americas and Australasia. Another remnant comprises the production of sugar, tropical tree crops (cocoa, coconut, rubber, palm oil, bananas) and beverages (tea, coffee) through quasi plantation systems of production – in other words large scale, monocultural, capital intensive farms, employing hired workers, with management and supervision by expatriates but now often under the control of agribusiness corporations.

The second food régime (1947–1970s)

The term 'productionist' (or 'productivist') has been used to summarize the second food régime as a period characterized by an intensive form of capitalist production relations and involving the modernization and industrialization of farming. This account of the régime incorporates the following key processes:

- The restructuring of agricultural sectors by agro-food capitals to supply mass markets.
- The development of durable food and intensive meat commodity complexes.
- Extension of the state system to former colonies (decolonization).
- Organization of the world economy under US geopolitical hegemony.
- Strong state protection for agriculture.

Starting with agro-food capitals, they are implicated in the restructuring of agriculture within the metropolitan states. Agri-inputs (chemicals and farm machinery) and food processing capitals instigated an 'industrialization' of agriculture that is discussed fully in Chapter 5. Capital replaced land and labour as the primary factor of production and almost all food reaching the consumer became subjected to some form of 'value added' processing. The results for the farm sector included a rapid increase in average farm size, rural depopulation and fewer people employed on the land. Friedmann and McMichael (1989, 103) observe that such restructuring blurred the distinction between sectors in national economies as large industrial capitals began to dominate both ends of the agro-food chain and farm products became inputs to manufactured foods rather than products for final use.

Agri-industrial capitals were also implicated in the growing incorporation of former colonies into the global agro-food system. The extension of food trade between metropolitan nation states and newly independent post-colonial states was led by nationally based agribusinesses from North America and Europe (e.g. Coca-Cola, Del Monte, Heinz, Kellogg, Nabisco, Pepsi and Unilever) as they sourced their raw materials through production contracts on a global basis. While these agribusinesses became increasingly important buyers of tropical raw materials, they also encouraged their substitution by temperate crops (e.g. sugar beet for cane sugar, soy and rapeseed oil for palm and coconut oil) and by generic sweeteners (e.g. high fructose corn syrup). Soybean oil also displaced other vegetable oils, both temperate and tropical, as the basis of animal feed cake. The increasing dependence of the newly independent states on imported food grains, such as wheat, from developed countries to feed their burgeoning populations, was paid for by the greater production of industrial crops (e.g. cotton) and speciality, high-value tropical and temperate food crops, with damaging consequences for traditional, subsistence-oriented agricultural systems.

These trends contributed to the further decline in the terms of trade for agricultural commodities from developing countries, with the result of a growing food import dependence amongst developing countries and a decline of their traditional agro-food exports (e.g. sugar, oils).

Turning to 'commodity complexes', the following account follows Friedmann (1994) in recognizing the development of wheat, durable foods and livestock complexes. *The wheat complex* drew its strength from national regulation, such as the farm price support programmes of America and the EU, with surpluses disposed of by food aid programmes and commercial trading. Newly independent states either received US wheat in the form of foreign aid or purchased wheat imports at the expense of domestically grown grains. For example, the share of world wheat imports by developing countries grew from 19 per cent in the late 1950s to 66 per cent in the late 1960s (Friedmann 1994). *The livestock complex* expanded greatly after World War II as income growth allowed northern consumers to eat meat and dairy products. Production intensified on pig, poultry and beef feed lot farms, a trend facilitated by the international trade in cheap US maize and soy-based feedstuffs. European agricultural protection against wheat imports, under the CAP, was accepted by the USA in return for the exclusion of soybeans and soy meal from import duties; this facilitated US exports of animal feed to the EU through the operations of North American grain corporations, such as Continental and Cargill. *The durable food complex* increased in step with the rising proportion of processed and manufactured ingredients in the diets of consumers in northern countries; indeed the mass consumption and mass production of standardised products underpinned the complex in the 1950s and 1960s. Under agri-industrialization, farm producers were faced by increasingly oligopolistic relations with corporate buyers rather than local markets.

These economic developments in food production and consumption were interrelated with global political developments, particularly the extension of the state system to former colonies (decolonization) and the organization of the world economy under US geopolitical hegemony. For example, decolonization broke up the colonial trading blocs and enabled the new states to import food, including food aid, to facilitate their industrialization. This development led to the displacement of traditional foods and the borrowing of capital to finance food imports, with consequences for the growing foreign debt burden of most developing countries. At the time, the USA was searching for new markets for its surplus wheat, including the use of grain exports/food aid to politically stabilize state systems in former colonial dependencies. Thus the spread of the state system diffused a dependence on wheat exports from the USA, although European states protected their agricultures against this development.

As regards state intervention, the second food régime developed under two international agreements: the 1945 Bretton Woods Agreement governing the stability of exchange rates between national currencies (based on the dollar/gold standard); and the 1947 GATT rules on international trade. The former underpinned the international diffusion of the national model of economic growth; the latter excluded agriculture from more liberal trading practices and instead facilitated the further development of national systems of state protection for agriculture. In the metropolitan states, such protection had its origins in the economic depression of the 1930s and the food production exigencies of World War II. National import controls and export subsidies first fostered and then regulated the international disposal of food surpluses from developed countries, initially to the advantage of US economic hegemony through the export of grain and other commodities. The states of the EU were subsequently drawn into subsidized food exports as their own surpluses began to emerge in the 1970s. The rapid expansion of agribusiness corporations during the second food régime facilitated these developments in food trade, as supported by state protection and the subsidy of national agricultures.

The third food régime (1980s–present)

The crisis of capitalist accumulation that ended the second food régime can be traced to the oil and food crises of the early 1970s (Friedmann 1994), comprising global recession, the collapse of Bretton Woods, soaring grain prices, the excessive costs of national farm support programmes, and the antagonism between the national regulation of agriculture and the growing commercial power of globally organized corporations. Goodman and Redclift (1989), Goodman (1991) and Friedmann (1994) account for the ending of the second food régime in these terms:

- The EU began to take on the status of an equivalent power bloc to the US, resulting in a decline in the geopolitical hegemony of America.
- Agricultural export competition under subsidies between the US and the EU threatened a trade war.
- As traditional patterns of trade with developing countries were disrupted, a crisis point was reached in 1973 with US–USSR grain deals eliminating the American wheat surplus from the international market.
- The increasing commercial power of agro-food corporations amplified the tension between nationally organized economies and transnational capital.
- The contradictions of institutionalized food surpluses, especially the economic cost of farm subsidies, emerged in the early 1980s to underpin the political desire amongst states to wind down state support for agriculture.

The final form of the third food régime, which has been emerging from the international farm crisis of the 1970s, is still far from certain but a number of often contradictory structures and processes have been identified (Figure 3.1). Le Heron (1993) usefully summarizes such transitional features under five headings: (1) increased global trading of food; (2) consolidation of capital in food manufacturing; (3) new biotechnology; (4) consumer fragmentation and dietary change; and (5) declining farm subsidies (deregulation).

These five headings are examined in more detail in the following chapters, but they are introduced here for completeness of the food régimes account. Looking first at the global trading of food, McMichael (1992) argues that the activities of the IMF and the WTO, the successor to the GATT, are central to the emergence of a new *global* regulatory structure. Under these new conditions, more liberal trading policies are bringing increased global competition to bear on those farming regions, food processors and food retailers in developed countries, which, for many decades, have been protected by national regulatory measures. At the same time new centres of food production have been emerging, termed 'new agricultural countries' (NACs) (Friedmann 1994). The new centres include Brazil (e.g. intensive livestock based on nationally produced grain and soy), Thailand, Chile, Kenya and Mexico (e.g. fruit and vegetable exports). New commercial relations of production have been penetrating further into agricultural economies historically characterized by their subsistence orientation. Table 3.1, for example, shows the increasing market share of a number of non-traditional agricultural exports from Latin American countries, for instance grapes from Chile and fresh pineapples from the Dominican Republic. Another outcome has been a reduction in the global dominance of the USA in food trade: for example, inputs to animal feed have been sourced globally by agribusinesses (e.g. cassava from Thailand) and extensive grazing on the frontier lands of central America has produced low cost, low quality beef for processing industries, supermarkets and fast food outlets (e.g. as hamburgers, frankfurters and canned products). The absorption of peripheral countries into the transnational livestock complex is but one example of an increasing globalization of food production, trade and consumption.

The processing, marketing and retailing agribusinesses that were forming during the second food régime have become major players in this new competitive trading of food. The concentration of corporate power is massive in some food sectors, for example beverages, with corporations diversifying horizontally as well as integrating vertically. With the organization of their operations now taking place at a global scale, they have been termed 'transnational corporations' (TNCs). The deployment of capital by TNCs has become both sophisticated and flexible, enabling them to react quickly to structural shifts in the world economy and to changes in consumer preferences. TNCs in the USA have led the process of transnational accumulation, which is undercutting the ability of individual states to regulate their domestic agriculture and trade. In addition, by limiting state farm support programmes, the WTO is extending the corporate power of

Table 3.1 Market shares of selected Latin American non-traditional agricultural exports (%)

Country	Product	1978	1990
Chile	Grapes	6.7	22.9
Mexico	Fresh tomatoes	23.8	21.0
Dominican Republic	Fresh pineapples	<1.0	3.9
Honduras	Fresh pineapples	10.3	8.7
Costa Rica	Fresh pineapples	<5.0*	20.1
Brazil	Orange juice	53.8	>60.0*
Colombia	Cut flowers	5.6	>7.0*

Note: *Estimate
Source: adapted from McMichael (1994, 194).

global agribusinesses relative to national (public) power; the WTO is also supervising new forms of reregulation arising out of the contest between nation states, the TNCs and popular movements (e.g. consumers, environmentalists), with the latter two groups not formally present at negotiations. Nevertheless, while the regulatory framework for trade is provided by the WTO, there remain tensions between the main trading nations regarding non-tariff barriers to 'free trade' (e.g. the EU ban on beef imports from the USA in the late 1990s because of the use of growth hormones in cattle production).

Turning to new biotechnology, McMichael (1992, 359) interprets the emerging régime in terms of agri-commodity production 'characterized by the reconstitution of food through industrial and bio-industrial processes via flexible global sourcing of generic crops'. The genetic engineering of crops and livestock is at present the most extreme form of this trend. But running in parallel with biotechnology and the advanced processing of food is a move away from mass production towards 'fresh' and organic fruit and vegetables for a global market (Le Heron and Roche 1995); it implies a reorientation from 'basic' foods and export crops to a more important role for the supply of inputs for 'elite consumption in the north' (McMichael and Myhre 1991, 100). This divergent trend in food production has been traced to the changing characteristics of the global labour force, which is fracturing the demand for food. On the one hand is the growth of new 'Fordist' consumers in NICs, served by large, specialized agri-industry farm complexes. On the other hand is the development of 'green' consumerism within social elites in developed countries, served by smaller speciality producers. Kenney *et al.* (1989) also show how the diversity of demand for food products is driven by different ethnic and social groups: the former exert a varied demand for non-traditional foods; the middle class in the latter enjoy a culture where food has become an experience rather than a sustenance, a trend that resonates with the health and environmental movements.

A contest is developing between these food sub-systems as corporate capital attempts to penetrate the market for high quality, speciality foods.

The emergence of new production and consumption trends in the third food régime has been termed 'post-productionist' by some researchers (Ilbery and Bowler 1998; Lowe *et al.* 1993). With the staged reduction of protectionist commodity programmes and the separation of state farm income supports from production, farmers in many developed countries are being asked to fulfil new and often contradictory roles. For instance, in the EU a re-regulation of agriculture includes the production of environmental goods, 'sustainable' farming practices (the 'greening of agriculture') and foods whose origins and quality can be traced and guaranteed so as to ensure 'food health'. In some ways this represents a relocalization of the agro-food system (Arce and Marsden 1993). The result in Western Europe is a patchwork of productionist and post-productionist trends that are often present to different degrees in neighbouring farms, farming regions and countries. Thus, while new developments in food production, distribution and consumption can be identified, many productionist processes and structures persist from the second food régime and are still being diffused globally.

Conclusion

The utility of the food régimes concept is not universally accepted and in this conclusion a summary is presented of claims and counter-claims so that the reader can form an independent judgement. First, the claims of advocates of the concept, such as Le Heron (1993) and Friedmann and McMichael (1989), are summarized. These writers contend that food régimes offer an organizing concept for interpreting the historical development of global food production, distribution and consumption under capitalism. In particular the concept brings a further dimension to the classical theorization of world trade in terms of comparative advantage by introducing those technological and organizational developments which drive restructuring and give or take away comparative advantage to or from national and regional agricultures (Le Heron 1993, 76). The advocates also claim an explanation of national economic development in terms of (a) sectoral changes within and between nations via the emergence of a collection of liberal nation states; and (b) the globalization of food production and consumption. Despite the variation in national contexts, food régimes offer a common framework of analysis of agriculture at the nation-state and international levels and transcend simplistic international comparisons and country case studies.

Second, food régime theorists claim an explicit consideration of a broad spectrum of agri-commodities within an historico-political context that links regulation to geopolitics, business economics, farming and supply industries, processing, retailing and consumption patterns. Central to the interpretation are the linkages forged between non-agricultural industries, especially TNCs, and agriculture, resulting in a political economy in which the two have become inseparable.

Third, the food régimes concept explicitly includes aspects of the spatial reorganization of production and consumption based on spatial patterns of investment within each régime. As Le Heron (1993, 76) argues, agriculture in any nation state can be assessed in terms of its insertion or non-insertion into the mainstream developments characterizing the prevailing food régime. Here significance is given to production-consumption links centred, at the production end, on various commodity complexes and, at the consumption end, on the mass and differentiating markets of the main industrialized nations. The varied national experiences of food régimes thus become a function of: national regulation, social norms, resources of countries and farming systems, the organization of farm enterprises, the historically contingent shape of agri-commodity chains, and the level of ability of rural producers to influence political processes.

The boldest criticism of the food régimes concept has come from Goodman and Watts (1994) and Moran *et al.* (1996). They counter-claim that theorists have concentrated on too few commodities and countries for a complete reading of global food development; as a result the regulationist food régime concept fails to accommodate the differentiated experiences of nation states in the way that farmers are being integrated into the world food system. In their view the concept seriously underplays national variations in: (a) social agency, (b) contestation and (c) regulation. On social agency, the critics observe that family farms still dominate production in most countries and there is limited evidence of the subsumption of family farms by non-farm capitals, as in the USA. Agricultural development cannot be explained using concepts drawn from the analysis of industry because of the land-based character of food production. In addition, individual farmers retain a degree of autonomy (agency) in deciding what to produce and how to market their output. Pritchard (1996) also observes that there is no place in the food régimes concept for endogenous development as an organizing vehicle for capital in food sectors at the national, regional, local or farm levels.

On contestation, the political and social influence of social movements and interest groups tends to be ignored in the food régimes concept. In countries such as France and New Zealand, for instance, farmers have had different experiences to those described in the food régimes literature in which empowerment of TNCs is a central theme (Moran *et al.* 1996). Rather farmers have been able to exert enough political power to ensure state sanctioning of producer co-operative control of the processing and marketing stages of several agri-commodity chains. In the member states of the EU, the winding down of farm supports under the CAP has been so resisted by farm groups that claims of a liberal trading régime appear premature.

On regulation, rural regional and national political organizations continue to influence law making but the food régimes literature tends to downplay the continuing importance of national and, particularly, local regulatory processes in their accounts of international regulation. Understanding national and sub-national legislation remains essential in interpreting the international variability in the organization of national food systems. Rather regulation is treated unproblematically by the food régime concept, whereas in reality it is contested and constructed through alliances and opposition. In Pritchard's (1996) terms, instead of comprising an accepted set of rules, participants in regulation play margin games with the aim of challenging established institutions. As cases are decided, the meaning of regulation affects and in turn changes the behaviour of social agents. Thus regulation is more correctly treated as contestable social practice in which outcomes are varied and contingent: nationally specific regulatory contexts emerge and regulatory influences from the past appear more persistent than is implied in the food régime literature.

Le Heron (1993, 75–6) is also critical of food régimes as an 'accumulationist account' that relies heavily on explaining international regulation in terms of US hegemony, yet fails to specify the rules accepted as part of that hegemony and how they elevated America into a position of regulatory dominance. In addition, he argues that food régime accounts gloss over the ways in which both crisis tendencies and their resolution vary by country. For instance, Friedmann (1994), McMichael (2000) and McDonald (2000) show how, at least until the 1980s, the history of Japan's agri-food sector was quite different from the US. There was little TNC investment to integrate Japan into the US agro-food model and Japanese agriculture retained much of its national character, for instance in the rice sector. Thus the timing and sequence of change in individual countries varies from the model advanced by the food régime concept and national production and distribution systems remain differentiated. Robinson (1997) amplifies this line of criticism. He argues that the third food régime contains many features of the second régime and consequently each régime cannot 'be clearly delimited from earlier ones'; nor can 'these general concepts advance understanding of the tensions between unique occurrences in particular locations'. History has been given priority over geography, while the high level of abstraction is problematic for understanding the historical experience of particular nations and regions.

Thus, on the one hand, the food régimes concept offers an integrated, structuralist reading of the recent *history* of global food production and consumption under capitalism; on the other hand, and despite claims to the contrary, the concept remains weak in explaining the wide variety of national and regional (i.e. *geographical*) experiences in the development of food systems. With the final dimensions of the third food régime still uncertain, the outcome seems likely to be influenced by the contest between private global regulation and democratic public regulation (Goodman and Watts 1994). In other words, the third food régime will be shaped, firstly, by the extent to which international institutions are developed to regulate rather than facilitate the globalizing activities and accumulation tendencies of TNCs. Secondly, the outcome will depend on the relative success of national and local regulatory power in reconnecting and redirecting national/local food production and consumption in resisting the development of global food networks.

Further reading and references

There is no single text that can be relied upon for a full account of food régimes. Rather, readers are referred to the work of four key writers: Friedmann, Goodman, Le Heron, and McMichael.

Aglietta, M. 1979: *A theory of capitalist regulation*. London: New Left Books.

Arce, A. and Marsden, T. 1993: The social construction of international food: a new research agenda. *Economic Geography* **69**, 293–311.

Friedmann, H. 1982: The political economy of food: the rise and fall of the postwar international food order. *American Journal of Sociology* **88**, S248–86.

Friedmann, H. 1987: Family farms and international food régimes. In Shanin, T. (ed.) *Peasants and peasant societies*. Oxford: Blackwell, 247–58.

Friedmann, H. 1993: The political economy of food: a global crisis. *New Left Review* **196**, 29–57.

Friedmann, H. 1994: The international relations of food: the unfolding crisis of national regulation. In Harriss-White, B. and Hoffenberg, R. (eds) *Food: Multidisciplinary Perspectives*. Oxford: Blackwell, 174–204.

Friedmann, H. and McMichael, P. 1989: Agriculture and the state system: the rise and decline of national agriculture, 1870 to the present. *Sociologia Ruralis* **29**, 93–117.

Goodman, D. 1991: Some recent tendencies in the industrial organization of the agri-food system. In Friedland, W., Busch, L., Buttel, F. and Rudy, A. (eds) *Towards a new political economy of agriculture*. Boulder, CO: Westview, 37–64.

Goodman, D. and Redclift, M. (eds) 1989: *The international farm crisis*. London: Macmillan.

Goodman, D. and Watts, M. 1994: Reconfiguring the rural or fording the divide? Capitalist restructuring and the global agro-food system. *Journal of Peasant Studies* **22**, 1–49.

Ilbery, B. and Bowler, I. 1998: From agricultural productivism to post-productivism. In Ilbery, B. (ed.): *The geography of rural change*. London: Longman, 57–84.

Kenney, M., Kloppenburg, J. and Cowan, J. 1989: Midwestern agriculture in U.S. Fordism. *Sociologia Ruralis* **29**, 131–98.

Le Heron, R. 1993: *Globalized agriculture: political choice*. Oxford: Pergamon Press.

Le Heron, R. and Roche, M. 1995: A 'fresh' place in food's space. *Area* **27**, 23–33.

Lipietz, A. 1987: *Miracles and mirages: the crises of global Fordism*. London: Verso.

Lowe, P., Murdoch, J., Marsden, T., Munton, R. and Flynn, A. 1993: Regulating the new rural spaces: the uneven development of land. *Journal of Rural Studies* **9**, 205–22.

McDonald, M.G. 2000: Food firms and food flows in Japan 1945–98. *World Development* **28**, 487–512.

McMichael, P. 1992: Tensions between national and international control of the world food order: contours of a new food régime. *Sociological Perspectives* **35**, 343–65.

McMichael, P. (ed.) 1994: *The global restructuring of agro-food systems*. Ithaca: Cornell University Press.

McMichael, P. 2000: A global interpretation of the rise of the East Asian food import complex. *World Development* **28**, 409–24.

McMichael, P. and Myhre, I. 1991: Global regulation vs. the nation-state: agro-food systems and the new politics of capital. *Capital and Class* **43**, 83–105.

Moran, W., Blunden, G., Workman, M. and Bradley, A. 1996: Family farmers, real regulation and the experience of food régimes. *Journal of Rural Studies* **12**, 245–58.

Noel, A. 1987: Accumulation, regulation and social change: an essay on French political economy. *International Organization* **41**, 303–33.

Peet, R. 1969: The spatial expansion of commercial agriculture in the 19th century: a von Thunen interpretation. *Economic Geography* **45**, 283–301.

Pritchard, W. 1996: Shifts in food régimes, regulation and producer cooperatives: insights from the Australian and US dairy industries. *Environment and Planning A* **28**, 857–75.

Robinson, G. 1997: Greening and globalizing: agriculture in 'the new times'. In Ilbery, B., Chiotti, Q. and Rickard, T. (eds): *Agricultural restructuring and sustainability: a geographical perspective*. Wallingford: CAB International, 41–53.

4

Globalization and food networks

Introduction

Globalization has been identified as one of the principal formative processes in the contemporary development of the capitalist world order. According to McMichael (1994, 277), globalization 'generally refers to the world-wide integration of economic process and of space', including 'a shift of power from communities and nation-states to international institutions such as transnational corporations (TNCs) and multilateral agencies'. Goodman (1994, cited in Burch *et al.* 1996, 28–9), however, is more specific in his definition of globalization and distinguishes between the following terms:

- *Internationalization:* economic development based on the logic of commodity exchange in world markets; national production systems in an expanding trading system.
- *Multinationalization:* corporations with countries of origin but with international operations as an extension of the national; central co-ordination and ownership of relatively autonomous overseas affiliates.
- *Transnationalization:* centrally controlled, vertically-integrated transnational production systems; footloose locational decisions driven by internal firm-specific criteria.
- *Globalization:* non-market forms of inter-firm collaboration; mutual reciprocity between regional innovation systems and global networks; technological convergence through collaborative research and design networks.

As interpreted by Goodman, therefore, globalization describes a future yet to be achieved; rather he recognizes transnationalization as a more accurate description of the contemporary régime of accumulation based on globally sourced and integrated production. Most writers, however, continue to employ the term 'globalization' rather than transnationalization and Buttel (1996, 23) is in the mainstream in identifying the following key features:

- Cross-border exchange by economic agents from different countries.
- Globally-integrated financial markets.
- Development of transnational corporations (TNCs) and cross-border intra-firm division of labour.
- Predominance of footloose, stateless, deterritorialized corporations.
- Global sourcing of raw materials through subsidiaries.

Globalization resonates, therefore, with an increasing socio-economic homogeneity, a dominant market culture and an erosion of national social, economic and political institutions. Indeed the restructuring of national institutions is viewed as a necessary response to stabilize capitalism under globalization, given the associated uncertainties of accumulation and legitimation. Nevertheless, McMichael (1994, 277–8) warns against claims for globalization as a linear economic or political trend; rather it should be interpreted as a formative process in a particular transitional but historically specific phase in the development of world capitalism.

Globalization in the context of the third food régime can be interpreted, therefore, in terms of the wider, contemporary global restructuring of economic activity, as experienced in the finance and manufacturing sectors. In this chapter we examine, first, the characteristics of globalization in the agro-food system, followed by a consideration of the role claimed for TNCs. Then the problem of linking varied local/regional/national responses to globalization is addressed, drawing on the concept of 'food networks'. In the final part of the chapter we review the growing evidence of local/regional resistances to the process of globalization through endogenous developments in agriculture. The overall theme is that globalization, like food régimes, can frame the historical analysis of case studies, with the concept of 'food networks' offering a potential for analysing the spatial or geographical outcomes of globalization. Indeed globalization, rather than exercising a homogenizing influence, will be shown as enhancing spatial diversity in food production, distribution and consumption.

The globalization of the agro-food system

The globalization of the agro-food system includes, in Le Heron's (1993, 17) words, 'agriculture's continued incorporation into the general dynamics of capitalist accumulation'. Expanding on this view, writers such as Bonanno *et al.* (1994) and Goodman and Watts (1997) have interpreted globalization for the agro-food system in terms of: the increasingly liberal international trading of agricultural and food products; the dominance of that trade by corporately restructured agro-food capital (i.e. food processing and retailing); the emergence of a new international division of agro-food labour; and the replacement of national by international institutions to regulate trading relations for agro-food products.

Looking first at trade, a number of writers have already observed that the trading of agro-foods across national boundaries is not a new phenomenon. But as national restrictions on food trade have been gradually reduced over recent years under the WTO, so evidence of a more liberal trading régime is slowly emerging. In addition, while the involvement of large agribusinesses in international trade also is not new, the increasing global scale of organization of some multinational companies has raised their status to TNCs, for example ConAgra based in the US, Gruppo Ferruzzi in Italy and Nippon Meat Packers in Japan. These TNCs are defined by the global sourcing of their supplies, the centralization of strategic assets, resources and decision making, and the maintenance of operations in several countries to serve a more unified global market. Unilever (an Anglo-Dutch TNC), for example, was a pioneer in global sourcing of raw materials, drawing upon tropical oilseeds from Asia and Africa as well as domestic European oilseeds and grains. Other agro-food TNCs are listed in Table 4.1 according to the size of their sales in the mid-1990s and, in Table 4.2, according to their value on the UK stock exchange in the late-1990s.

New dimensions of global trading patterns in food reflect, on the one hand, the emergence of a number of new agricultural countries (NACs) within the global economy and, on the other hand, the restructuring of the

Table 4.1 Top twelve transnational corporations in food manufacturing and processing by total sales in 1994

Corporation	Sales (US $b)	Country	Sector
Philip Morris	53.3	US	Multi-product
Cargill	50.0	US	Cereals
Nestlé	40.2	Switzerland	Multi-product
PepsiCo	28.5	US	Beverages and soft drinks
Unilever	26.2	Netherlands	Multi-product
Coca-Cola	23.8	US	Beverages and soft drinks
Conagra	23.8	US	Multi-product
RJR Nabisco	15.4	US	Multi-product
Danone (BSN)	12.8	France	Multi-product
Anheuser Bush	11.4	US	Beer
Grand Metropolitan	11.3	UK	Multi-product
Snow Brand	10.6	Japan	Dairy products

Source: adapted from FAO (1998).

agro-food sector in developed countries. Such developments are based on the relocation of certain types of agro-food production to areas where the factors of production, particularly labour, are less costly. This can be interpreted as the emergence of a new international division of labour in the agro-food system. For example, labour-intensive, high-value, crop and livestock product exports from developing countries, such as Thailand, Brazil, Chile, Kenya and Mexico, have become a major growth sector in international agricultural and food trade. Jaffee (1994), for instance, points out that in 1989 the trade in high value foods (HVF), such as fresh fruit, vegetables, poultry, dairy products and shell fish, was equivalent to five per cent of the value of world commodity trade – roughly equivalent to that of crude petroleum. Taking Kenya as an example of an NAC, together with its attendant spatial division of labour, Barrett *et al.* (1999) show how exports of horticultural crops increased by 58 per cent in the 5 years between 1991 and 1996, and accounted for 10 per cent of the total export earnings of the country (behind tea and coffee). Of total horticultural exports, cut flowers (roses, statice – a bouquet filler – and carnations) comprised 42 per cent by weight, green beans 18 per cent, avocado 12 per cent, and the remaining crops each less than seven per cent. Of the fresh horticultural produce, 93 per cent left Kenya by air through Nairobi airport, the remainder passing through the port of Mombassa using refrigerated containers. Horticultural produce has been sent to the UK (30 per cent), the Netherlands (29 per cent), France (16 per cent) and Germany (10 per cent). Bonanno (1993) provides two further examples of NACs: first, the People's Republic of China, with its low labour costs, now manufactures and exports animal feed to Western markets, albeit under the control of TNCs; second, in poultry production, birds are raised in the US, deboned in Mexico, and exported to Asian markets. Taken together, evidence of this type has provoked claims of an increasing agricultural specialization at regional and farm levels within the global economy, with agricultural production providing inputs to the manufacturing system rather than immediate consumption.

In developed countries, a restructuring of the agro-food system is also underway, but with a greater impact on the food processing compared with the farm sector. Australia is an example of a country which, under an 'economic rationalist' philosophy of market forces and efficiency, has had to shift from a nationally coherent to a globally competitive economy. This process began in the 1970s, a decade ahead of other countries, as Australia lost its traditional market in the UK with that country's membership of the EU. Australian agriculture, followed after 1984 by New Zealand's, has had to undergo a profound reorientation in its food trade away from Europe and towards South East Asia, as well as a restructuring of its agro-food sector. Within the dairy sector, for example, Pritchard (1998) has traced the withdrawal of state government regulations and the resulting locational shifts in dairy processing capacity, with less profitable regions losing out to

Table 4.2 Food companies with shares quoted on the London Stock Exchange in 1999

Company	Market capital (£m)*
Beverages	
Diageo	21,493.6
Allied Domecq	5,943.8
SA Breweries	3,877.7
Coca-Cola	1,296.4
Food producers	
Unilever	17,218.6
Cadbury Schweppes	8,315.0
AB Foods	3,251.2
Tate & Lyle	1,988.4
Kerry Group A	1,225.8
Unigate	940.7
United Biscuits	851.7
Northern Foods	767.4
Food retailers	
Tesco	11,568.4
Sainsbury	7,594.4
ASDA	6,714.4
Safeway	2,409.6
Morrison	2,346.7
Somerfield	1,099.8
Iceland Group	475.6

*Market capital > £400m.
Source: compiled from *The Guardian* (November 1999).

higher-productivity regions in Australia's south. Deregulation of regional milk processing co-operatives after 1986, including the dismantling of the market pooling system (fresh and manufactured milk prices) and milk zones, contributed to a reduction in dairy co-operatives from 44 to 27 between 1983 and 1993 and dairy companies from 65 to 31. Implicated in the restructuring of the dairy processing sector have been: (a) mergers and take-overs, so that two companies, Bonlac and Murray-Goulburn, account for 70 per cent of Australian dairy exports; (b) strategic alliances for market access between milk producers and dairy processors; (c) the restructuring of transport and distribution to serve supermarkets; and (d) the equity financing of co-operatives leading to the demise of regional co-operatives and the growth of a few large co-operatives. These changes took place against the background of an increase in dairy production by 40 per cent and, significantly for the contested definition of globalization, within the context of national and not global capital. In the wheat sector, some deregulation took place in the early 1990s but the Australian Wheat Board retains a monopoly over exports (85 per cent of Australian wheat is exported). Here a diversification of cereal production can be observed in unregulated domestic feedgrains (barley, oats, sorghum) to meet the growth in demand from domestic dairy and beef feed-lot enterprises. Beef production in particular has attracted inward investment, in this case from the US and Japan, especially into New South Wales and Queensland. Pritchard (1998) concludes that new regional commodity complexes are emerging within Australia and are expressive of a global shift in food regions.

The replacement of national by international institutions (e.g. WTO, NAFTA, Asia Pacific Economic Co-operation, EU) to regulate trading relations for agro-food products is contributing to these developments. National deregulation is taking place to accommodate the internationalization of production and financial capital, as reflected in the growth of multinational and transnational food companies and their global food production and processing operations. This reduction of national regulation threatens the alliance between the state and farmers that has existed since World War II in most developed countries and is discussed further in Chapter 18. Recent reforms of the CAP, for example, reflect the decline of the political power of farm groups and the fragmentation of agricultural lobbies both in individual member states and supranationally within the EU. Symptomatically, farm groups in developed countries were unable to block the conclusion of the Uruguay Round in 1992 with its promise of reduced restrictions on agro-food trade.

The role of transnational corporations

Transnational corporations (TNCs) form a third stage in the continuing reorganization of capital within the agro-food system. Agribusiness first emerged as a feature of the second food régime – that is vertically integrated businesses involved in the supply of agri-inputs (e.g. fertilizers and animal feeds), the processing of agri-outputs (e.g. freezing, canning, grading produce) and the distribution of food products (e.g. retailing) (Wallace 1985). With some notable exceptions, the type of company listed in Table 4.3, for US broiler (poultry meat) production, was essentially national rather than international in business organization, sourcing any non-domestic raw materials through conventional bilateral trading relations. During the 1980s, however, in parallel with other economic sectors, agribusinesses became involved in take-overs and mergers in a second-stage process termed 'conglomerate integration'. The result was a concentration of productive capacity in a relatively few, but still nationally oriented corporations, some with foreign subsidiaries so that the term 'multinational corporation' could be applied. Rogers (2000, 14), for example, cites concentration ratios for the four largest food firms in the USA in 1992 varying from 85 per cent for cereal breakfast foods and refined cane sugar,

Table 4.3 The top ten broiler (poultry meat) firms in the US in 1989*

Rank	Firm	Rank	Firm
1	Tyson Foods	6	Continental Grain
2	ConAgra	7	Hudson Foods
3	Goldkist	8	Foster Farms
4	Perdue Farms	9	Townsends
5	Pilgrim's Pride	10	Seaboard Farms

* The top four firms accounted for 45 per cent of production.
Source: adapted from Bonanno *et al.* (1994, 33).

to 22 per cent for fluid milk and 19 per cent for fish. For Thailand, Burch *et al.* (1996) show how Thai-based, but globally-oriented, agribusiness corporations drew their raw materials from domestic producers before exporting them on to international markets (e.g. frozen poultry, prawns, canned and frozen fruit and vegetables, snack foods and fast food products).

By the 1990s, some of these companies had begun to develop a global structure through international scale direct investment, take-overs, mergers and joint partnerships. Rickson and Burch (1996, 80) for Australia, for example, cite the case of Pacific Dunlop whose food division was the largest Australian-owned agro-food company before 1995. In that year, brands such as Edgell-Birds Eye and Herbert Adams were sold to J.R. Simplot (US), with the remaining brands, such as Peters Ice Cream and Yoplait, sold to Nestlé (Switzerland). Other companies embarked upon a strategy of 'multiple sourcing', drawing their raw materials competitively from different countries and regions. Thus in the third stage of transnational development, the largest transnational corporations (TNCs) began to assume control over the whole of a production and distribution chain – from agri-inputs, including farmland, to distribution and retailing – and with interests spread over both a range of food products and countries. Under global organization, individual farmers, food processors and food retailers are being thrown into global rather than national competition and all are required, for economic survival, to restructure their businesses into increasingly large, economically efficient units. However, the organization and ownership of corporate capital is complex. Tables 4.4 and 4.5, for example, show how joint ventures/partnerships, rather than full ownership, are used by many large corporations to gain access to global raw materials and markets.

For Bonanno *et al.* (1994, 2), TNCs are 'the central defining element behind (the) configuration of new capital accumulation spaces', but Watts (1996) cautions against exaggerating the development of TNCs. He observes that most agro-food businesses remain multinational rather than transnational in their organization. That is to say, they operate 'multidomestic' strategies rather than sourcing through centralized, global intra-firm

Table 4.4 Joint ventures by selected TNCs importing into Japan

TNC	Joint venture partner	Partner countries	Exporting country
Tyson	C. Itoh	US, Japan	Mexico
Cargill	Nippon Meat Packers	US, Japan	Thailand
Mitsubishi	Perdigao	Japan, Brazil	Brazil
Mitsui	Malayan Flour Co.	Japan, Malaysia	Malaysia
Ajinomoto	d'Osat Ajinomoto Alimentos	Japan, Brazil	Brazil

Source: adapted from Bonanno *et al.* (1994, 43).

production systems. Consequently the development of TNCs should be interpreted as an emerging tendency, with the multinational corporation (MNC) recognized as more significant for the contemporary organization of the agro-food sector.

To illustrate this point, one of the principal mechanisms of control exercised by MNCs is through production contracts rather than the ownership of farmland on an international basis (Rickson and Burch 1996). Production contracts are legally binding agreements between independent farm and food processing or retailing businesses for the sale and purchase of agricultural produce. On the one hand, farm producers gain the security of a guaranteed buyer at an agreed price before production takes place; on the other hand, processors/retailers gain access to a raw material with a guaranteed quality and volume, at a fixed price, to be delivered at an agreed date. However, an unequal power relation exists between producer and buyer. Not only are there more producers than buyers but produce that fails to meet the contract specifications in terms of quality, quantity or timeliness

Table 4.5 Agro-food capital structures in Thailand, 1990

Sector	No. of companies	Percentage of total capital investment		
		All Thai	Joint venture	All foreign
Rubber processing	94	61	39	0
Frozen/canned fruit and vegetables	100	60	39	1
Frozen/canned seafood	48	51	47	2
Milk/dairy produce	12	27	57	16
Animal feed	84	28	71	1

Source: adapted from Burch *et al.* (1996, 328–9).

can be declined by the purchaser. In addition the farming practices of the producer are commonly specified by the food processor/retailer, for instance as regards crop variety, pesticide use and planting date, and not surprisingly farm producers under contracts have been termed 'landed labourers'.

Production contracts have been used in agriculture throughout the second food régime and they are not in themselves novel. For instance, Bonanno *et al.* (1994, 35) found that in the US by 1980, 25 per cent of all farm output was under contract, with a further 69 per cent of production from independent farms and six per cent from corporations. Contracts were most prevalent in the production of poultry meat, eggs and fresh vegetables. Friedmann (1994, 264) cites the use of contracts by McCain Foods, one of the largest corporate empires in the world, to assure a steady supply of genetically standard potatoes into its processing factories for the production of frozen french fries. One outcome was the reorganization of traditional agricultural communities in eastern Canada to create a monocultural region of potato production under contracts. A further example of the use of contracts is the Bud Antle corporation in the US in securing year-round supplies of lettuce and tomatoes using machine harvesting and long-distance shipping. The western US has been most affected, including the employment of non-unionized labour. But contracts are now increasingly used by MNCs to secure farm produce at a global level from competing countries (Little and Watts 1994). From one perspective, when contracts are placed in developing countries, an unequal power relation is imposed not just between buyer and producer but also between northern buyers and southern producers. From another perspective, however, Miller (1996) argues that contracts form stable relationships between growers and buyers, with mutual benefit gained from long-term production outcomes, grower and buyer loyalty, and brand-name quality.

Food networks – linking the local to the global

Conventional political economy analyses of globalization tend to explain national and regional outcomes in terms of contingent processes, for example the varied local/regional penetration of the food system by non-farm capitals, differences in the national deregulation of the agro-food sector, or spatial variations in the operation of the labour process. Indeed this tradition underpins the account given in the previous sections where precedence has been given to structure over agency. Buttel (1996, 18), however, challenges this uncritical interpretation of the term 'globalization': he draws attention to the sectoral and local specificities of farm structure and agri-ecology. But in addressing such specificities, the tendency has been to interpret 'the local' and 'the global' as separate spatial domains – the problem remains of identifying how global processes of change are incorporated by social actors in localities, firms and institutions (Arce and Marsden 1993).

One approach lies in the analysis of *commodity systems* (sometimes termed 'production chains', 'food systems' or 'filières') (Bowler 1992, 12; Friedland 1984). This form of analysis, taking one food product at a time (e.g. tomatoes, lettuce, oranges), traces production from the first agri-inputs, through farm production to food processors, wholesalers, retailers and consumers, and includes the labour process, technology and state policies. Research in Australia by Burch and Pritchard (1996) for processed tomatoes, and by Hungerford (1996) for sugar, is illustrative of this mode of analysis. The previously noted study by Barrett *et al.* (1999), on Kenyan horticultural produce, also falls into this tradition (Figure 4.1). Two commodity chains are identified: one based on the traditional wholesale chain and the other on the supermarket chain. All the 'actors' within each commodity system are identified, from those operating at the local level (e.g. individual small, medium and large growers, middlemen and agents), through actors operating at regional levels (e.g. freight/cargo agents, warehousing and coldstores), through national level actors (e.g. charter/scheduled airlines and UK importers), to actors significant at the international level (e.g. independent retailers: *see* Chapter 7). Using fresh horticultural produce as the centre of the analysis, the local response can be traced through to the actions of actors at the global level.

More recently the concept of *food networks* has been advanced as a means of addressing the global-local dichotomy on the one hand, and the inter-relationships between human and non-human intermediaries in the social construction of food on the other. For Arce and Marsden (1993), food networks offer the potential for showing how actors shape, and are shaped by, the political, cultural and social environment – in other words, how commodities are both produced and consumed in social contexts. A food network is conceptualized as a hybrid, that is comprising the inter-relationships between the human actors in a commodity chain but extended to include the non-human intermediaries that bind the actors together in power relationships. Examples of non-human intermediaries include the contracts that link farmers to processors, the regulations that link processors and farmers to national politicians, and the international agreements that link MNCs to the WTO.

The theoretical underpinning of this interpretation lies in actor-network theory (ANT), which itself was introduced into the English language literature on the sociology of science and technology by writers such as Latour (1986), Callon (1986) and Law (1986). ANT has been extended to the geographical literature by Marsden *et al.* (1993, 140–47), Murdoch and Marsden (1995) and Thrift (1996, 23–26). These writers have developed the spatial concepts of 'network lengthening' and 'acting at a distance'. The former deals with the ways in which additional actors and intermediaries are enrolled into a network; the latter addresses the means by which one actor, separated by space, has the capacity to influence other actors, for instance in defining the production and valuation of food. A production

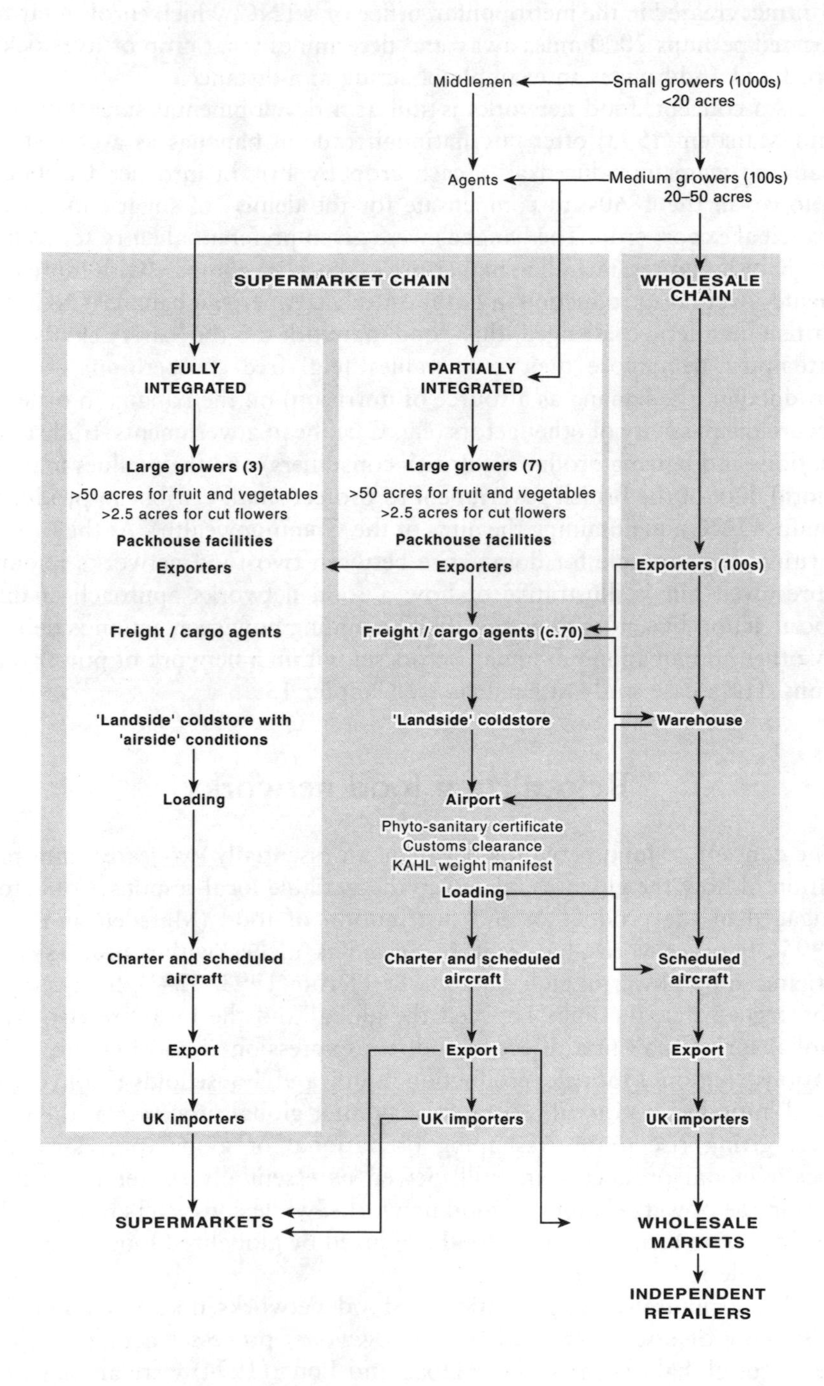

Fig. 4.1 The Kenyan marketing network for the export of fresh horticultural produce to the UK. *Source*: based on Barrett *et al.* 1999.

contract created in the metropolitan office of a TNC, which enrols a farmer located perhaps 2000 miles away and determines what crop or livestock is produced and how, is an example of acting at a distance.

As a concept, food networks is still at a developmental stage but Arce and Marsden (1993) offer international trade in bananas as a case study. Bananas were introduced as a cash crop by Britain into her Caribbean colonies in the 1860s, to compensate for the demise of sugar cane as the principal export crop. The bananas were given preferential entry terms into the British market, including higher prices, so as to compete with more efficient, lower cost production in Latin America. American banana TNCs and British retailers challenged this food network in the early 1990s and attempted to impose their own values (e.g. free competition; efficient production; the banana as a source of nutrition) on the banana in order to secure the passivity of other actors (e.g. Caribbean governments, trade associations and banana producers; British consumers) and their values (e.g. the moral duty of the British government to protect small Caribbean producers against TNCs; maintaining the aims of the Commonwealth). At the time of writing, this struggle for dominance between two food networks remains unresolved but is illustrative of how a food networks approach enables social action by actors to be specified, including how such action is defined by other human and non-human actors set within a network of power relations. For a case study of bananas *see* Chapter 13.

Relocalizing food networks

The concept of food networks began as an essentially *exogenous* interpretation of how the global is related to the variable local responses of actors engaged in the production and distribution of food (Marsden and Arce 1995). In other words, locales were viewed as 'niches' within a food system organized by powerful global actors. Le Heron (1993, 32–3), for example, has argued that the links between the global and the local are two way: global forces may find direct or indirect expression in local changes (i.e. nations, regions, locales, production units and households); conversely, local initiatives and local resistance may alter global processes at the local level giving rise to the reshaping of national or global processes. But local/regional producers are still viewed as essentially subservient actors within the power relations of food networks, while consumers remain relatively passive recipients of the food produced by globalized food processors and retailers.

This structuralist interpretation of food networks does not admit the possibility of essentially local (i.e. *endogenous*) processes acting independently of globalization. Van der Ploeg and Long (1994) were amongst the first to argue against the exaggerated claims of the power and homogenizing capacity of global macro-structural forces in accounts of the agro-food system. Observing different 'styles of farming', they found little evidence of an homogenization of technology, structure and management in world agricultures; rather, they argued that farmers are not passive recipients of global economic forces but develop locally-based adaptive strategies rooted in culture, local agri-ecology and farm household resources. The themes of this approach include localism, the recovering of local knowledges, the avoidance of marginalization within unpredictable global markets and rural diversity as a livelihood strategy. Clusters of similarly developing farm businesses, linked to local food processing industries and distributors, have the capacity to create 'agro-food districts', similar in concept to the industrial districts identified in contemporary industrial organization. Claims for the identification of these 'agro-food districts' have been made mainly by researchers in Italy, and include cereal production in the Colline Pisano–Livornesi of Italy (Iacoponi *et al.* 1994), Parmesan cheese production in the provinces of Parma and Reggio Emilia, processed pork and ham in Modena, and poultry in Verona and Forli (Fanfani 1994).

Farm-based accounts of farming styles and agro-food districts, however, have paid insufficient attention to the necessary relationship between consumption and production in the development of adaptive strategies. Evidence is mounting of an increasing consumer resistance to the globalizing tendencies of food networks, as expressed through various social 'movements', which form the economic base for *alternative food networks*. Buttel (1996, 35) identifies the following range of social movements:

- environment
- sustainable agriculture
- community-supported agriculture
- consumer and health
- genetic resources conservation
- animal rights
- rural social justice
- consumer preference (e.g. organic food)
- farmers' markets
- non-traditional medicine
- ethnic cuisine.

Watts (1996) terms such social/consumer movements 'the politics of collective consumption' and evaluates them only as counter-tendencies; for Bonanno *et al.* (1994, 11), they are evidence of an emerging 'flexible consumption'; while Friedland (1994) interprets them as part of the development of a 'post-modern diet'. However explained, consumer resistance to 'industrialized' food products is growing, with new political alliances and marketing networks emerging between consumers and farmers in different parts of the world. These issues are considered in more detail in Chapter 7 but are mentioned here for completeness of the argument. On the one hand, increasing numbers of consumers are prepared to pay a premium price to assure the welfare conditions under which meat, eggs and milk are produced, as well as guarantee chemical-free fruit and vegetables. On the other hand, awareness of the health risks attached to food has increased. Risk is now interpreted not only in terms of how different foods and qualities of food impact indirectly on individual health, for example as regards heart disease and obesity, but also directly in terms of salmonella poisoning, pesticide residues, *E. coli* infection and new variant Creutzfeldt–Jacob disease (nvCJD from bovine spongiform encephalopathy-infected meat – BSE) (see Chapter 16). When taken together, these consumer behaviours support alternative food networks that are situated outside conventional food networks.

A study by Whatmore and Thorne (1997) illustrates the potential for investigating alternative food networks using ANT, in their case based on the social movement associated with rural social justice and international trade in coffee. Two networks are compared: one a conventional 'commercial' coffee network, the other a 'fair trade' network (Figure 4.2). In the former network, commercial imperatives form the 'mode of ordering', in which an unequal power relationship exists between numerous small-scale producers of coffee and a relatively few dealers, and where the economic returns from brand-related, mass produced coffee favour the end producer and retailer. In the latter network, however, the mode of ordering is based on partnership, alliance, responsibility and fairness, thereby empowering the marginalized actors, in this case the small-scale producers working within local co-operatives. They are paid a guaranteed minimum premium price, but otherwise have to work within the disciplines of quality control, marketing deadlines and so on. This fair trade network is one of many that have grown over the last three decades out of the work of non-governmental organizations, such as Oxfam in the UK. A number of trading companies, Cafédirect in the case study, have emerged as key actors in connecting local producers, for instance in Latin America, with consumers in developed countries who are prepared to pay a higher price for a fair-trade product (e.g. Cafédirect ground and freeze-dried coffee) compared with conventionally marketed produce.

A case can be made, therefore, that alternative food networks are in the making based on a range of social movements, with a resulting relocalization of food based on endogenous development processes. At present consumer resistance within alternative networks is confined to social élites in developed countries; the question remains if such resistance will spread to other social groups and locations.

Conclusion

This chapter has examined claims of a globalization of agro-food systems under the third food régime. Globalization is revealed, on the one hand, as

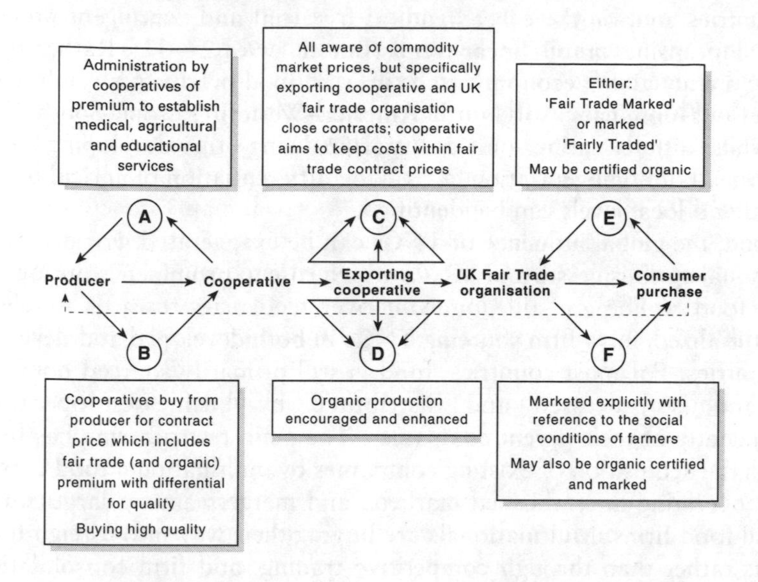

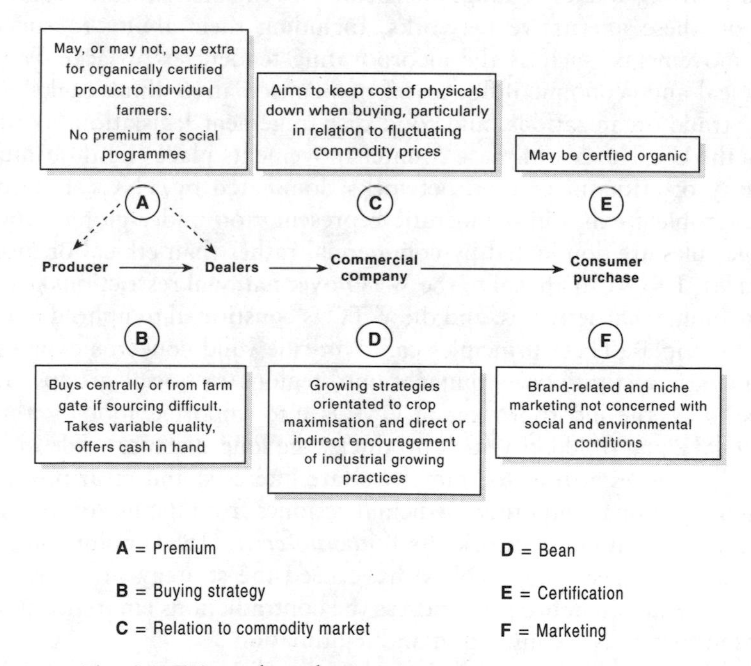

Fig. 4.2 Fair trade and conventional food networks.
Source: redrawn after Whatmore and Thorne 1997.

a contextual variable of restructuring within both developed and developing countries and, on the other hand, as a partial and contingent process rather than an inexorable linear trend (Buttel 1996, 24–32). Rather than creating a truly global economy, trade in agro-food products is reinforcing the previous dominance of North America–Western Europe–South East Asia, whilst differentiating some and marginalizing other developing countries. Instead of homogenization, a growing differentiation of agriculture at regional and local levels can be identified.

Second, the global influence of TNCs can be exaggerated. For example, the national agribusinesses and MNCs which rose to prominence during the second food régime are still dominant over tendencies towards stateless, deterritorialized, intra-firm sourcing TNCs in both developed and developing countries. For most countries, food is still primarily sourced domestically and is processed and distributed by businesses operating independently of their parent companies. The main restructuring has been through the acquisition of existing companies by multinational food corporations operating in established markets, and mergers among large multinational food firms. Multinationals are buying their way into foreign food markets rather than through competitive trading, and firm consolidation remains the most crucial mechanism of structural change in the food industry.

Third, the analysis has identified evidence of alternative food networks that are growing out of a range of social movements. More research is needed on these alternative networks, including their ability to survive counter-movements, such as the incorporating tendencies of large TNCs, the political and economic influence of neo-conservative 'think-tanks' and industry trade organizations, and food disparagement legislation, as now exists in the US. The last of these counter-movements places legal obstacles in the way of criticism of food networks dominated by TNCs. Indeed a growing problem exists in democratic representation under globalization: economic rules are dominated by commercial rather than ethical or moral imperatives. TNCs can appeal to the WTO over national restrictions placed on their commercial activities, and the WTO is constituted to uphold liberal trading principles. These principles can over-ride valid concerns expressed by democratic institutions within nations, regions or social groups over such issues as: the incorporation of GMOs into imported foods without adequate labelling to identify such products; the long-distance trade in live animals that causes affront to animal welfare interests; and meat products containing hormonal and other medicinal residues from the use of drugs to promote weight gain in livestock. As Bonanno *et al.* (1994) points out: 'In the transitional phase, they (TNCs) have used the strategy of by-passing domestic state action in order to address the contradictions embodied in the relationship between accumulation and legitimation'.

Thus the dismantling of national regulatory policies removes food policy further from public scrutiny (Robinson 1997, 46) but brings 'greening' and 'environmentalism' closer (Buttel 1992). McMichael (1994, 283) observes that 'the issue is whether and to what extent states surrender economic sovereignty to supranational institutions – some say the surrender has begun and defines the current transition'. Or will new claims be made on national states for protection against market forces, with supranational regulation empowering locally based political movements? Regional blocs are already forming to help regulate currency movements and trade imbalances among member states at a multinational level. But new forms of regulation for the world food economy are still unclear in terms of what they will be and what directions they will take.

Further reading and references

There are several edited collections on the theme of globalization that have papers on food. The most helpful for the general reader are probably Bonanno *et al.* (1994), Goodman and Watts (1997), and McMichael (1994). Le Heron (1993) and Watts (1996) are also good value. Thompson and Cowan (2000) regret the neglect of certain broad regions, especially Asia, in this literature, which as a result cannot be called truly global in scope.

Arce, A. and Marsden, T. 1993: The social construction of international food: a new research agenda. *Economic Geography* **69**, 293–311.

Barrett, H., Ilbery, B., Browne, A. and Binns, T. 1999: Globalization and the changing networks of food supply: the importation of fresh horticultural produce from Kenya into the UK. *Transactions of the Institute of British Geographers* NS **24**, 159–74.

Bonanno, A., Busch, L., Friedland, W., Gouveia, L. and Mingione, E. (eds) 1994: *From Columbus to ConAgra: the globalization of agriculture and food.* Lawrence: University Press of Kansas, Lawrence.

Bowler, I. (ed.) 1992: *The geography of agriculture in developed market economies.* London: Longman.

Burch, D. and Pritchard, W. 1996: The uneasy transition to globalization: restructuring of the Australian tomato processing industry. In Burch, D., Rickson, R. and Lawrence, G. (eds) *Globalization and agri-food restructuring*. Aldershot: Avebury, 107–126.

Burch, D., Rickson, R. and Lawrence, G. 1996: Globalization and agri-food restructuring. Aldershot: Avebury.

Buttel, F. 1992: Environmentalization: origins, processes and implications for rural social change. *Rural Sociology* **57**, 1–28.

Buttel, F. 1996: Theoretical issues in global agri-food restructuring. In Burch, D., Rickson, R. and Lawrence, G. (eds) 1996: *Globalization and agri-food restructuring*. Aldershot: Avebury, 17–44.

Callon, M. 1986: Some elements of a sociology of translation. In Law, J. (ed.) *Power, action, belief: a new sociology of knowledge*? London: Routledge and Kegan Paul, 196–233.

Fanfani, R. 1994: Agro-food districts: a new dimension for policy making and the role of institutions. In Centre for Rural Research (ed.) *Restructuring the agro-food system: global processes and national responses*. CRR, University of Trondheim, 81–9.

Food and Agriculture Organisation 1998: *The state of food and agriculture*. Rome: FAO.

Friedland, W. 1984: Commodity systems analysis: an approach to the sociology of agriculture. In Schwarzweller, H. (ed.) *Research in rural sociology of agriculture*. Greenwich, CT: JAI Press, 221–35.

Friedland, W. 1994: The global fresh fruit and vegetable system: an industrial organization analysis. In McMichael, P. (ed.) *The global restructuring of agro-food systems*. Ithaca: Cornell University Press, 173–89.

Friedmann, H. 1994: Distance and durability: shaky foundations of the world food economy. In McMichael, P. (ed.) *The global restructuring of agro-food systems*. Ithaca: Cornell University Press, 258–76.

Goodman, D. 1994: World-scale processes and agro-food systems: critique and agenda. *Working Paper* 94–11, Centre for the Study of Social Transformations, University of California, Santa Cruz.

Goodman, D. and Watts, M. (eds) 1997: *Globalizing food: agrarian questions and global restructuring*. London: Routledge.

Hungerford, L. 1996: Australian sugar in the global economy: recent trends, emerging problems. In Burch, D., Rickson, R. and Lawrence, G. (eds): *Globalization and agri-food restructuring*. Aldershot: Avebury, 127–38.

Iacoponi, L., Brunori, G. and Rovai, M. 1995: Endogenous development and the agroindustrial district. In Ploeg, J. van der and Dijk, G. van (eds): *Beyond modernisation: the impact of endogenous rural development*. Assen: Van Gorcum, 28–69.

Jaffee, S. 1994: Exporting high-value food commodities: success stories from developing countries. *Discussion Paper* 198. Washington, DC: World Bank.

Latour, B. 1986: The powers of association. In Law, J. (ed.) *Power, action and belief: a new sociology of knowledge?* London: Routledge and Kegan Paul, 264–80.

Law, J. 1986: On the methods of long-distance control: vessels, navigation and the Portuguese route to India. *Sociological Review Monograph* **32**, 234–63.

Le Heron, R. 1993: *Globalized agriculture: political choice*. Oxford: Pergamon Press.

Little, P.D. and Watts, M. 1994: *Living under contract: contract farming and agrarian transformation in sub-Saharan Africa*. Madison: University of Wisconsin Press.

Marsden, T. and Arce, A. 1995: Constructing quality: emerging food networks in the rural transition. *Environment and Planning A* **27**, 1261–79.

Marsden, T., Murdoch, J., Lowe, P., Munton, R. and Flynn, A. 1993: *Constructing the countryside*. London: UCL Press.

McMichael, P. (ed.) 1994: *The global restructuring of agro-food systems*. Ithaca: Cornell University Press.

Miller, L. 1996: Contract farming under globally-oriented and locally-emergent agribusiness in Tasmania. In Burch, D., Rickson, R. and Lawrence, G. (eds) *Globalization and agri-food restructuring*. Aldershot: Avebury, 203–218.

Murdoch, J. and Marsden, T. 1995: The spatialization of politics: local and national actor-spaces in environmental conflict. *Transactions of the Institute of British Geographers* **20**, 368–80.

Pritchard, W. 1998: The emerging contours of the third food régime: evidence from Australian dairy and wheat sectors. *Economic Geography* **74**, 64–74.

Rickson, R. and Burch, D. 1996: Contract farming in organizational agriculture. In Burch, D., Rickson, R. and Lawrence, G. (eds) *Globalization and agri-food restructuring*. Aldershot: Avebury, 173–202.

Robinson, G. 1997: Greening and globalizing: agriculture in 'the new times'. In Ilbery, B., Chiotti, Q. and Rickard, T. (eds) *Agricultural restructuring and sustainability: a geographical perspective*. Wallingford: CAB International, 41–53.

Rogers, R.T. 2000: Structural change in US food manufacturing, 1958 to 1997. Paper presented to a conference organized by the Economic Research Service, United States Department of Agriculture on 'The American Consumer and the Changing Structure of the Food System', Washington, DC, May 4–5th.
Thrift, N. 1996: *Spatial formations*. London: Sage.
van der Ploeg, J. and Long, A. (eds) 1994: *Born from within: practice and perspectives of endogenous rural development*. Assen: Van Gorcum.
Thompson, S.J. and Cowan, J.J. 2000: Globalizing agro-food systems in Asia. *World Development* **28**, 401–407.
Wallace, I. 1985: Towards a geography of agribusiness. *Progress in Human Geography* **9**, 491–514.
Watts, M. 1996: Development–iii: the global agrofood system and late 20th-century development (or Kautsky redux). *Progress in Human Geography* **20**, 230–45.
Whatmore, S. and Thorne, L. 1997: Nourishing networks: alternative geographies of food. In Goodman, D. and Watts, M. (eds) *Globalizing food: agrarian questions and global restructuring*. London: Routledge, 287–304.

5

Transformation of the farm sector

Introduction

Thus far in our account of the political economy of food we have considered mainly the 'external' relations of the farm sector as regards non-farm capitals. In this chapter we turn in more detail to the 'internal' transformation of farming brought about by those relations during the second and third food régimes. To begin with we examine the theoretical perspectives that interpret the restructuring of agriculture; then we turn to empirical evidence on how farming has been changing over the last five decades in producing the raw materials of food.

Theorizing the transformation of farming

Contemporary political economy theorizations of agrarian transformation can be traced to the writings of Marx, Lenin, Chayanov and Kautsky (Newby 1987). Many reviews of this literature are available, for example Goodman and Watts (1994), Marsden *et al.* (1986a), Vandergeest (1988) and Watts (1996), and only a brief summary is produced here following the interpretation by Bowler (1992).

The term 'commoditization' can be used to describe the process of agrarian transformation from non-market (i.e. subsistence) to market forms of production. As farm businesses and farm households are drawn into a dependency on non-farm goods and inputs purchased in the market, they are compelled to produce agricultural commodities with an exchange value in order to obtain a cash income. Economically unsuccessful, usually smaller, farm businesses become marginalized through competition in commodity markets and owners of more successful, usually larger, businesses are able to purchase their land and so further enlarge their farm sizes. Theorists predict a polarizing trajectory in the farm-size structure towards fewer but larger farms, with medium-sized farms least able to resist the economic pressures of marginalization.

For neo-Marxists, the process of capitalist accumulation unleashed by the logic of commoditization also produces a polarized agrarian class structure: on the one hand lies a bourgeoisie (capitalist land-owning class), occupying or owning the large-scale, wage-labour farms; on the other hand can be found a proletariat comprising the marginalized and ultimately landless peasant class that supplies the wage-labour for capitalist farms. At any point in time there exists a continuum of 'transitional' states between pure 'peasant' farming at one extreme, as found within many developing countries, and large-scale, capitalist (wage-labour), corporately-owned farming at the other, as exists in regions favoured by climate, soil or location for farm production in developed countries. An analytically separate category is recognized by some writers, namely simple or petty commodity production (SCP); in other words, a stratum of medium and small-sized family farms based exclusively on family labour and with no pressures for the enlargement of the business (i.e. expanded reproduction) outside demographic and cultural factors (Friedmann 1986).

This narrow interpretation of SCP has been criticized by Goodman and Redclift (1985) as comprising an historically contingent phenomenon and not a theoretical concept. For example, SCP assumes no class relations within the farm business, whereas empirical research has shown that even within family labour farms there is a need to hire workers at early and late stages in the farm-family demographic cycle. Such hiring of workers on to small and medium-sized farms blurs the distinction between family-labour and capitalist farms, while recent research has shown that many large-scale, capital-intensive farm businesses can be owned and operated almost exclusively by multi-generational family labour (Marsden and Symes 1984). A debate also continues over evidence of the 'disappearing middle' of medium-sized farms and the development of a polarized (dualist) farm-size structure. Outside the USA, the resistance of medium-sized farm businesses to marginalization has been greater than assumed by theorists and this condition is exemplified in Table 5.1 for the member states of the EU. While marked national differences exist between countries in their farm-size structures, medium-sized farms are retaining their importance despite the increasing area of land controlled by large farm businesses.

Some neo-Marxists still explain the survival of SCP as 'arrested marginalization', but an increasing number of writers within the 'commoditization' school is prepared to recognize the persistence of family labour farms as a structural feature of agriculture under capitalism. Following Kautsky, a number of writers explain such persistence in terms of the 'self-exploitation' of farm families, that is their willingness to accept low economic returns on their labour and invested capital (Watts 1996). Other writers, such as Moran *et al.* (1993), emphasize the importance of farmer co-operatives in the defence of small and medium-sized farms. Table 5.2 is a

Table 5.1 Variations in farm-size structure in selected European countries (1993)

Country	Percentage holdings in each farm-size group (ha)				
	1–5	5–10	10–20	20–50	>50
Germany	31	15	18	23	13
Greece	75	15	7	2	0.4
Italy	77	11	6	4	2
Spain	57	16	11	8	7
UK	14	12	15	24	33
(EU 12)	(58)	(13)	(10)	(11)	(7)

Source: abstracted from agricultural statistics of the European Commission.

reminder of the significance of co-operatives in countries such as Denmark and the Netherlands, and in the production of certain farm products, for instance milk and cereals. By contrast, Kautsky, as early as 1899, emphasized the role played by off-farm work (other gainful activity – OGA) by the farmer and/or the farm family, including work on capitalist farms. OGA generates alternative sources of income and allows small and medium-size farms to persist, even if as part-time farms. The term 'pluriactivity' is now applied to farm households that survive by means of income from OGA or on-farm diversification into non-agricultural activities, such as tourism and horse-riding (Evans and Ilbery 1993). In this interpretation farm households are not viewed as passive receptors of market forces but active agents entering into a variety of relations with non-farm capitals in order to ensure their survival (reproduction). Non-commoditized social relations within the farm family also support the farm business, for example non-wage work by women and children and the inter-household exchange of labour. While pluriactivity can be interpreted as a form of proletarianization of the rural workforce, it enables ownership of the means of production, especially land, to remain in the hands of farm families. For Goodman *et al.* (1987), however, the structural separation between the farm sector and agri-industry is a function of the natural production process which confronts non-farm capitals and over which they can exercise only partial control.

Commoditization theory thus raises a theoretical concern on how control is exerted over the means of production. The previous two chapters have described the ways in which non-farm capitals have increasingly penetrated and subsumed the agricultural labour process, for example through production contracts and control of marketing chains. A distinction between the 'formal' and 'real' subsumption of labour by capital has been theorized to explain the process. With the labour process acting as the main mechanism for creating surplus value, non-farm capitals can seek the 'real' subsumption of the process by direct ownership of the means of production, especially land, through vertical integration within the agro-food

Table 5.2 Agricultural marketing through farm co-operatives: selected products and West European countries (%)

Product	Denmark	France	Netherlands	UK
Pigmeat	98	78	24	17
Eggs	60	25	16	18
Milk	91	50	84	4
Cereals	47	75	65	19
All fruit	90	45	77	21

Source: abstracted from agricultural statistics of the European Commission.

system. For example, large agribusinesses can purchase farmland so as to control the whole labour process from agri-inputs, through farm production to processing and even retailing. This agri-industrial form of organization has been a feature of economies under state socialism and corporately-owned farming in regions such as the south eastern and south western states of the USA (Gregor 1982).

In general, however, capitalist agriculture has not followed this form of agrarian development and consequently the concept of 'formal' subsumption of the labour process has been developed. In this conceptualization, non-farm capitals extract surplus value from the farm sector indirectly, for instance through interest paid on farm finance (e.g. mortgages and credit facilities) and obtaining raw materials for food processing by production contracts with individual farms rather than through the open market. Table 5.3, for example, shows the varying significance of production under contracts in Western Europe: sugar beet and peas, together with France and the Netherlands, have the highest incidence of this type of formal subsumption. As Marsden *et al.* (1986a, 512) write, 'The legal ownership of the farm business and land remains (with the farm family), whilst they become increasingly separated from effective control, as management depends on external technical and economic factors governed by monopoly industrial and finance capitals'. From this perspective, only relatively unprofitable, residual activities are left in the farm sector, for example those agricultural activities in which 'production time' exceeds 'labour time': where labour time is less than production time for a particular crop or livestock, the average rate of profit is reduced and external capitals find direct investment inefficient (Mann and Dickinson 1978). However, where the difference between production and labour time can be closed through the industrialization of agriculture, for instance in pig, poultry and beef feed-lot production, real subsumption can be more fully developed.

Until recently, commoditization theory of agrarian transformation has had relatively little to say about the roles of international non-farm capitals or state intervention. However, looking first at international non-farm

Table 5.3 Agricultural production under contract in selected West European countries (%)

Product	Belgium	France	Netherlands	UK
Pigmeat	55	30–32	35	70
Poultrymeat	90	45–50	90	95
Eggs	70	15–20	50	70
Milk	–	1	90	98
Sugar beet	100	100	100	100
Peas	98	90	85	100

Source: abstracted from agricultural statistics of the European Commission.

capitals, as the previous chapters have shown, monopoly capital organized within national boundaries has increasingly yielded to global capital, with agribusinesses becoming both vertically and horizontally integrated and having the capacity to seek out the lowest-cost agricultural raw materials on the international market. The 'new political economy of agriculture' is attempting to incorporate this dimension by extending the concepts of real and formal subsumption from a national to an international scale of operation (Buttel and McMichael 1990). In an alternative conceptualization, the terms 'appropriationism' and 'substitutionism' have been applied to the development of new agri-technologies by corporate, non-farm capitals (Goodman *et al.* 1987). Appropriationism describes the partial but persistent transformation into industrial activities of certain farm-based labour processes, and their subsequent reintroduction into the food system in the form of purchased farm inputs and processed food outputs. The appropriations are partial in the sense that not all of the production processes on farms have been affected. The replacement of farm animals by corporately produced tractors and the processing of milk into cheese and butter in factories, rather than on the farm, provide two examples of appropriationism. Facets of the Green Revolution in many developing countries can also be interpreted in terms of appropriationism, for instance the replacement of seeds native to individual regions by commercially grown 'miracle' strains of hybrid rice and wheat produced under the control of non-farm capitals based in developed countries.

Substitutionism, on the other hand, occurs when non-farm capitals replace agricultural *outputs* by chemical and synthetic raw materials and so eliminate the rural base of agriculture (Goodman *et al.* 1987, 4). This process is most advanced in the substitution of natural fibres, such as cotton and wool, by synthetic fibres, such as rayon and nylon. Another recent example is the replacement of natural cork by synthetic bottle stoppers in the wine sector. With processed or manufactured foods also increasingly fabricated from reconstituted generic food components (e.g. starch, glucose and vegetable protein) and chemical additives, the traditional constraints of

Table 5.4 Substitutionism in the potato commodity chain

Labour processes			
Agricultural labour	Agricultural labour	Scientific labour	Scientific labour
Potato	*Potato*	*Modified potato seed*	*Modified potato seed*
Household labour	Industrial labour	Agricultural labour	Agricultural labour
Cooked potato	*Potato crisp*	*Engineered potato*	*Engineered potato*
		Industrial labour	Scientific labour
		Processed potato	*Potato starch feedstock for protein culture*
			Industrial labour
			Synthetic 'meat' products

Source: adapted from Whatmore (1995, 43).

land, space and biological reproduction within the agricultural production process, including the social division of labour, are being removed (Goodman *et al.* 1987, 58). Table 5.4 shows the application of these tendencies within the potato commodity chain: the potato is increasingly transformed from a raw food, through a range of processed food products, to a generic raw material (starch). Nevertheless, the 'uniqueness' of agriculture in terms of its natural production process – nature as the biological conservation of energy, as biological time in plant growth and animal gestation, and as space in land-based rural activities – places biological limits on the ability of non-farm capitals to impose a unified transformation (vertical disintegration) on agriculture.

Turning to intervention by the state, there are three implications for agrarian transformation. First, the agricultural policies of developed countries have provided protection for their farm sectors from international competition, while price subsidies on farm produce have assisted in the persistence of small and medium-sized farms. Farm subsidies in general have not prevented the working of capitalist accumulation and concentration, but the speed and scale of the processes have been reduced. A more liberal market system, with a farm sector more open to international competition, would undoubtedly have accelerated the demise of small and medium-sized farm businesses. Individual farm subsidies, however, have also supported the transformation of agriculture, for example farmer retirement pensions, land and farm consolidation grants, farm modernization grants, subsidies on fertilizers, land drainage and land reclamation grants, and subsidies on the cost of borrowing capital. Second, the state has maintained an economic and political environment favourable to competition between, and mergers/take-overs amongst, non-farm capitals. These twin processes have ensured both the emergence of agro-food businesses with a global reach and the continuous development and application of new technologies in the agro-food sector. Third, the state has funded the research and development of new farm and food technologies and assisted in their diffusion throughout the farm sector, for example through grants and subsidies to farm businesses, the provision of agricultural education, the maintenance of extension services and the funding of experimental and demonstration farms. In sum, the state has been, and remains, deeply implicated in the technological transformation of agriculture, including the relationship between the farm sector and non-farm capitals, and this theme is taken up in Chapter 18.

Attempts to apply these theoretical propositions at the farm level have revealed the heterogeneity, or absence of a unilinear pattern of development, within agriculture. One result has been a variety of typologies of farm businesses, each recognizing diversity in the relationship between farms and non-farm capitals as regards *technology* (e.g. manufactured inputs, advice, technical assistance), *finance* (e.g. mortgage and credit facilities) and *marketing* (e.g. linking production to processing and market outlets). Marsden *et al.* (1986b), for example, recognize a fundamental difference between 'survivalist' and 'accumulationist' farm businesses. The former category includes (I) 'marginal closed farms' (i.e. sub-marginal farms reliant on pensions, savings and insurance schemes for their survival, including holdings occupied by hobby and retired farmers). Occupiers of these holdings keep financial debt to a minimum, with little investment in new farm technology. Survivalist behaviour is also evident in a category of farms termed (II) 'transitional dependent farms': the category includes pluriactive farm households with diversified activities on the farm and OGA off the farm. These households often have a debt relationship with finance capitals, such as banks, for the development of their diversified businesses. In contrast, accumulationist behaviour characterizes (IV) 'subsumed farm units': this category includes corporately-owned businesses that are heavily indebted, apply the most advanced agri-technologies available, and are commonly contracted to produce for food processors and retailers. Capitalist farms in farm family ownership exhibit many of the same characteristics. Full-time, family labour farm businesses are more difficult to categorize. Some comprise (III) an 'integrated farm' category, incurring debt so as to purchase new technologies, perhaps to enter into production contracts with food processors, and so place themselves on an accumulationist trajectory of development. Others minimize their debt and maximize their efficiency within traditional farming technologies and thereby exhibit the survivalist behaviour of 'transitional dependent farms'.

Estimating the relative significance of typological groups and their trajectories within a national agriculture is difficult, not least because suitable agricultural census data are not collected. However, using random samples of farms within small study regions, Whatmore *et al.* (1987) have made estimates for the UK (Table 5.5). Farms have been categorized on their scores of formal subsumption by external capitals (e.g. debt level, importance of production under contract) and real subsumption as measured by the diffusion of control away from a single person within the farm business (e.g. ownership of farm capital, type of land ownership, control over business management and labour relations). Table 5.5 shows the relative importance of typological groups for three study areas within the UK (west Dorset, east Bedfordshire and metropolitan London). Setting aside the problem of assuming a linear relationship between the scores of real and formal subsumption, the analysis shows a significant level of subsumption on only 20 to 28 per cent of the farms surveyed. Rather 'transitional dependent' farms are most significant, that is traditional family labour farms, especially in the west Dorset study area. When measured for the period 1970 to 1985, 64 per cent of the farms in the survey experienced no change in the degree of subsumption. However, with 28 per cent of farms showing increasing levels of subsumption, evidence is available to support the contention of an increasing polarity in the trajectory of individual farm business development.

From a commoditization perspective, therefore, the uneven spatial and temporal transformation of agriculture can be interpreted as a function of regional variation in the relationship between agriculture and non-farm capitals. The latter penetrate those farming systems where the greatest

Table 5.5 A typology of the internal and external relations of the farm business as applied to samples of farms within England* (percentage in each category)

		Internal relations (real subsumption)				
External relations (formal subsumption)	*Score*	4–9	10–16	17–25	26–32	Total
	0–3	16 [I]	12	1	0	29
	4–6	12	26[II]	4	1	43
	7–9	2	10	8[III]	1	21
	10–12	0	2	3	2[IV]	7
	Total	30	50	16	4	100

I: Marginal closed farms; II: Transitional dependent farms; III: Integrated farms; IV: Subsumed farms.* See text for explanation.
Source: adapted from Whatmore *et al.* (1987).

financial returns can be obtained, as mediated by farm type and farm size. As limitations are placed on the maximization of surplus value in a 'crisis of accumulation', for example between food régimes, so external capitals switch their investments between regions to produce phases or waves of capitalization (Marsden *et al.* 1987). Some writers have reversed this argument by claiming that differential degrees of commoditization result from the resistances to transformation (survival strategies) devised by specific categories of family-farm households. In this view, local and regional responses are reduced to cultural and historical diversity, enabling other writers to claim them as 'historically superfluous'. In general though, both 'land' (its productive capacity) and 'space' are viewed as constraints/opportunities in the process of commoditization, helping to explain both the differential transformation and durability of the family farm under capitalism.

Empirical observations on the transformation of farming

The broad transformation of the farm sector has been towards farming systems having characteristics normally associated with manufacturing industry, namely the creation of scale economies at the farm level (larger farms), a reliance on agri-inputs manufactured in other sectors of the economy (e.g. farm plant and machinery, agri-chemicals, fertilizers), resource substitution (capital for labour), specialization of labour, and mechanization of production methods. Significantly, this 'industrial model' of agricultural

Table 5.6 Yields of selected crops in the European Union, 1980–86 (100 kilogram/hectare)

Crop	1980	1986
Common wheat	41.9	50.2
Durum wheat	23.0	26.1
Rye	28.7	29.1
Barley	37.3	38.5
Oats	29.0	29.1
Maize	54.6	64.6
Sugar	66.0	71.1
Rapeseed	27.3	29.1
Soya beans	20.3	32.2

Source: abstracted from agricultural statistics of the European Commission.

development can be found in capitalist and (former) socialist farming systems alike, as well as more selectively on large farm units within developing countries. However, the industrialization of farming is best interpreted as an ongoing process rather than an achieved structure. In other words 'industrialization' describes a particular trajectory of development, with modern pig, poultry, beef feed-lot and greenhouse horticultural production probably representing the most industrialized sectors of agriculture. This trajectory of farm development can be usefully summarized by the following five terms: intensification, concentration, specialization, diversification and extensification (Bowler 1992, 13–16; Ilbery and Bowler 1998, 70–71).

Intensification in the farm sector

The term 'intensification' can be used to describe the rising level of purchased agri-inputs (e.g. farm machinery, agri-chemicals, fertilizers) and the related increases in output per hectare of farmland. These trends are readily observable in the rising per hectare application of fertilizers and the rising average yields of most crops and livestock (Table 5.6). One secondary result of this 'capitalization' of agriculture has been the displacement of labour from agriculture by machinery and other purchased inputs – initially the hired workers and more recently the owner-occupiers (farm families) –

Table 5.7 Increasing specialization within the farm sector of the Irish Republic, 1960–1980

		Percentage holdings with tillage or livestock type	**Coefficient of localization by rural district**	**Percentage in districts with location quotients >1.5†**
Tillage	1960	82	0.25	28
	1980	52	0.39	54
Cows	1960	80	0.17	18
	1980	63	0.17	34
Cattle	1960	88	0.09	0
	1980	79	0.09	0
Sheep	1960	30	0.34	44
	1980	17	0.43	62
Pigs	1960	38	0.27	44
	1980	5	0.43	46
Poultry	1960	78	0.17	13
	1980	35	0.55	54

Notes: * Larger values = greater localization; †>1.5 = greater localization.
Source: adapted from Gillmor (1987).

leading to rural unemployment and depopulation. Farmers cannot avoid the 'treadmill' created by the technology that drives intensification: when a new farm technology is made available, those farmers who adopt it first (innovators) gain an advantage in the market (lower unit production costs or higher output per unit input); other farmers must subsequently adopt the technology if they are not to become uncompetitive in the market; the increased output by all producers then has the effect of over-supplying the market, reducing product prices, and thereby stimulating innovative farmers to seek out and apply a further round of new, cost-reducing or output-increasing technology.

Concentration in the farm sector

The competitive market process that drives the least economically successful farm businesses from agriculture also enables their land to be purchased by the remaining, more successful businesses. As already explained, those with large land holdings tend to be able to out-compete smaller farms when bidding for land in the market; farmland becomes concentrated into fewer but larger farm businesses and the number of small and medium-sized farms declines. Concentration can be measured by the proportion of total productive resources (land, labour, capital) or output (crops, livestock) located in farms of various sizes, especially larger farms. Another factor behind concentration is the search for economies of scale by individual farmers. On larger farms, the fixed costs of production, for example in land, buildings, machinery and labour, can be spread over greater volumes of production, so reducing production costs per litre of milk, per dozen eggs, or per tonne of wheat.

Specialization in the farm sector

Another way of gaining economies of scale is to limit production to fewer products on the farm and so concentrate the costs of production on a narrow range of items. In addition, in an increasingly sophisticated technological environment, there are management advantages in learning and applying specialist skills and knowledge. This process of specialization on a narrow range of crops and livestock, when aggregated for groups of farms, produces the increasingly specialized agricultural regions observable throughout the world. Specialization can be seen in the functions of the labour force, the types of farm equipment employed, and in the resulting tendency towards monocultural land use and livestock systems. However, the processes of enterprise and regional specialization are uneven as between types of farm production, as illustrated in Table 5.7 for the Irish Republic between 1960 and 1980. The processes of specialization are more

Table 5.8 Types of diversification in farming

Agricultural diversification: new crops (for example linseed, teasels, evening primrose, borage and fennel); new livestock (for example, goats, deer, buffalo, rabbits and ostriches); and value-added activities on the farm (for example, farm shop, pick-your-own, delivery round, cheese making, ice-cream making, yoghurt making, potato packing).
Non-agricultural diversification: farm tourism (for example, bed-and-breakfast, accommodation, holiday cottages); farm recreation (for example, camping, sport shooting and fishing, horse riding, farm visits, farmhouse refreshments); alternative resource use (for example, farm woodland, farm building conversion for industry or rural craft centre, fish farming, environmental conservation).

Source: adapted from Slee (1989, 62).

evident in pig and sheep production, for example, compared with (beef) cattle and (milk) cows.

Diversification in the farm sector

Empirical evidence of a turnaround in intensification/concentration/specialization tendencies within farming under the third food régime is fragmentary and uneven. Only writers in western Europe have reported sustained signs of post-productivism within the farm sector in terms of diversification rather than specialization and extensification rather than intensification (Ilbery and Bowler 1998). The term 'diversification' is normally used to describe the introduction of a non-traditional enterprise into a farm business. Long lists of such enterprises have been produced, but they can be usefully categorized into agricultural and non-agricultural groups (Table 5.8).

The long-term economic viability of many of these enterprises is uncertain, with relatively few farmers involved in their production for 'niche' markets. Where a form of diversification becomes successful, for example deer farming or farm accommodation, the increased numbers of farmers attracted into production can rapidly over-supply the market so that the same problem of 'surplus' production can emerge as in agricultural commodities. In these circumstances, only the economically successful survive; but for those that do, the diversifying enterprise can provide a significant contribution to the total profits of the farm business.

Extensification in the farm sector

Turning now to 'extensification' within the farm sector, this implies lower inputs to and outputs from the farm business. Two processes appear to be at work. Firstly, the state is attempting to solve the problem of over-production by providing financial stimuli and regulations to reduce farm inputs. For example, set-aside of cropland has been introduced into the USA and the EU as one policy mechanism for reducing the amount of arable land in production. In addition in the EU, stocking density limits (quotas) have been placed on the number of livestock (sheep and beef cattle) eligible to receive subsidy on each farm, and designated groups of farmers are being compensated financially for reducing the amount of nitrate fertilizer they apply to their land or for continuing to farm in traditional (low input) ways. Critics of these policy mechanisms claim, with some justification, that the measures tend to prevent further intensification rather than induce extensification in agriculture.

A second process, however, reflects the decision by certain farmers, and groups of farmers, to produce crops and livestock by sustainable, post-productivist, 'ecological' farming methods. A range of 'alternative' farming systems is being developed, including low input–output farming, permaculture, alternative agriculture, organic farming, regenerative farming and biodynamic farming. The differences between these and 'conventional' agricultures are outlined in Table 5.9 under three bi-polar dimensions: centralization–decentralization, individualism–community, and specialization–holism. At a higher level of abstraction, a choice is available between the presently dominant scientific paradigm of conventional agriculture and a radical reconception of science (Beus and Dunlap 1990).

Looking just at organic farming, this system uses fewer purchased inputs compared with conventional farming, especially agri-chemicals and fertiliz-

Table 5.9 Alternative agricultural paradigms

Conventional agriculture	Alternative agriculture
Centralization	*Decentralization*
Fewer farms	More farms
Concentrated resources	Dispersed resources
National/international marketing	Local/regional marketing
Individualism	*Community*
Self-interest	Co-operation
Reduced labour	Retained labour
Farming as a business	Farming as a way of life
External costs ignored	All costs considered
Material success	Non-material values
Specialization	*Holism*
Farming reduced to individual components	Farming as a system
Standardized production	Diversification

Source: adapted from Beus and Dunlap (1990, 598–99).

ers, and consequently produces less food per hectare of farmland. For early entrants to organic farming (the innovators), the decision was as much based on environmental and food ethics as economics. Indeed a similar concern with ethics was required by consumers who had to pay a considerably increased price for their organic food to support the method of production. More recent entrants to organic farming, however, have been attracted by the financial profits to be made as retailers have responded to the growing consumer demand for 'healthy' food. Indeed organic producers have tended to polarize: on the one hand are those who remain small in scale, retain the original ecological philosophy of the first organic farmers about the simplicity of food production, distribution and consumption, and who supply largely local markets through small retail outlets, farm shops or vegebox deliveries direct to consumers. On the other hand are the large, commercial producers, supplying produce in volume to distant consumers through large retail outlets or wholesale markets, and even agribusinesses (Campbell 1996). These producers are motivated mainly by commercial considerations and may not even have all of their farmland under organic production. Locationally, the small producers tend to cluster either around their markets in the large urban conurbations or on marginal land where entry costs to organic production are relatively low. In contrast, the larger producers tend to locate in farming regions traditionally associated with the production of the crop or livestock concerned, where organic production opens up an alternative market compared with conventional production. This mixture of locational processes is evident on Figure 5.1, which shows the distribution of certified organic growers in New Zealand. Canterbury and Bay of Plenty form two areas of concentrated development in response to the business activities of Watties Frozen Foods Ltd. (a New Zealand-based subsidiary of H.J. Heinz & Co.) and the Kiwifruit Marketing Authority.

Evidence to date suggests that organic farming is not without its own environmental problems, for example nutrient run-off, but the problems are less than in conventional agriculture. Also organic farming, because it is more labour intensive, supports more jobs per hectare of farmland and thereby contributes to the social stability of the farm population and rural society. Evidence on the economic sustainability of organic farming is more mixed: comparisons of organic with conventional farms do not reveal systematic differences in economic returns. Organic farming produces lower outputs per hectare but is compensated by higher output prices. Much depends on the relative efficiencies and sizes of the individual farms being compared and the market conditions for their respective produce at the time of the survey. In general, organic farms are capable of producing economic returns at least equal to those of conventional farms. A main economic problem lies in generating farm income during the transition from high input–output to low input–output farming. While state assistance is emerging for the transitional period, including certification schemes to

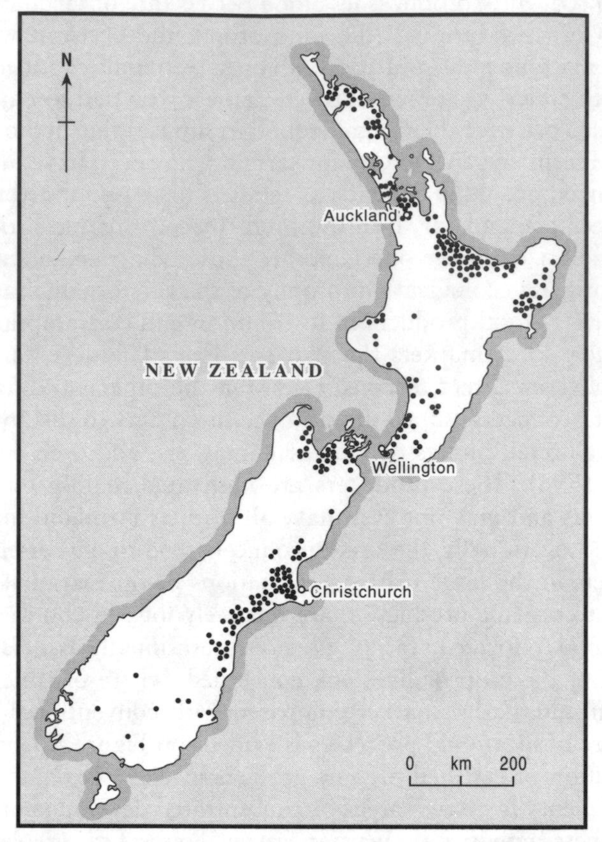

Fig. 5.1 The distribution of transitional and fully-certified organic growers in New Zealand. *Source*: after Campbell (1996, 156).

validate the 'organic' status of farm produce, organic farming remains a small component of both agriculture and food consumption even in Europe.

Organic agriculture, however, is only one option: a range of farming principles and practices compatible with a sustainable 'ecological' agriculture are widely available for implementation, including diversified land uses, minimum soil cultivation, the integration of crop and livestock farming, crop rotations, nutrient recycling, low energy inputs, low inputs of agri-chemicals, and biological pest and disease control. A strong case can be made, for example, for the greater development of diversified land-use systems on the grounds of reduced ecological and economic risk, improved sustainability in the medium term, and the higher efficiency with which natural resources are used. Organizational procedures, such as integrated farming systems (IFS) – integrated crop management (ICM), integrated livestock management (ILM) and integrated pest control (IPC) – incorporate many of these principles and practices. Such systems lie midway on the continuum between the stringent limitations imposed by organic farming, on the one hand, and the damaging environmental practices of conventional farming on the other hand. ICM, for example, involves the systematic use of crop rotations, the carefully targeted use of agri-chemicals and fertilizers, the application of biological controls on pests, and the management of field margins to create habitats for the predators of crop pests, for example beetles. Two problems with IFS mean that at present implementation is fragmentary. One problem lies in its conflict with the conventional wisdom of the industrial model of development, for instance intensification, monocultural specialization and the employment of purchased agri-inputs. A second problem is the level of information, knowledge and management skills needed to implement the more complex and risky farming practices of IFS. While 'networks' of agencies providing the necessary information exist in most countries, the problem remains of recruiting a sufficiently large number of farmers to participate in acquiring new knowledge and skills, some would argue the rediscovery of lost knowledge and skills, so as to be able to implement and diffuse IFS. The state is implicated in the development of ecological farming through the international commitment given to 'sustainable development' and the development of agri-environmental policies.

Conclusion

The industrial model of agriculture is now a global phenomenon, although with varying degrees of development within most counties and farming systems. However, the industrializing process in the farm sector has differential spatial and structural (i.e. farm-size) effects and three broad 'paths of farm business development' (Bowler *et al.* 1996) can be observed: (1) large-scale, specialized farm units following the industrial model of farm development and drawn into globally networked, urban-centred food complexes (Marsden *et al.* 1996); (2) small-scale, pluriactive family farms (i.e. diversified or OGA), dependent on local or regional markets; and (3) medium-sized, traditional farms increasingly under pressure to join one of the first two groups. The result is a rich heterogeneity of farming trends within agriculture leading to multiple rural spaces within national economies. In Western Europe, for example, a distinction is emerging between areas with dynamic, intensive farming systems (e.g. Paris basin, East Anglia, Emilia Romagna, southern Netherlands), which produce the majority of farm output, and those areas suffering increasing marginalization (e.g. mountain regions), from which a minority of farm production emanates.

Evidence of a further transformation of the farm sector towards sustainable, post-productivist, 'ecological' farming systems is more fragmentary, although most widely documented within European agriculture. At present the level of consumer support for the food produced by these farming systems is insufficient to promote their widespread development. Similarly, state support for sustainable farming systems is equivocal, with more financial support still being directed towards farms on the industrial trajectory of development, and with agri-environmental grants and subsidies 'bolted-on' to conventional agricultural policies (Robinson 1991). Nevertheless, a transformation towards more sustainable farming systems appears to be a likely characteristic of the emerging third food régime.

Further reading and references

Those wishing to pursue this topic further have a rich and varied literature to call on. Bowler (1992) is one possible starting point, and major authors include David Goodman, Terry Marsden and Michael Watts.

Beus, C. and Dunlap, R. 1990: Conventional versus alternative agriculture: the paradigmatic roots of the debate. *Rural Sociology* 55, 590–616.

Bowler, I. 1992: *The geography of agriculture in developed market economies.* London: Longman.

Bowler, I., Clark, G., Crockett, A., Ilbery, B. and Shaw, A. 1996: The development of alternative farm enterprises: a study a family labour farms in the northern Pennines of England. *Journal of Rural Studies* 12, 285–95.

Buttel, F. and McMichael, P. 1990: New directions in the political economy of agriculture. *Sociological Perspectives* 3, 89–101.

Campbell, H. 1996: Organic agriculture in New Zealand: corporate greening, transnational corporations and sustainable agriculture. In Burch, D., Rickson, R. and Lawrence, G. (eds): *Globalization and agri-food restructuring.* Aldershot: Avebury, 153–69.

Evans, N. and Ilbery, B. 1993: The pluriactivity, part-time farming and farm diversification debate. *Environment and Planning A* 25, 945–59.

Friedmann, H. 1986: Family enterprises in agriculture: structural limits and political possibilities. In Cox, G., Lowe, P. and Winter, M. (eds) *Agriculture: people and policies.* London: Allen and Unwin, 41–60.

Gillmor, D. 1987: Concentration of enterprises and spatial change in the agriculture of the Republic of Ireland. *Transactions of the Institute of British Geographers* 12, 204–16.

Goodman, D. and Redclift, M. 1985: Capitalism, petty commodity production and the farm enterprise. *Sociologia Ruralis* 25, 231–47.

Goodman, D. and Watts, M. 1994: Reconfiguring the rural or Fording the divide?: capitalist restructuring and the global agri-food system. *Journal of Peasant Studies* 22, 1–49.

Goodman, D., Sorj, B. and Wilkinson, J. 1987: *From farming to biotechnology.* Oxford: Blackwell.

Gregor, H. 1982: Industrialization of US agriculture: an interpretative atlas. Boulder, CO: Westview.

Ilbery, B. and Bowler, I. 1998: From agricultural productivism to post-productivism. In Ilbery B (ed.) *The geography of rural change.* London: Longman, 57–84.

Mann, S. and Dickinson, J. 1978: Obstacles to the development of a capitalist agriculture. *Journal of Peasant Studies* 5, 466–81.

Marsden, T. and Symes, D. 1984: Land ownership and farm organisation: evolution and change in capitalist agriculture. *International Journal of Urban and Regional Research* 8, 388–401.

Marsden, T., Munton, R., Whatmore, S. and Little, J. 1986a: Towards a political economy of capitalist agriculture: a British perspective. *International Journal of Urban and Regional Research* 10, 489–521.

Marsden, T., Whatmore, S., Munton, R. and Little, J. 1986b: The restructuring process and economic centrality in capitalist agriculture. *Journal of Rural Studies* 2, 271–80.

Marsden, T., Whatmore, S. and Munton, R. 1987: Uneven development and the restructuring process in British agriculture: a preliminary exploration. *Journal of Rural Studies* 3, 297–308.

Marsden, T., Munton, R., Ward, N. and Whatmore, S. 1996: Agricultural geography and the political economy approach: a review. *Economic Geography* 72, 361–75.

Moran, W., Blunden, G. and Greenwood, J. 1993: The role of family farming in agrarian change. *Progress in Human Geography* 17, 22–42.

Newby, H. 1987: Emergent issues in theories of agrarian development. In Thorniley, D. (ed.) *Economics and sociology of rural communities*. Norwich: GeoBooks, 7–22.

Robinson, G. 1991: E.C. agricultural policy and the environment: land use implications in the U.K. *Land Use Policy* 8, 95–107.

Slee, B. 1989: *Alternative farm enterprises*. 2nd edn. London: Farming Press.

Vandergeest, P. 1988: Commercialization and commoditization: a dialogue between perspectives. *Sociologia Ruralis* 28, 7–29.

Watts, M. 1996: Development – iii: the global agrofood system and late twentieth-century development (or Kautsky *redux*). *Progress in Human Geography* 20, 230–45.

Whatmore, S. 1995: From farming to agribusiness. In Johnston, R., Taylor, P. and Watts, M. (eds) *Geographies of global change*. Oxford: Blackwell, 36–49.

Whatmore, S., Munton, R., Little, J. and Marsden, T. 1987: Towards a typology of farm businesses in contemporary British agriculture. *Sociologia Ruralis* 27, 21–37.

6

Food processing and manufacturing

Introduction

In this chapter we explore in more empirical detail the food processing and manufacturing industry that lies 'downstream' from the farm sector in the food supply system. Whereas many writers use the terms interchangeably, here we use 'food processing' to imply the manipulation of agricultural raw materials into food products which retain many of the characteristics of the original materials. The freezing and canning of vegetables, the slaughter, evisceration, deboning and packaging of poultry, and the pasteurization and bottling of milk are illustrative of food processing. By 'food manufacturing' we imply the transformation of agricultural raw materials into food products that have lost many of the characteristics of the original materials. The production of bread, cakes and biscuits from flour, of meat pies from pork, beef and poultrymeat, and of butter, cheese and yoghurt from milk provide examples of food manufacturing. The term 'agribusiness' is also employed by some writers to describe the totality of food processing and manufacturing, but here we use the term 'food industry' (Burns *et al.* 1983).

Industrial capital and the manipulation of food

The companies comprising the food industry are increasingly severing the traditional links between agricultural raw materials and food products through a process termed 'substitutionism'. Summarizing the argument of Goodman *et al.* (1987, 58), and as introduced in Chapter 5, substitutionism involves the progressive reduction of agricultural products to simple industrial inputs; these inputs include proteins, carbohydrates and fats derived from either food or non-food vegetable matter by biotechnologies. Such technologies are enabling the agricultural production process to be eliminated either by utilizing non-agricultural raw materials or developing industrial substitutes for food. The resulting food products can be based on generic food components and an increasing technological control of food production, for instance through chemical additives. Food producers are diversifying their input sources to achieve greater interchangeability, while trends have developed in 'product fractionating' and the production of 'fabricated foods' (Goodman 1991). These methods, which allow agricultural products to be broken down into generic intermediate food ingredients, are resulting in reconstituted, or manufactured, foods with a longer shelf-life or convenience in preparation.

Mass production in flour milling, sugar refining and oilseed pressing were early influences on substitutionism by providing standardized, homogeneous inputs that included flour, edible vegetable oils, animal fats, sugar and powdered milk. Margarine was one of the first recognizably manufactured foods to be reconstituted from cheaper intermediate ingredients, in this case as a substitute for butter. Another example has been the production of low calorie sweeteners from corn starch (high fructose corn syrups) by biocatalysis. Food manufacturers further 'downstream' have been given increased scope to differentiate the form, composition and packaging of their agricultural products, while many new foods have a physical form and appearance that disguises their industrial origins and allows them to compete with 'natural' foods.

With 'nature' in agriculture – as biological time, production space and land – placing limits on the appropriation of the agricultural production process (*see* Chapter 5), food processing and manufacturing capital has turned its attention to increasingly sophisticated forms of substitutionism. Mechanical skills and scientific knowledge have been applied to food production, so that food itself has become more heterogeneous with specific properties created by processing techniques, product differentiation and merchandising. Innovations in the preservation of food have also been significant, for example canning, freezing, freeze-drying, chilling and dehydration. New products have been created, including a wide range of ready-to-eat meals and 'fast' foods, with additional attributes such as ease of handling and storage and longer shelf-life. The outcome has enabled processors and manufacturers to capture a greater proportion of the 'value added' in food products at the expense of the farm sector.

However, the level of substitutionism within the food industry is uneven, and in practice a range of food products can be identified, varying in their 'natural' as compared with 'industrial' composition. At the 'natural' end of the range can be found processed foods such as frozen vegetables, including peas, beans, carrots and broccoli, graded and pre-packed beverages such as tea and coffee, and animal meats that have been butchered, packaged and frozen for consumer convenience. At the 'industrial' end of the range lie products such as reformed meats, for example chicken nuggets, meat substitute products based on soya, canned 'fruit' drinks with their high content of artificially-introduced chemicals, and soft-form ice-creams.

Between these extremes lie the majority of processed and manufactured foods, such as 'ready to eat' chilled/frozen meals, milk, egg and potato powders, pastas and pizzas, each with their varying content of food preservatives and stabilizers, colour additives, flavour enhancers, and supplementary water, vegetable oils, animal fats, starch and sugar.

Substitutionism, however, does not take place just within an industrial plant; food companies can also ensure that it takes place on the farm, and a case study of this process has been provided by Baldwin (1999) for hyperimmune milk. Summarizing her study, hyperimmune milk is produced by dairy cows that have been immunized with a sterile vaccine for human pathogens, which induces the formation of specific antibodies in the milk. The milk can then be processed into a milk powder, as produced by Stolle Milk Biologies International Incorporated based in Ohio (USA) but operating in New Zealand in partnership with the New Zealand Dairy Board. Health benefits for humans are claimed for the product, for example in respect to infectious bacteria, which reside in the gastrointestinal tract of rheumatoid arthritis sufferers. However, the company is careful to advertise the milk powder as a food rather than a medicine, so as to avoid having to meet more stringent medical regulations. The passive immune protection offered to humans who consume the milk on a regular basis raises issues of food safety, the meaning of 'natural' foods, the burden of scientific proof, and the blurring of the boundary between the nutritional and medical content of food.

The varied national development of food processing and manufacturing

Attention has been drawn already in Chapter 4 to the contemporary global restructuring of the food industry. Indeed that discussion identified a fourfold development sequence, namely from (1) individual national food companies, through (2) integrated national food companies, to (3) integrated multinational food corporations, and lastly to (4) integrated transnational food corporations. This four-stage sequence of development can be elaborated in the following way, with developing countries now more prevalent in the first two stages (Athukorala and Sen 1998) and developed countries more prevalent in the last two stages:

1 Food processing and manufacturing based on national capital (e.g. flour milling, oil pressing, dairy products and cheese) – usually relatively small-scale business activity with low value-added and without significant international industrial or marketing linkages.
2 Fuller development of national food processing and manufacturing, commonly including some international capital (e.g. food preserves, fruit

juice, canned fruit and vegetables) – stronger links are created with non-food industries for inputs such as chemicals, glass, aluminium and paper, and marketing services to create channels for the flow of goods and a demand amongst consumers.

3 International agri-industry involved in large-scale processing and manufacturing of standardized but ever more sophisticated products – greater use of finance capital for growth and technology for efficiency and innovation.
4 Food products processed and manufactured with the most advanced machinery to supply high-income economies internationally – complex food system linkages guarantee quality and reduce the costs of inputs to a minimum; farmers are contracted to produce under conditions that are highly prescriptive, with strong links downstream to supermarket chains.

Chapter 4 has already examined the international reorganization of the food industry through the activities of transnational corporations. Here we focus attention on national-level food processing and manufacturing, drawing on case studies in developed countries where these industrial sectors are most fully developed. Looking first at the EU, in global terms the member states taken together have a greater value of food, drink and tobacco production than either the USA or Japan – the other two dominant producer countries. Over three quarters of agricultural production, by value, in the EU is transformed by the food industry before it reaches the consumer (European Commission 1983), while the sector employs approximately 2.4 million people (ranked fourth amongst industrial sectors). The relative importance of

Table 6.1 The significance of sub-sectors within food processing and manufacturing in the European Union, 1994

Sector	Production value (m ECU)	Employment	Number of firms*
Meat	81,223	436,000	–
Dairy products	75,277	242,000	–
Baking	28,086	449,412	27,272
Brewing and malting	26,572	117,986	616
Chocolate	23,824	158,407	1,180
Soft drinks	19,564	–	–
Soft drinks	17,917	89,921	1,166
Sugar	16,622	–	–
Spirits	13,150	31,946	461
Wine	10,537	46,000	–
Pasta	7,188	33,978	317
Other foods	35,363	181,204	2,205

ECU: European Currency Unit. * More than 20 employees. – No data.
Sources: adapted from European Commission (1997a and 1997b).

the various sub-sectors within EU food processing and manufacturing is shown in Table 6.1, based on partial surveys of all possible sub-sectors (European Commission 1997a, 1997b). Comparable data for individual countries across the EU are not available. However, the dominance of the meat and dairy products sub-sectors, together with the significance of the baking (bread, cakes, biscuits), brewing and chocolate sectors is evident in production values, although there is more variation in terms of employment and the number of firms where data are available to show this. However, estimates of levels of employment by food sector in the EU are as follows: meat products 436,000; dairy products 242,000; baking 449,000; pasta manufacture 34,000; grain milling 36,000; and wine production 46,000. Amongst the branded foods and drinks with the highest sales values within the EU are Coca-Cola, Barilla pasta, Nescafé coffee, Langnese ice-cream, Knorr soups, Nutella spread, Kellogg's cornflakes, Mulino biscuits and Walker's crisps.

Table 6.2 shows the value added by the food, drink and tobacco industries in the national food processing and manufacturing sectors of the member states of the EU. The size of the food industry in each country is proportional to the domestic population it serves, so that Germany, the UK and France are the three most important countries in terms of value added by production. These three countries are also the most active in the formation of mergers and acquisitions (M&A) – both within their boundaries and with other countries – and the formation of alliances and joint ventures (AJV). M&A activity *within* individual member states accounts for between 40 and 70 per cent of all EU merger activity in a given year, while cross border M&A *between* the member states (i.e. integration) varies from 40 to 50 per cent of all world mergers in most years. The main objective of M&A is to increase the market share of individual companies. AJV activity is greatest in the alcoholic beverages and soft drinks sectors, for example by companies such as Carlsberg, Tate & Lyle, Danone, Sodiaal and United Biscuits. To summarize this complex variation in corporate strategy, four broad distinctions can be drawn: (1) continued centralization of control by companies in commodities where local market competition is intense; (2) diversification into other sub-sectors by food companies whose origins lie in one food sub-sector; (3) diversification into food production by companies whose origins lie in other sectors of the economy; and (4) the globalization of capital investment by individual companies, including the building of international alliances.

These structural changes have produced high 'concentration ratios' in the food industries of the EU, as measured by the combined value of production of leading companies, although there are wide variations in concentration across the Member States and between products (Martin 1995). For example, in France 4-firm concentration ratios vary from over 90 per cent of the total value of output for soups and breakfast cereals, through 79 per cent for mineral waters, to 32 per cent for chocolate. Even in countries with less developed food and drinks industries, such as Portugal and Greece, 4-firm concentration ratios are commonly above

Table 6.2 The relative importance of food, drink and tobacco industries within the member states of the European Union, 1994

Country	Value added (m ECU)	% EU AJVs*	% EU M&A†	
			National	Cross border
Germany	23,876	14	15	13
United Kingdom	21,631	24	20	18
France	18,169	12	21	13
Spain	13,910	9	11	16
Italy	10,937	7	12	12
Netherlands	7,686	4	7	6
Denmark	4,107	12	2	6
Ireland	3,611	9	5	1
Belgium and Luxembourg	2,400	6	4	8
Greece	1,147	2	2	4
Portugal	1,137	5	1	2
EU total number	108,608	191	278	108

ECU: European Currency Unit. *Alliances and Joint Ventures. † Mergers and Acquisitions in 1992.
Source: adapted from European Commission (1997b).

70 per cent in a majority of sub-sectors. These high concentration ratios indicate the oligopolistic market power held by food companies (Bhuyan and Lopez 1997) and the barrier against entry into the food industry for new companies. Table 6.3 lists the largest of the existing food, drink and tobacco companies in the EU by value of their turnover. Unilever emerges once again as the largest company on all criteria, while the importance of the UK and France is emphasized, together with the significant employment that the food companies provide in their national economies. Nestlé, a Swiss-owned company and thus not shown in Table 6.3, is the second largest company in Western Europe. Together with Unilever, Nestlé supplies markets at an international level, whereas the remaining companies tend to have less international diversification and are strongest within single national markets (Martin 1995).

So as to complete this part of the analysis, some evidence is presented on the growth of the food industry in developing countries, drawing on the work of Athukorala and Sen (1998). Their study shows the emergence of significant food processing rather than manufacturing in countries such as Bangladesh, Bolivia, Chile, Indonesia, Korea, Malaysia and Thailand. Many of these countries also have dynamic manufacturing sectors, and the common role of permissive public policy régimes is an important factor in both types of development. The main sub-sectors are in processed fish,

Table 6.3 The 15 largest food, drink and tobacco companies in the EU (mid-1990s)

Company	Country	Turnover (m ECU)	Net profit (m ECU)	Employment ('000)
Unilever	Netherlands–UK	38,299	2,012	304
British American Tobacco	UK	15,062	1,555	173
Hanson	UK	14,069	1,383	74
Ferruzzi Finanziaria	Italy	11,955	–528	39
Groupe Danone	France	11,679	536	68
Montedison	Italy	10,723	–183	32
Grand Metropolitan	UK	9,054	584	64
Eridania Beghin-Say	France	7,721	184	22
ABF	UK	5,859	400	50
Hillsdown Holdings	UK	5,499	131	40
Tate & Lyle	UK	5,289	222	15
Cadbury Schweppes	UK	5,199	347	41
Bass	UK	5,103	453	76
Tomkins	UK	4,874	272	46
Société au Bon Marché	France	4,827	193	21

ABF: Associated British Foods.
Source: adapted from European Commission 1997b.

preserved vegetables, animal oils and vegetable oils. By 1994, 41 per cent of the non-manufactured exports of all developing countries were in processed foods (developed countries 35.3 per cent), with national figures varying from 79 per cent in Senegal, through 50 per cent in Nicaragua, to 35 per cent in Peru. There is a mixture of national and international capital underpinning these developments, but both sources of capital draw developing countries into the global food processing and manufacturing sector.

The national organization of food processing and manufacturing

The national organization of food processing and manufacturing can be examined by looking at one country and one product; here the chicken filière in the USA is used to illustrate a number of characteristics of the recent restructuring of food companies. Drawing on the account by Boyd and Watts (1997), the raising of chickens to produce eggs, with meat as a by-product, has been a traditional farmyard activity; by the 1920s it was widely developed on small farms around the north-east manufacturing belt's relatively affluent, urban consumers. Over three quarters of flocks had fewer than 100 birds and less than two per cent were in flocks of over 2500 birds. As chicken production expanded to meet growing consumer demand, so egg and meat production became specialized into sub-sectors, with independent farmers increasingly contracted to processors as 'growers' of broilers (young chickens) and occupying a residual location within the filière (Figure 6.1). Economies of scale were achieved by independent companies specializing in each sector of the production process (e.g. hatching eggs, feed, processing plant, distribution), but these companies in their turn were subjected to the economic logic of integration into large corporations.

By the 1960s, broiler production in the USA had fallen under the oligopolistic control of a few massive, vertically integrated corporations (e.g. Tyson Foods, Perdue, Lane Poultry and Holly Farms) and by the 1990s the top four corporations accounted for nearly 45 per cent of USA broiler output. In addition, average flock size had increased, production had been moved from the farmyard into buildings housing thousands of birds, while advances in breeding genetically uniform broilers, capable of high feed conversion ratios (i.e. feed into liveweight gain), permitted a bird to reach market weight in as little as 40 days. In addition, the location of broiler production had changed significantly and was located less widely across southern and south eastern states (especially Georgia, Alabama and Arkansas). Four factors are commonly advanced to explain this locational shift: (1) the early development of broiler production under contract in the American South; (2) the existence of small marginal farms requiring an alternative source of income to cotton; (3) a pool of surplus rural labour for the processing plants; and (4) a tradition of merchants and finance capital in extending credit to small farmers. Contracted growers in the American South now feed their raw material (live chickens) into industrial processing and manufacturing units that use every body-part of the chicken – from meat for human consumption to head, feet and offal wastes ground up and reincorporated as protein into feed for following generations of chickens. A wide array of products are created from the chicken meat, including the standardized chicken portions sold through fast-food chains, frozen and fresh whole-chickens, standardized packages of chicken breasts, legs and wings, and manufactured 'chicken nuggets', often formed from reconstituted carcass meat. Boyd and Watts (1997), for example, record 4600 different chicken products from Tyson Foods.

The restructuring of food processing and manufacturing, however, can involve the reorganization of existing operational and business practices rather than the relocation of investment. This theme has been developed by Burch and Pritchard (1996) for the Australian tomato processing industry and the following summary follows their account. Australia is a minor producer of tomatoes in global terms, accounting for only 0.6 per cent of

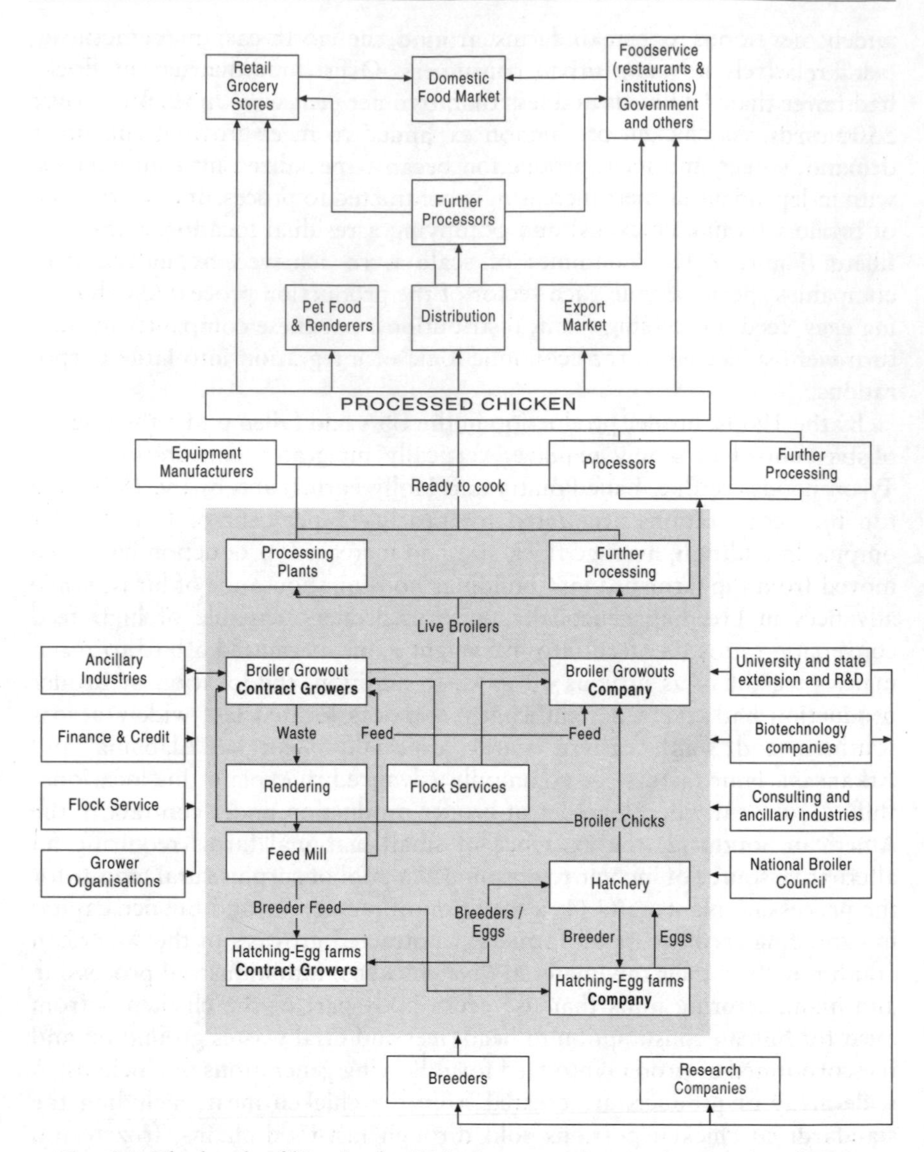

Fig. 6.1 The broiler filière in the USA. *Source*: redrawn after Boyd and Watts (1997, 205).

world production, of which approximately half is processed rather than consumed fresh. Processed tomato production is concentrated in the Murray River basin regions of Victoria, New South Wales and South Australia, where there are nine processing plants owned by companies such as Unifoods (Unilever), Heinz, Leggo's (JR Simplot) and the Sheppparton Preserving Company. 80 per cent of the processed tomatoes are converted into paste, which is then sold for use by other 'downstream' food manufacturers, such as Campbells and McCains. With the deregulation of the Australian market for processed tomatoes in the 1990s, the processing companies adopted three strategies. First, some companies, such as Unilever, expanded their Australian operations as a base for supplying the market in Asia. Research and development functions were relocated from California to Sydney and investment was made to upgrade the plant at Tatura, Victoria. Second, companies such as Heinz and Unilever began to offer more diverse pricing in their contracts with growers, including bonuses for high quality and early/late season production. These contracts favoured larger growers with the result that many smaller growers gave up production and so bore the brunt of the costs of restructuring. Third, Heinz, with production capacity in both Australia and New Zealand, allowed competition to take place between its plants, with resulting productivity and efficiency gains within the existing structure of processing plants.

The regional influence of food processing and manufacturing

A common theme in food processing and manufacturing is the concentration of capacity into a few companies and processing plants and the creation of agro-industrial districts or agribusiness complexes. In USA poultry production, for instance, such concentrations can be identified in north west Arkansas, north Alabama and north Georgia. Within each district can be found the full range of sub-sectors comprising the broiler industry, from egg hatcheries, through feed production units and chicken growing farms to processing plants. Their locational clustering promotes efficient integration in the movement of raw materials through the filière, where 'just-in-time' provision of materials for the chicken 'assembly line' is almost as important as in industrial manufacturing (Boyd and Watts 1997). Similar locational concentrations in national broiler industries have been identified elsewhere, for instance within southern Ontario – in five counties stretching from Huron in the west to Lincoln in the east (Bowler 1994), and in the East Midlands of the UK (White and Watts 1977). Agri-industrial districts have also been identified for a range of other products, for instance cheese and pork production within Italy (Iacoponi *et al.* 1995) and horticultural production under glass in the Netherlands (Maas and Cardol 1984). In the latter case, for example, the authors found that between 64 and 85 per cent of specific inputs to production originated in the local area, but through independent businesses rather than large, vertically integrated companies. In the literature on former socialist economies, agricultural districts are termed 'agri-industrial complexes' (Enyedi 1976). These complexes, under the former command economies of socialist states, were based on the production of a range of agricultural raw materials from large collective or state farms, which were themselves vertically integrated into large food processing/manufacturing, distribution and retailing organizations, sometimes called combines or 'kombinats'. Many of these agri-industrial complexes and combines are struggling to survive international competition in the post-socialist era.

Food processing and manufacturing plants, set in their agricultural districts, are supported by farms producing raw materials under contract, as described in Chapter 4. Initially, food companies were drawn to two types of location in the supply of their raw materials. The first type was the port city where imported agricultural raw materials were landed and processed close to the quayside. The processing of sugar, grain and oilseeds, for example, has been a traditional port industry for many decades, particularly in food importing countries such as Japan, Germany and the UK. The second type of location was the farming region producing the required type of agricultural raw material, for example wheat, milk or beef cattle. Everitt (1993), for example, has examined the history of flour milling on the Canadian Prairies where, in the late nineteenth century, a large number of locally-owned, small mills were developed to process wheat from localized 'tributary areas' and serve local regional markets. Later a few larger mills, owned by interests outside the region, came to control the export of wheat from the Prairie Provinces for milling elsewhere, leading to the demise of flour milling on the Prairies. Similarly the distribution of dairies producing butter and cheese has closely matched the prior location of dairy farms (e.g. Maas and Wever 1986), while the distribution of abattoirs has reflected the distribution of cattle and sheep raising. Contracts placed by food processors and manufacturers with producers in these regions took three types of spatial distribution (Hart 1978): clustering close to the production plant, so as to minimize transport costs – very evident for sugar beet and sugar cane refining, for instance; a random distribution reflecting the prior location of favoured larger farms in a region, as found in vegetable freezing; and clustering in localities with natural advantages for the production of the particular agricultural raw material – a particular soil type or frost-free location – for example in the production of soft fruit.

However, as food processors and manufacturers have restructured their processing and manufacturing plants, first at a national scale and more recently at an international scale, so individual abattoirs, dairies and canneries have been closed down in some agricultural districts and relocated or expanded in others. Rather than following the prior location of production of their raw materials, industrial capitals are now able to influence the location of that production through the relocation of their factories and the placing of contracts. The example of the changing location of dairying in Australia, following the relocation of dairy processing and manufacturing plant, has already been considered above.

The 'downstream' influence of food processing and manufacturing

Food processing and manufacturing firms invest heavily in advertising their brand names; the aim is to develop 'product recognition' amongst consumers when they shop for household food. Similarly, brand names or trademarks are important for marketing intermediate food products to the catering trade, where a brand name may become associated with a guarantee of product quality, taste, texture, reliability or price. Brand names, and associated copyrights, patents and logos, become 'intangible assets' of a company, and these assets, representing access to a segment of the market, rather than tangible assets such as real estate and processing plant, are often the objective of the mergers, acquisitions and joint ventures identified earlier in this chapter. Pritchard (1999) gives the example of Danone – a French-owned transnational food company – whose intangible assets comprise 86 per cent of its net asset worth. A comparable figure of 78 per cent is quoted for Grand Metropolitan, while the brand name of Coca-Cola is estimated to be worth US$43 billion. Pritchard also cites the protracted take-over battles between the Australian-New Zealand transnational Goodman Fielder Wattie, and the British transnational Rank Hovis McDougall plc, during 1988–89, as evidence of corporate strategies to capture and trade intangible assets internationally.

All types of food business have initially sought dominance with a few branded products and then increased their product range and differentiation so as to challenge other market leaders. The new lines embody certain of the characteristics of other products without actually replicating them, and enable companies from one country to penetrate markets in other countries. Indeed many well-known brands are manufactured by subsidiaries or overseas agents of the parent company. Coca-Cola syrup, for example, is transported around the world and bulked with water and sugar and bottled locally. Table 6.4 shows how the Nestlé brand name is deployed across its numerous products to be recognizably present in international

Table 6.4 The international presence of Nestlé brand names

Product	Product brand names in over 34 countries
Milk	Carnation, Coffee Mate, Nestlé, Nido
Confectionery	Crunch, Kit Kat, Nestlé, Smarties
Coffee	Nescafé, Decaf, Classic
Infant formula milk	Cerelac, Nan
Other dry goods	Libby
Petfoods	Friskies

Source: adapted from Pritchard (1999).

markets, but where 'local' brand names have been acquired so as to increase market penetration, often employing imported food products.

Increasingly, however, branded products have to compete for shelf space with the 'own-label' brands of large supermarket chains. It is not uncommon for the same product coming off a production line in a food manufacturing plant to be variously packaged with the advertising of competing producer-brand and own-label brands. We return to this topic in the following chapter, which deals with the retailing of food.

Conclusion

This chapter has shown how food processing and manufacturing capital increasingly defines the identity of food and thereby captures a rising proportion of the final monetary value of consumer food. It has also sought to give more empirical flavour to the 'lived experience' of restructuring in the food industry at the regional level. The resulting concentration of production capacity into fewer companies organized at the national and, increasingly, international level has had the following consequences. First, smaller producers of agricultural raw materials for the food processing and manufacturing sector have either been excluded from production contracts or have had their farming practices increasingly defined for them under tight price margins. Second, so as to remain economically competitive, food processing and manufacturing has become associated with low wage, low skill labour, while communities where plant has been closed down have suffered from the loss of employment. But, thirdly, consumers have benefited from food products that have fallen significantly in real price over the last five decades, while national economies have benefited from the growing export trade in processed and manufactured foods (*see* Chapter 13).

As we have shown for hyperimmune milk, the new biotechnologies developed by the food industry have the capacity to engineer the outputs from agriculture in ways more closely required by food processors and manufacturers (custom made to meet processing and nutritional requirements). The new technologies are overcoming species barriers to genetic variation allowing plant breeders to introduce genes derived from any plant, animal or micro-organism into crop varieties (Shaw 1984). Genetic engineering is also increasing the efficiency of converting different feed stocks into human food. Food processors and manufacturers have had to acquire plant genetic research firms and seed companies so as to have control over their raw material inputs, although a social movement is developing to resist the absorption of genetically modified food into the food chain (*see* Chapter 17).

Resistance has also emerged amongst agricultural producers, food retailers and food consumers against the increasing market dominance of large food processors and manufacturers. Amongst producers, alternative ways are being sought in marketing farm produce directly to consumers, including pick-your-own, farm shops, farmers' markets and vegebox schemes for organic produce. Amongst retailers, the emergence of large, financially powerful supermarket chains has brought a countervailing power into the market for food. Consumers, in their turn, are increasing their demand for 'natural' organic foods and speciality foods whose quality, content and authenticity can be guaranteed. These developments in food marketing, retailing and consumption form the focus of the next chapter.

Further reading and references

The following references are only a small sample of the literature that is available on food processing and manufacturing. There is an especially rich vein of market research reports, many of which are publicly available (see Key Note, Mintel and other websites). Also there are thousands or perhaps tens of thousands of publications on specific industries and on particular countries, the majority of which are not in English.

Athukorala, P. and Sen, K. 1998: Processed food exports from developing countries: patterns and determinants. *Food Policy* **23**, 41–54.

Baldwin, C. 1999: Food safety and the New Zealand dairy industry: the politics of Stolle hyperimmune milk. In Burch, D., Goss, J. and Lawrence, G. (eds) *Restructuring global and regional agricultures*. Aldershot: Ashgate, 189–202.

Bhuyan, S. and Lopez, R. 1997: Oligopoly power in the food and tobacco industries. *American Journal of Agricultural Economics* **79**, 1035–43.

Bowler, I. 1994: The institutional regulation of uneven development: the case of poultry production in the Province of Ontario. *Transactions of the Institute of British Geographers* NS **19**, 346–58.

Boyd, W. and Watts, M. 1997: Agro-industrial just-in-time. In Goodman, D. and Watts, M. (eds) *Globalising food: agrarian questions and global restructuring*. London: Routledge, 192–225.

Burch, D. and Pritchard, B. 1996: The uneasy transition to globalisation: restructuring of the Australian tomato processing industry. In Burch, D., Rickson, R. and Lawrence, G. (eds) *Globalization and agri-food restructuring*. Aldershot: Ashgate, 107–126.

Burns, J., McInerney, J. and Swinbank, A. 1983: *The food industry: economics and policies*. London: Heinemann.

Enyedi, G. 1976: *Agrarian-industrial complexes in the modern agriculture*. Budapest: Research Institute of Geography, Hungarian Academy of Sciences.

European Commission 1983: The Common Agricultural Policy and the food industry. *Green Europe* **196**. Luxembourg: Office for Official Publications of the European Communities.

European Commission 1997a: *Impact on manufacturing: processed foodstuffs*. Luxembourg: Office for Official Publications of the European Communities.

European Commission 1997b: *Panorama of EU industry* (Vol. 1). Luxembourg: Office for Official Publications of the European Communities.

Everitt, J. 1993: The early development of the flour milling industry on the Prairies. *Journal of Historical Geography* **19**, 278–98.

Goodman, D. 1991: Some recent tendencies in the industrial organisation of the agri-food system. In Friedland, W., Busch, L., Buttel, F. and Rudy, A. (eds) *Towards a New Political Economy of Agriculture*. Boulder, CO: Westview, 37–64.

Goodman, D., Sorj, B. and Wilkinson, J. 1987: *From farming to biotechnology*. Oxford: Blackwell.

Hart, P. 1978: Geographical aspects of contract farming: with special reference to the supply of crops to processing plants. *Tijdschrift voor Economische en Sociale Geografie* **69**, 205–15.

Iacoponi, L., Brunori, G. and Rovai, M. 1995: Endogenous development and the agroindustrial district. In Ploeg, J. van der and Dijk, G. van (eds) *Beyond modernisation: the impact of endogenous rural development*. Assen: Van Gorcum, 28–69.

Maas, J. and Cardol, G. 1984: The agribusiness complex in its spatial context. In Smidt, M. and Wever, E. (eds) *A profile of Dutch economic geography*. Assen: Van Gorcum, 150–67

Maas, J. and Wever, E. 1986: Dairy complexes in the Netherlands. In Hottes, K., Wever, E. and Wever, H-U. (eds) *Technology and industrial change in Europe*. Bochum: Universitat Bochum, 18–38.

Martin, S. 1995: European Community food processing industries. *European Review* **3**, 147–57.

Pritchard, B. 1999: Switzerland's billabong? Brand management in the global food system and Nestlé Australia. In Burch, D., Goss, J. and Lawrence, G. (eds) *Restructuring global and regional agricultures*. Aldershot: Ashgate, 123–40.

Shaw, C. 1984: Genetic engineering of crop plants: a strategy for the future and the present. *Chemistry and Industry* **23**, 11–19.

White, R. and Watts, H. 1977: The spatial evolution of an industry: the example of broiler production. *Transactions of the Institute of British Geographers* NS **2**, 175–91.

7

Food marketing and the consumer

Introduction

Processors and manufacturers occupy only one sector in the complex marketing chain for food between agricultural producers and consumers. Until recently political economy accounts of agriculture and food were relatively silent on the non-food industry sectors, such as food retailers, relying on an assumed relationship between mass food production, on the one hand, and class-based food consumption on the other hand. A more critical appreciation of the varied relationships between food consumers and food retail capital has emerged over the last two decades, however, perhaps best illustrated by the writings of Fine *et al.* (1996), Ritson *et al.* (1986) and Wrigley and Lowe (1996). Fine *et al.* (1996, 13–26), for example, show how a multidisciplinary approach is needed to investigate the increasingly fragmented marketing chain, in which food has to be interpreted as both a physical and social object once consumers are entered into the analysis. These authors offer the concept of 'systems of provision' (SoP) to describe the vertically integrated marketing chain, but differentiated by commodity and national context, between farm gate and consumer household. Indeed their argument relies on consumer differentiation through features such as gender, ethnicity, household structure and age of consumer, rather than social class, to explain the varied relationships between consumers and the food they eat.

We begin by reviewing the development of the marketing chain under the second food régime, emphasizing the varied ways in which food can be moved between producer and consumer. Then we focus on the significance of the restructuring of the retail sector, including the emergence of 'fast food' restaurants. We conclude with an examination of emerging trends in the third food régime, which we interpret as growing resistance by consumers in developed countries to mass-produced and marketed food. These trends, which are examined further in Chapter 15, include alternative ways of more directly reconnecting food consumers with agricultural producers, and the concern of consumers with food quality rather than food quantity.

The marketing chain for food

The marketing chain for food can be examined from an evolutionary viewpoint (Hart 1992), beginning with simple face-to-face trading between agricultural producers and food consumers. Clusters of stalls selling farm produce can still be found in open village squares and covered city markets throughout developed and developing countries alike where face-to-face trading continues, although more commonly in the latter context. With the increasing separation of rural and urban economy and society, wholesalers emerged to accumulate production from individual farms and trade with urban retailers, as investigated by Parsons (1996) in an Australian context, sometimes through organized metropolitan wholesale and auction markets that persist to the present day. A recent trend has been for electronic markets to threaten these long-established 'live' wholesale markets for the buying and selling of agricultural products, including livestock, fruit, vegetables and flowers (Blandford and Fulponi 1997). As shown in the previous chapter, food processors and manufacturers have developed within the last four decades to act as an alternative market for agricultural producers, with many farmers grouping into marketing co-operatives so as to gain some measure of 'countervailing power' in the negotiation of forward contracts. But with the continuing uneven balance of power between

Table 7.1 The largest 10 grocery corporations, 1997

Retailer	Country	Turnover (Euro bn)	Countries with stores	
			Europe	Non-Europe
Wal-Mart	USA	130	1	8
Metro	Germany	35	17	5
Rewe	Germany	34	9	0
Edeka	Germany	33	5	0
Tengelman	Germany	30	8	1
Promodès*	France	29	7	6
Carrefour*	France	28	6	15
Auchan	France	25	7	3
Tesco	UK	25	6	2
Ahold	Netherlands	22	5	9

* Carrefour acquired Promodès in 1999.
Source: adapted from *The Guardian*, June 15th 1999, 22.

agriculture and the food industry, the state has commonly intervened to regulate the marketing of farm produce (*see* Chapter 11), including marketing boards, intervention agencies, surplus food storage, and the disposal of those surpluses on international markets using food subsidies.

More recently, large retail capital (i.e. supermarket chains) in developed countries has revolutionized the retailing of food by eliminating many small, independent grocers and green grocers, becoming a powerful negotiator of supply contracts with both individual farm businesses and food processors and manufacturers. Table 7.1 lists the largest of these grocery corporations in the late 1990s by total sales. Apart from Wal-Mart in the USA, French and German food retailers head the list. Fast-food chains from the USA, such as Burger King and McDonald's, have emerged as the most recent innovators in how food is retailed to largely urban consumers, placing particular contractual demands on the suppliers of their raw materials.

This condensed and simplified account of the evolving marketing chain for food covers the period we have previously described as the second food régime; we leave for consideration until the last part of this chapter those contemporary changes in food marketing that characterize the present third food régime. However, the overall impact of these evolutionary developments has been to lengthen the marketing chain (or network) between farmers and consumers, with separate sectors of the chain adding value to the food produced, packaged, transported or resold.

The restructuring of food retailing

We look now at a particularly dynamic sector of the marketing chain for food, namely retailing. In most countries of the world, food is still purchased from relatively small independent grocers and green grocers, often on a daily basis. In many developed countries, however, large food retailers have emerged since the 1960s to capture a majority of food sales through their supermarket and hypermarket chains (Agra Europe 1994); indeed a number of food retailers have diversified into other products, such as clothes, furniture and entertainment products, to become multiple retailers. In western Europe, for example, supermarkets now account for over 70 per cent of purchases of packaged foods, soft drinks and cheese, and over 40 per cent of fresh fruit, vegetables and wine. Specialist stores retain a significant share of purchases of only fresh bread and meat. However, discount grocery retailers, with their limited but low cost product lines, are a developing threat to the dominance of the corporate retailers, particularly in Germany (25 per cent of food sales in 1992), Norway (31 per cent) and Austria (26 per cent). Kwik Save and Lo-cost are the relevant discount grocers in the UK, Wal-Mart in the USA, with Aldi and Netto in Denmark, Belgium and Germany.

Table 7.2 Food sales through supermarkets in western Europe, 1997

Country	Market share of top retailers*	Percentage own-label sales
Norway	89 (4)	–
Germany	80 (5)	24
Austria	73 (4)	–
Finland	72 (2)	–
Belgium	68 (4)	–
Sweden	61 (3)	–
UK	61 (5)	28
France	59 (5)	20
Ireland	59 (3)	–
Denmark	57 (2)	–

* Number of retailer in brackets. – No data.
Source: adapted from *The Guardian*, June 28th 1999, 20.

Table 7.2 shows the level of concentration in food retailing at a national level within western Europe, although, as noted by Dawson and Burt (1998, 161), individual corporate food retailers tend to have their strength in particular regions – for instance, Eroski in the Basque country, Cora in Alsace, Savia in Silesia, Kethreiner in Bavaria and Sainsbury in southern England. By the late 1990s Norway, Germany, Austria and Finland had more than 70 per cent of their food sales through four or five food retailers. The national scale of retail organization can be illustrated in more detail by taking the UK as an example. As late as the 1950s, most food sales in the UK were made by independent retailers or in chains of self-serve food shops that, by present-day standards, were small in size and largely organized on a regional rather than national basis. However, consumers began to demand a wider choice, small organizations carried diseconomies of scale, and small retailers lacked new marketing and managerial skills. Through the processes of acquisitions, mergers and new investment, four corporate retailers had come to dominate nearly half of food sales by the late-1990s, namely Tesco, Sainsbury, Asda and Safeway (Table 7.3; *see also* Table 4.2). The date has to be specified because the process of mergers and acquisitions continues; for example, Somerfield merged with Kwik Save in 1998, thereby threatening Safeway's position as fourth largest supermarket chain.

The demise of small, independent food retailers has been relatively rapid and on a large scale in developed countries. In Spain, for example, food stores decreased in number by 34,000 between 1988 and 1995, while between 1971 and 1991 the UK lost 120,000 food shops, (west) Germany 115,00, and France 105,000 (Dawson and Burt 1998, 171). The process continues in southern Europe, although Italy has national legislation to preserve small, non-self service shops. For the remaining independent retailers, they have

Table 7.3 The largest four food retailers in the UK

Retailer*	**Share of grocery market (%)**		**Cumulative share (%)**	
	1988	**1998**	**1988**	**1998**
Tesco	9.0	15.6	9.0	15.6
Sainsbury	10.2	12.6	19.2	28.2
Asda	4.7	8.7	23.9	36.9
Safeway	6.9	7.7	30.8	44.6

* Includes associated companies.
Source: adapted from *The Guardian*, 19th March 1999, 24.

found some competitive refuge by joining one of a number of national buying groups, such as Spar in the UK and Leclerc in France. Through collaboration, independent retailers gain access to cheaper food supplies and access to improved stock management and marketing services.

Within the last 5 years, the national structure of food retailing has been subjected to the same types of international acquisition and merger as characterized the food industry two decades earlier. For example, the American Wal-Mart corporation (annual sales £86b) acquired Asda in the UK (annual sales £8b) in 1999, while in Spain, the three main food chains of Pryca, Continente and Alcampo are French owned, and the French corporations of Carrefour, Promodès and Auchan have acquired outlets in Portugal through mergers and acquisitions (Table 7.1). Carrefour has a market presence in Argentina, Brazil and Taiwan, while Sainsbury has expanded into the North American market through investments in Shaw's Supermarkets (1983) and Giant Foods (1994) (Shackleton 1998). These international developments are driven in large part by the ownership of the large food retail corporations by financial institutions. Share prices and dividends rely on profits that are driven by continual increases in sales and growth in the number of retail outlets. Some national governments have put limits on the domestic growth of supermarkets, in part to protect the remaining stratum of smaller food retailers, in part to retain the coherence of city centre shopping, and in part to limit urban development onto out-of-town, greenfield sites. As a result the larger food retailers have tended to looked globally for new locations for capital investment and the further diffusion of the supermarket way of selling food. In addition national retailers are entering into cross-border alliances (buying groups) at a European scale of organization; the largest, with turnovers exceeding 60 billion ECU a year, include European Marketing Distribution, Eurogroupe, Associated Marketing Services, Deurobuying and Inter-Co-op. Deurobuying, for example, brings together Asda, Asko, Carrefour and Metro.

Harrison *et al.* (1997) have traced the source of power amongst large food retailers to four 'regulatory domains', namely (1) competition and pricing policies; (2) planning and environment; (3) food law; and (4) food quality. Here we deal with the first two of these domains, beginning with competition policy, leaving the latter two to subsequent chapters (Chapters 18 and 15 respectively). Wrigley (1992) and Hughes (1996) have shown how antitrust (competition) legislation and its interpretation have been more restrictive in some countries, for instance the USA, compared with others, such as the UK. Thus in the USA, in the 1970s, the leading four retailers accounted for only 18 per cent of the grocery market (e.g. Kroger Company, American Stores Company, Safeway Inc. and Great Atlantic and Pacific Tea Company) (Wrigley and Lowe 1996, 101). However, following the relaxation of corporate regulation in the 1980s, a restructuring of retail capital brought about a degree of concentration more in line with that developed in the UK (Wrigley 1999).

On pricing, five dimensions can be identified. First, with price determining a large part of food purchasing behaviour, larger retailers are able to obtain price discounts from their suppliers in the food industry when negotiating supply contracts for their branded goods. In addition extended supply contracts tie suppliers to a supermarket chain on a long-term basis. Second, large retailers can also levy a charge on the same suppliers for giving their products access to their supermarket shelves. Product ranges are tightly controlled so as to balance maximizing consumer choice against limiting products to those with the highest sales potential. Third, because of the volume of their sales, large retail corporations are able to discount the selling price of their goods and so undercut the prices of both independent retailers and, from product to product, some of their large competitors. Indeed discounting is probably the most dynamic force of restructuring within the retail sector; for instance, in the USA, Sam's/Wal-Mart and Target/Dayton-Hudson grew mainly by this process rather than through value-added production. Fourth, by introducing cheaper, 'own-brand' labels, large retailers have taken an increasing share of the retail turn-over in competition with 'producer-brand' label products (Table 7.2). This aspect extends to large supermarket chains now operating their own baking and butchering divisions within individual stores. Fifth, the 'just-in-time' restocking of individual supermarkets from centrally-located depots (warehouses) reduces the operating costs of leading food retailers. This development has been made possible by new technology, such as centralized computer-controlled sales records and stock management that employs bar-code scanning systems at points of sale. Indeed the largest retailers have subsumed the wholesale and distribution functions previously provided by independent companies. Supermarket chains now have their own road transport divisions so as to regulate the efficient delivery of food products to individual supermarkets. This level of control reduces the costs of inventory (i.e. stock) at individual supermarkets and depots, minimizes delay in the delivery of fresh products, and ensures the maximum turnover from each metre of retail sales shelving through continual restocking.

One outcome of the power of corporate retailers in developed countries is their ability to source materials globally through forward contracts. 'Fresh' fruit and vegetables, for example, are grown under direct contract with individual, large farms in developing countries (or a number of smaller farms acting co-operatively), where the same tightly-drawn requirements of crop variety, method of husbandry, quantity, quality and timeliness of production are imposed as in contracts with producers in developed countries. Under national buyer programmes, supermarkets are developing long term, reliable, supply relationships with carefully selected 'preferred' suppliers (i.e. trust-based partnerships – Doel 1999), which bypass conventional wholesale markets. One retail company, Sainsbury, in its quest to assure supplies of high-quality, organic, tropical fruit, has gone so far as to propose placing contracts with so many producers on the Caribbean island of Grenada as to dominate, in neo-colonial form, the economy and society of that island (*The Guardian*, 19th January 2000, 2). Such global reach by retail capitals involves the destruction of seasonality in food consumption, as products are offered all the year round as a marketing strategy. Fresh strawberries, mangoes and runner beans, for instance, can now be bought throughout the year by consumers in the northern hemisphere.

The restructuring of national food retailing faces consumers with, on the one hand, less choice in the source of their food purchases; on the other hand the range of food products on offer continues to expand and there is considerable inter-regional and international variation in the availability of food and drink products. Indeed Dawson and Burt (1998, 167) claim that individual supermarkets are differentiated to suit local market conditions: product range, price policy, promotions and opening hours are adjusted to local consumer demand. Equally the international sourcing of food by large retailers has brought about a marked increase in the 'food miles' (Paxton 1994) element of different food products. The term 'food miles' concerns the transport cost component in consumer food prices, where various constituents of a food item, including its packaging, may have been assembled at a location from a variety of sites of production throughout the world, then processed or manufactured together at that location, before being transported onwards to a range of retail (or consumption) sites. The accumulated food miles (distance) costs to consumers can be considerable for relatively simple products, such as packaged and cooled but unprocessed mange-tout peas, as well as complex products, such as pre-cooked, boxed and chilled, ready-to-eat meals.

Turning now to the second regulatory domain – planning and environment – we begin by noting that the structural changes in food retailing in developed countries have been supported by changes in consumer demand, attitudes and behaviour. For example, average consumer age is increasing and older consumers demonstrate high levels of brand and store loyalty but often expect high levels of service (Dawson and Burt 1998, 162). In addition, a greater proportion of women now work outside the home, prompting an increased demand for convenience, 'ready-to-eat' foods, as well as once-a-week, one-stop shopping, with the private car as the means of transport and the whole family often involved in shopping as a leisure activity. This change in behaviour has enabled large supermarkets to exercise their influence within national planning systems to permit reinvestment from restricted city-centre locations to suburban and out-of-town locations (Nayga and Weinberg 1999). These new locations have permitted the construction of large car parks, and wide aisle spaces, children's entertainment areas and restaurants within stores. New retail parks, with their clusters of large multiple retailers, have also been favoured locations, with the 'shopping mall' (shopping centre) concept, as developed in the USA, becoming an international phenomenon. The 'land take' on green-field sites from out-of-town retail developments, together with their disruptive impact on both traffic movements and city-centre retailing, has proved sufficiently contentious that in some countries, for example the UK and Italy, limits have been placed on their growth.

This focus on the restructuring of retail capital, however, should not obscure the continuing wide range of retail outlets for food. Table 7.4 is a reminder that individual consumers still vary their food purchasing behaviour – typically visiting three to five different types of store in a three month period in western Europe (Dawson and Burt 1998, 163). Indeed with time now seen as a commodity, large food retailers are starting to move into internet or 'tele-shopping'. In this system, a shopping order can be placed from the home to a supermarket using a personal computer, a modem link to the supermarket's internet site, an interactive menu of food products, and e-credit card payment; the shopping order is subsequently delivered direct to the home of the consumer. In addition, if the trends outlined in the last section of this chapter deepen, then the sites of food purchase could diversify further.

Table 7.4 Types of retail outlet for food in western Europe

Type of purchase	Type of retail outlet
Major food replenishment	Food superstore, hypermarket, warehouse club
Weekly food replenishment	Food superstore, hypermarket, discount store
Short-life and convenience food purchase	Town centre supermarket, discount store, neighbourhood supermarket, convenience store
Top-up, out-of-stock food	Convenience store, garage store, corner shop
Specialist food items	Specialist food shops, food halls in department stores, internet and mail order

Source: adapted from Dawson and Burt (1998, 162).

Fast food and the 'McDonaldization' of society

Fast food, as an American innovation, can be traced to the first franchising of the McDonald's hamburger chain in 1955. Indeed, with its growth in subsequent decades, Burch and Goss (1999) argue that fast food is a manifestation of Fordist economic organization and development. Other restaurant chains, specializing in a variety of fast foods, were subsequently established in the USA, notably Kentucky Fried Chicken (KFC), Burger King, Wendy's, Subway, Pizza Hut and Charley Chan's. By 1994 sales in fast-food restaurants in the USA exceeded those in traditional full-service restaurants, and the gap between them has continued to grow (Ritzer 1993, 2). These chains have now been franchised on an international basis; the first McDonald's outlet in China, for instance, was opened in Shenzhen province in 1990 and another in Beijing in 1992 (700 seats, 29 cash registers and 1000 employees); the first McDonald's restaurant in Moscow was opened in the same year. Indeed since 1993 over a third of McDonald's restaurants have been operating outside the USA.

As with corporate food retailers, the main impetus for the global spread of franchises has been market saturation in the USA and other developed countries. Developing countries, with their increasing wealth and urban populations, have offered a commercial opportunity for continued growth and capital accumulation (Burch and Goss 1999). For example, Pizza Hut opened its first outlet in Thailand, at Pattaya, in 1980 and by 1995 a total of 85 outlets were in operation. Other countries have also developed their own variants. For example, the Wimpey hamburger chain in the UK in the 1960s, the Nirula muttonburger chain in India, and the Charoen Pokphand (CP) Group in Thailand which operates Chester's Grill outlets. In Paris (France) four large restaurant chains have developed – Bistro Romain (39 outlets), Buffalo Grill (53), Leon de Bruxelles (26) and Hippopotamus (26). Each chain of Parisian restaurants is supplied from central processing plants located in suburban locations, where bulk raw materials are processed into restaurant meals and delivered to the outlets.

The success of fast food can be traced in the first instance to changes in society. Such changes include: the demand for 'casual dining out' as a form of entertainment by a wider spectrum of society, together with the discretionary income to express that demand; the increased pace of economic and social life, in which personal time is under increasing pressure; and the increased economic activity of women in society, as a result of which eating outside the home becomes an occasion for family interaction. From a commercial point of view 'McDonaldization' has many advantages and these have been summarized by Ritzer (1993) into four categories, namely (1) efficiency; (2) calculability; (3) predictability; and (4) control. His argument is that these categories are being applied to a wide range of commercial and service activities, so that a broader and inexorable McDonaldization of society is underway. Summarizing, fast food restaurants offer an *efficient* way for consumers to access and eat a meal, for food to be processed and delivered to the consumer, and for employees to have their work organized. On *calculability*, consumers know in advance exactly what they are purchasing in terms of price, quantity and service offered. To an extent quantity (e.g. large fries) has been made equivalent to quality. Indeed franchises operate to formalized, organizational rules and regulations, so that a uniform consumer experience can be achieved in any outlet of a chain. These features thus contribute to the *predictability* of the product and service on offer, at all times and in all places for that chain: as Ritzer (1993, 10) observes, it appears that food consumers prefer a world in which there are few surprises. *Control*, of the product, the employee and the consumer, is the fourth dimension of fast food. Non-human technology is substituted for human agency, for example in the display of the food menu, the placing of the order for food, the seating arrangements and the self-clear tables. Employees work to a fixed procedure and consumers are 'processed' through the restaurant.

The global spread of fast food attests to its social and economic success: the availability of nutritious food at affordable prices has been increased for a larger proportion of the population; people are able to eat what they want almost instantaneously; there is a high degree of convenience for the consumer; fast, efficient food and service is available to a population with fewer hours to spare outside work and leisure activities; and food consumption becomes a form of entertainment. On the other hand, according to Ritzer (1993), these advantages are achieved at three main costs, which are also observable in the 'McDonaldization' of other services. First, 'McDonaldization' is essentially dehumanizing. Humans become part of an assembly line from which there is no scope for deviation, either as a producer or consumer of food. The menu is limited and fixed: consumers are unable to control their (relatively high) intake of salt, fat and sugar from the food products. Fast food is not necessarily healthy food. Second, on careful examination, the main gains in efficiency from McDonaldization accrue to the supplier of the service rather than the consumer. For instance, consumers are required to wait their turn in a queue so as to order and receive their food efficiently; they are expected to clear their tables on completion of the meal; the price of the meal has a high convenience component and a low content (ingredients) component; employees work at relatively low wage rates. Third, the system is environmentally costly: for example, the packaging (polystyrene and paper) and eating utensils (plastic) associated with the meal are thrown away once the food is consumed; low cost beef is produced largely from intensive feed-lots with their associated hazards of environmental pollution; and much of the potato is wasted in creating uniform french fries. In sum, Ritzer accuses fast food chains of the 'irrationality of rationality' in offering only the illusion of efficiency and cheapness.

Fast food restaurants can be viewed as a sub-sector of the catering industry, which itself is of increasing importance for the consumption of food

Table 7.5 The top 10 UK hotel and catering groups, 1991

Group	Turnover (£m)
Forte	2,641
Hilton International	780
Bass Hotels	570
Queens Moat Houses	484
Rank Holidays and Hotels	437
Scottish and Newcastle Leisure	295
Scottish and Newcastle Retail	287
Compass Services	264
Mount Charlotte Investments	241
Rank Leisure	109

Source: adapted from British Hotels, Restaurants and Caterers Association (1991).

outside the home. A wide range of outlets can be identified, varying from independent cafés and restaurants, through full-service restaurant and hotel chains, to food outlets in educational, hospital, prison, armed services, industrial and office premises. Large catering groups again dominate the sector. Table 7.5, for example, shows data available for the early 1990s on the top 10 hotel and catering groups in the UK. These groups have been subject to the same processes of national and international acquisitions and mergers as found elsewhere in the food sector since the data were gathered.

Resistance and reaction to processed, manufactured and fast food

Observers of consumer behaviour have detected a number of changes in the last decade. According to Dawson and Burt (1998, 163) these include:

1. A search for individualism in lifestyle through products and services purchased – this results in a shift away from a uniform pattern of mass demand and an accentuation of local differences.
2. Higher quality expectations in product, service and shopping environment.
3. Non-price responsiveness to advertising and promotional methods.
4. Less direct price comparison with the increase in product range and differentiation.

For the present discussion on food, a number of trends can be identified which are illustrative of these changes, can be interpreted as a growing consumer resistance and reaction to processed, manufactured and fast food, and lend support to the contention that we are entering a third food régime.

In sum, alternative methods of producing and marketing 'non-industrial' foods are emerging in developed countries; these methods are being supported by an increasing proportion of 'elite' consumers who are willing to pay premium prices for such foods.

On alternative marketing methods, farmers are seeking to reconnect their production directly with consumers through a variety of business developments, which include 'pick-your-own' (PYO), farm shops, farmers' markets and vegeboxes. All of these methods attempt to by-pass the control exercised by the food industry and retail capitals over the marketing chain. Under PYO, for example, the consumer visits the farm and undertakes the harvesting of the crop, be it strawberries, potatoes or apples. The consumer can influence the quality and quantity of the product harvested and at a lower retail price; the farmer is saved the labour cost of harvesting the crop and achieves a higher profit margin on the product sold. Originating in the USA in the 1960s, this form of direct marketing spread to the UK in the 1970s and other West European countries in the 1980s (Bowler 1992, 192). Farm shops have developed in parallel and also involve the consumer in a visit to the farm. As with PYO, consumers at farm shops tend to be limited to car owners; so as to broaden this consumer base, farmers are also participating in farmers' markets located in urban centres. This form of marketing has been popular in the USA for several decades but is now spreading to western Europe. In vegebox schemes, consumers contract to receive a box of seasonal vegetables from a producer, delivered to their place of residence on a weekly basis. At the retail level, an increasing number of small, specialist food shops are being re-established in inner-city areas to provide high quality foods, along with more personal service, which together differentiate them from large retailers.

These 'supply-side' developments are associated with organic and speciality foods, so that an association is being formed between the differentiation of food product and the means of purchase by the consumer. While organic and speciality foods still comprise niche rather than mass markets, the development of a supportive consumer movement is underway as a reaction to the concerns over food health, food safety and food ethics associated with mass-produced foods. In particular, consumers are increasingly concerned about the 'traceability' of food back through the supply chain to its site of origin. The term 'food quality' is increasingly used to summarize these concerns of both producers and consumers (Bowler 1999) and this topic is given full consideration in Chapter 15. However, the point being made here is that, with the increased consumption of organic and quality foods providing a new commercial opportunity, supermarket chains are mobilizing to sell organic food, and food produced through quality assurance schemes, using their own national and international forward contracting systems. Campbell (1996) terms this development 'corporate greening', but the growth in international trade in organic food bears testimony to the outstripping of domestic supply by consumer demand in many developed countries (Campbell and Coombes 1999). Once again large corporate capitals seem poised to appropriate an emerging dimension of the food supply chain.

Conclusion

This chapter has outlined the struggle for control over food amongst a variety of actors, including food processors and manufacturers, food retailers and food consumers (Hallsworth and Taylor 1996). A number of 'sites' of struggle have been identified, including supermarkets, fast food restaurants and the farm gate itself. Exercising control over food is being made more complex by: (1) the growing fragmentation and segmentation of consumers into sub-groups, for example by age, ethnicity, household structure, and ethical values; (2) the increasing centrality of food in the construction of lifestyle; (3) the emphasis being placed on the quality and convenience of food rather than price or quantity; and (4) concern over food safety and food health. Taken together, these trends offer new market opportunities for both individual agricultural producers and food capitals.

Another theme has been the shifting locus of power within the marketing chain, first to food processors and manufacturers, and then to the retail sector, including fast food restaurants. For each sector in turn, an initial organizational structure within regional and then national economies has given way to restructuring at a global scale, a trend now evident in the production of organic food. However, a reaction to these developments amongst consumers can now be detected, with Dawson and Burt (1998) going so far as to suggest that a more fragmented retail structure may emerge to serve regional cultures, with regional food factories supplying regional distribution networks. However, as shown by the appropriation of organic food (Campbell 1996), large retail capital has the capacity to subsume emerging market trends, so that the development of local and regional market niches for quality foods will remain problematic in the face of the continuing globalization of food.

Further reading and references

Patrick Hart has written a useful general introduction to the marketing of agricultural produce, and Neil Wrigley's publications on retailing, only a few of which are listed here, are a valuable resource.

Agra Europe 1994: Domination of food retailing by supermarkets: a world wide trend. *Food Policy International* 3, 2–4.

Blandford, D. and Fulponi, L. 1997: Electronic markets in the agro-food sector. *OECD Observer* 208, 20–23.

Bowler, I. (ed.) 1992: *The geography of agriculture in developed market economies*. Harlow: Longman.

Bowler, I. 1999: 'Quality food' and the reproduction of small rural businesses. In Bowler, I., Bryant, C. and Firmino, A. (eds) *Progress in research on sustainable rural systems*. Lisbon: Centro de Estudos de Geografia e Planeamento Regional, 60–71.

British Hotels, Restaurants and Caterers Association 1991: *Wages and salaries survey*. London: BHRCA.

Burch, D. and Goss, J. 1999: An end to Fordist food? Economic crisis and the fast food sector in Southeast Asia. In Burch, D., Goss, J. and Lawrence, G. (eds) *Restructuring global and regional agricultures*. Aldershot: Ashgate, 87–110.

Campbell, H. 1996: Organic agriculture in New Zealand: corporate greening, transnational corporations and sustainable agriculture. In Burch, D., Rickson, R. and Lawrence, G. (eds) *Globalization and agri-food restructuring*. Aldershot: Ashgate, 153–69.

Campbell, H. and Coombes, B. 1999: New Zealand's organic food exports: current interpretations and new directions in research. In Burch, D., Goss, J. and Lawrence, G. (eds) *Restructuring global and regional agricultures*. Aldershot: Ashgate, 61–74.

Dawson, J. and Burt, S. 1998: European retailing: dynamics, restructuring and development issues. In Pinder, D. (ed.) *The new Europe: economy, society and environment*. Chichester, John Wiley, 157–76.

Doel, C. 1999: Towards a supply-chain community? Insights from governance processes in the food industry. *Environment and Planning A* **31**, 69–85.

Fine, B., Heasman, M. and Wright, J. 1996: *Consumption in the age of affluence: the world of food*. London: Routledge.

Hallsworth, A. and Taylor, M. 1996: 'Buying' power: interpreting retail change in a circuits of power framework. *Environment and Planning A* **28**, 2125–37.

Harrison, M., Flynn, A. and Marsden, T. 1997: Contested regulatory practice and the implementation of food policy: exploring the local and national interface. *Transactions of the Institute of British Geographers* NS **22**, 473–87.

Hart, P. 1992: Marketing of agricultural produce. In Bowler, I.R. (ed.) *The geography of agriculture in developed market economies*. Harlow: Longman, 162–206.

Hughes, A. 1996: Retail restructuring and the strategic significance of food retailers' own-labels: a UK–USA comparison. *Environment and Planning A* **28**, 2201–26.

Nayga, R. and Weinberg, Z. 1999: Supermarket access in the inner cities. *Journal of Retailing and Consumer Services* **6**, 141–45.

Paxton, A. 1994: *The food miles report: the dangers of long distance food transport*. London: SAFE Alliance.

Parsons, H. 1996: Supermarkets and the supply of fresh fruit and vegetables in Australia: implications for wholesale markets. In Burch, D., Rickson, R. and Lawrence, G. (eds) *Globalization and agri-food restructuring*. Aldershot: Ashgate, 251–70.

Ritson, C., Gofton, L. and McKenzie, J. (eds) 1986: *The food consumer*. Chichester: John Wiley.

Ritzer, G. 1993: *The McDonaldisation of society: an investigation into the changing character of contemporary social life*. California: Pine Forge Press.

Shackleton, R. 1998: Exploring corporate culture and strategy: Sainsbury at home and abroad during the early to mid 1990s. *Environment and Planning A* **30**, 921–40.

Wrigley, N. 1992: Antitrust regulation and the restructuring of grocery retailing in Britain and the USA. *Environment and Planning A* **24**, 727–49.

Wrigley, N. 1999: Market rules and spatial outcomes: insights from the corporate restructuring of US food retailing. *Geographical Analysis* **31**, 288–309.

Wrigley, N. and Lowe, M. (eds) 1996: *Retailing, consumption and capital: towards the new retail geography*. Harlow: Longman.

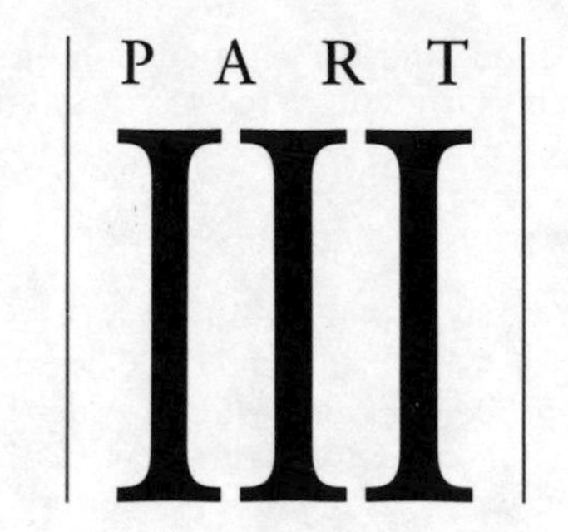

GLOBAL AND GEOPOLITICAL FOOD ISSUES

8 Introduction

Part III picks up the dictum 'grow globally, eat locally'. It argues in effect that the significance of global political and economic structures, and recently of global cultural trends, insists upon a broad view of food systems.

Probably the issue of most prominence for the general public is the world's ability to feed future generations. Chapter 9 addresses this by examining the claims of the pessimists and the optimists. Although the number of at-risk people in poor countries does not seem likely to fall in the near future, the evidence suggests that this is less the result of the exhaustion of production capacity than a serious structural problem of economic development. The main area of concern remains Africa south of the Sahara, where several countries have stagnant food production sectors, constrained by a mix of endogenous and exogenous factors.

Chapter 10 discusses malnutrition and famine. There are a number of theoretical contexts for this and we examine in particular Amartya Sen's use of 'entitlements' to food as an explanation of shortages. In the last 20 years social science has moved away from thinking in terms of 'natural disasters' such as drought in the Sahel of Africa, towards the social and political background of hunger. It is now recognized that poverty and war are the two most important causes.

The ironic and obscene counterpart of food shortage in the poor world is food surplus in the North. Chapter 11 may seem a strange juxtaposition to Chapter 10, but it represents an important reminder that food production is policy-led. Here we deal with the food mountains of the CAP of the EU and similar frameworks in other rich countries. The various devices to put an upper limit on production are examined, such as quotas and set-aside land, along with recent moves to an internationalized reduction of price subsidies under the aegis of the WTO.

Chapter 12 returns to the theme of shortage by studying food security. In particular we look at interventions such as famine early warning systems and food aid. Several detailed examples are offered along with some comments on the 'post-development' school of thinkers who would like international food aid to be abolished. Finally, the role of war in food shortage receives a brief treatment.

Violence is also a point of discussion in Chapter 13, which starts with food riots caused by the structural adjustment required for many LICs by the IMF. This has geopolitical implications of course, which are then developed in a discussion of the international trade in food.

9

Food production and population

Food scarcity will be the defining issue of the new era now unfolding, much as ideological conflict was the defining issue of the historical era that recently ended (Brown 1996, 19).

Introduction

The relationship between the world's growing population and our ability to feed it at an acceptable level of nutrition is one of the most important and also one of the most controversial current global issues. This chapter studies some of the arguments which have been put forward, but the literature is so vast and complex that we cannot do more than sketch them in outline. Caution is essential, because any student of the relationship between population and resources soon realizes that simplistic explanations are potentially misleading, and perhaps even dangerous, if they spawn inaccurate forecasts and misplaced policy responses.

Figure 9.1 is one conceptual approach to the relationship between population growth and food supply. It stresses the role of poverty rather than food supply, the implication being that wealth-creation would be of fundamental assistance in reducing the size of families. Not all commentators agree with this analysis and one purpose of this chapter is to alert the reader to the polarized nature of the debate between what we shall call the pessimists and the optimists.

The pessimists

1999 is a particularly apposite time to be writing this chapter because the United Nations' Fund for Population Activity has estimated that this year

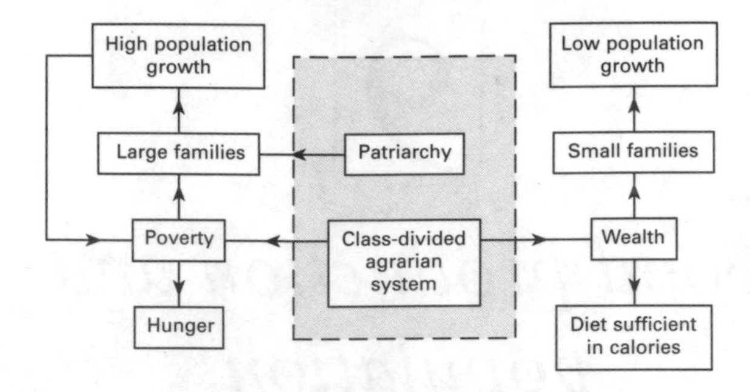

Fig. 9.1 Relationships between poverty, hunger and population growth. *Source*: based on Abraham (1991).

the world's population reached six billion. The number itself is almost impossible to comprehend but it represents a highly significant stage psychologically on the *via dolorosa* to what some writers have called the 'overpopulation' of the planet, with the accompanying potentially disastrous environmental and human consequences.

Thomas Malthus (1766–1834) was the most influential of the early writers in this area. His *Essay on the Principle of Population* (1798) and subsequent publications argued in essence that in hard times society is subjected to the 'checks' of increased death rates due to starvation and disease. These limit the numbers of the indigent classes when food is scarce or expensive. His analysis was not as crude as is sometimes portrayed, but it was nevertheless gloomy because he recognized a tendency of the population to expand at a faster (geometric) rate than the (arithmetic) growth in agricultural output.

As a theory, Malthusianism actually gained little ground in nineteenth century Europe because its predictions of hunger were undermined by the increased availability of cheap food. This was caused by the opening up of new lands such as the North American grain-growing prairies and also by improved yields from crop breeding and the use of fertilizers. Nevertheless an undercurrent of worry remained and resurfaced from time to time in the twentieth century as localized famines in Africa and Asia have highlighted the fragility of food security. Since the 1960s there has been a well publicized school of thought known as 'neo-Malthusianism' which has revived parts of the original model and has propagated its warnings about the consequences of rapid population growth.

Lester Brown, founder of the Worldwatch Institute, is perhaps the best known proponent of demographic pessimism. For over 30 years he has been alarmed by what he regards as the unsustainable use of the earth's resources and one of his recent books was subtitled a 'wake-up call for a small planet'. Brown identifies 16 dimensions of the population problem but here we will rehearse only those parts of his proposition which address the question of food. Table 9.1 summarizes the main points of the pessimists' thesis.

The first and most obvious point is the striking acceleration in the global population that took place in the late twentieth century (Figure 9.2). Although the developed countries have ageing populations and have reduced their fertility to replacement levels only, the poor nations have yet to complete this 'demographic transition' and in consequence their growth potential over the next few decades remains high. There are 78 million additional mouths to feed each year, 97 per cent of which are in low-income countries.

Set against this 'ticking population time-bomb' is the argument that the increase in food production is unlikely to keep pace. Lester Brown is in good company when he articulates the view that many aspects of modern farming systems are unsustainable because of their intensive use of limited resources such as cropland and irrigation water, and their reliance upon environment-damaging inputs from the agro-chemical industries. An additional point in the 1990s has been the prediction that Global Warming will disrupt agricultural production because of shifting weather patterns and the flooding of fertile coastal land by rising sea levels.

In its most extreme form, neo-Malthusianism foresees widespread famine amongst poor people in the least developed countries. Ehrlich, Ehrlich and Daily (1995) remark that 250 million people have died from hunger-related causes since 1970, with the implication that this is a trend which will become more pronounced in future if population is not controlled.

Table 9.1 A summary of the pessimists' case

- Population growth, coupled with income growth and a recognition of 'the right to food', has meant a rapid acceleration in demand.
- Food-producing resources are reaching their limit and there is evidence of a reduction in the availability of agricultural land and water per head.
- Inappropriate human activity is leading to environmental degradation, including reduced yields and the ruination of some formerly productive cropland and pasture.
- Global warming and other secular changes in the environment will impact negatively on productivity in some densely populated parts of the developing world.
- Improvements in the technology of production and marketing will not be enough to offset the negative factors. Food availability will decline and prices will rise. Food security is under threat.
- There will be an increase of hunger, malnutrition and famine, coupled with food riots and possibly wars between nations over scarce resources such as irrigation water.

Principal sources: Brown 1994, 1996; Ehrlich and Ehrlich 1990; Hardin 1993.

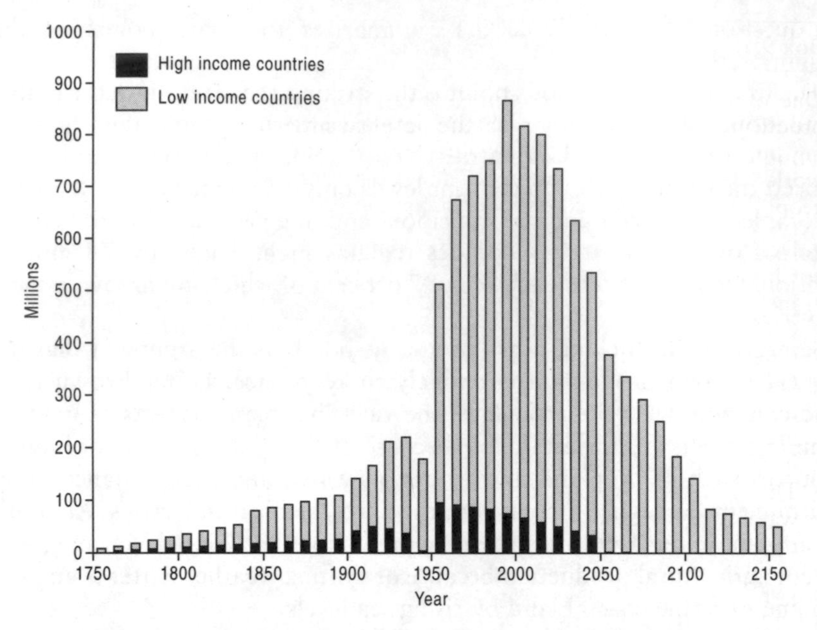

Fig. 9.2 Estimates and projections of decennial world population growth. *Source*: Bongaarts (1995).

The optimists

It is perhaps no coincidence that the pessimists and the optimists in the main have different disciplinary origins. Many of the neo-Malthusians have a scientific background and a knowledge of the ecological mechanisms which balance animal and plant populations with the natural environment. Lester Brown was originally trained as an agricultural scientist and Paul Ehrlich as an ecologist. In contrast, many optimists work in social science and especially in economics, where there is a long tradition of identifying the flexible responses that humans adopt to imbalances between the supply of goods and demand in the market. Without wishing to make too much of these divisions of intellectual genealogy, we can reasonably speculate that some of the disagreement between the two sides is a function of their conflicting constructions of knowledge.

Until his untimely recent death, Julian Simon was in the forefront of the demographic optimists. He saw human beings as our greatest asset and argued (Simon 1981, 355) that:

> The standard of living has risen along with the size of the world's population since the beginning of recorded time. And with increases in income and population have come less severe shortages, lower costs, and an increased availability of resources . . . And there is no

Box 9.1 Case study: who will feed China?

One of Brown's most interesting books (1995) concerns the problems of producing sufficient food for China's people. It is summarized in the following bullet points. The arguments are not new but using the context of the world's most populous nation is telling, especially when one considers the geopolitical implications of food shortages in this emerging super-power.

- Despite its efforts at containment, including the one-child-family policy, China's population will rise to 1.6 billion by 2030.
- Incomes have risen, increasing the demand for all food products. Consumers are now wanting more meat, especially pork, and this requires the use of scarce feeding stuffs such as grain.
- Grain land is being diverted from its traditional uses, especially diversification into higher value crops and loss to building land for new houses and industry. The availability of grain land per head will shrink from 0.08 ha in 1990 to 0.03 ha in 2030.
- The area under irrigation has stagnated in the last 20 years and groundwater reserves are over-exploited. Water must now be used increasingly for the rapidly growing urban population.
- Yield increases have slowed in recent years despite a heavy use of chemical fertilizers.
- China's grain production will fall by 20 per cent between 1990 and 2030.
- If population and income growth meet expectations, there will be a shortfall of 369 million tons a year, nearly double the world's entire grain exports at present. Large-scale imports are inevitable and these will drive up the world price of grain.
- Signs of strain in the economy have already appeared with several periods of high inflation due mainly to increases in food prices.
- Hunger is nothing new in China. During the Great Leap Forward (1959–61) approximately 30 million people died in a famine caused by climatic problems and administrative incompetence.

convincing economic reason why these trends towards a better life, and toward lower prices for raw materials (including food and energy) should not continue indefinitely.

Boserup (1965) was less concerned with skills of individuals than with the social response to the pressure of population upon resources. Contrary to the Malthusian concept of 'checks' through famine, she identified examples where changes in the organization of cultivation and the adoption of technological innovations had helped to increase food production through a gradual process of intensification, even in the most 'primitive' agricultural economies. In other words, population density might actually be a causal variable contributing to the upgrading of productivity. If this is correct then the idea of 'over-population' is redundant and famines are rather the result of short-term anomalies such as wars or climatic phenomena. In the longer term the balance between population and resources will be maintained.

One example of the Boserupian stance is given by Tiffen *et al.* (1994). In the Machakos District of Kenya these researchers studied an area which had been described in the 1930s during the colonial era as 'an appalling example of a large area of land which has been subjected to uncoordinated development . . . the inhabitants of which are rapidly drifting to a state of hopeless and miserable poverty and their land to a parching desert of rocks, stones and sand'. Since then, through the efforts of the Akamba people, and without significant aid from outside, the region has been transformed in 60 years into a model of sustainable agriculture. Soil improvement measures were widely adopted, including conservation tillage, contour farming and terracing. Coffee, fruit and horticultural crops were introduced and the rural infrastructure was improved. All of this was achieved despite, or perhaps because of, a six-fold increase of population between 1932 and 1989.

Other optimists, this time from science and technology, point to the food-producing potential of the new high-yielding varieties of crops which have been bred in the last 40 years, during the era of the Green Revolution, and also to the more recent developments in biotechnology which are promising 'designer crops' to suit particular circumstances of each region (Chapter 17). In view of the tremendous strides made in crop breeding and agronomy during the twentieth century, it would seem unreasonable to expect such advances to cease now, even if their impact may begin to slow.

The evidence

The first element of evidence relates to demography. Most of the developed world has already achieved very low levels of population growth, with replacement levels of fertility only. The 'South' is now following, with encouraging achievements of demographic control in the 1990s that have led to successive reductions in global population forecasts by the UN. In Figure 9.3 we can see the UN's predictions according to low, medium and high assumptions about fertility. It seems most likely that a stable population of about eleven billion will be reached in the next hundred years.

At this point it is worth reminding ourselves that high fertility is usually less to do with religious taboos on contraception or with cultural preferences for large families than with poverty (Figure 9.1). Poor households have to think of every means at their disposal to increase their income and the addition of more children is a strategic investment in the family's labour force. Where infant mortality rates are high, having many babies may be essential to ensure that some survive to the economically productive age. Economic development and better infant health then would make important contributions to reduced fertility, but also very important are the education of women and empowering them with control over their own

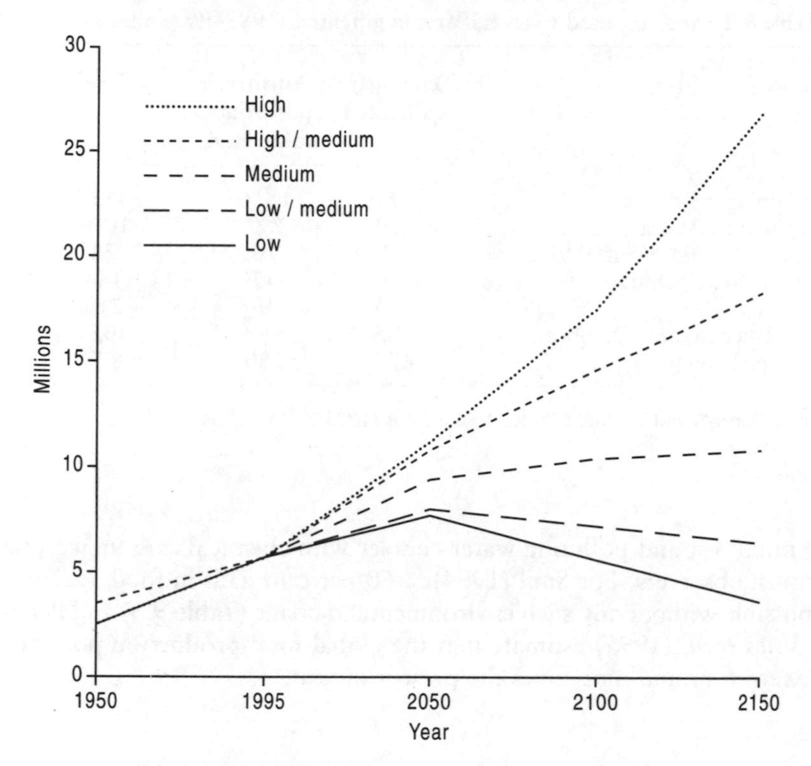

Fig. 9.3 World population projections. *Source*: UN.

bodies. This was one of the main conclusions of the World Population Conference at Cairo in 1994. Contraception is generally welcome where people genuinely wish to limit their family size but imposing it officiously can be both offensive and counter-productive.

Secondly, the evidence must answer the question 'are we running out of agricultural resources?' The pessimists would like us to think so, but optimists in the past have rejected such limitations and indeed 40 per cent of estimates made in the last 50 years of the earth's carrying capacity say that it could exceed 20 billions, over three times the present population figure.

In reality only 11.2 per cent of the earth's land area is presently exploited by arable agriculture, and much of that at relatively low levels of intensity. Mountainous topography, poor soils, short growing seasons near the poles, and shortages of irrigation water are all obvious constraints, yet there does seem to be plenty of scope for increasing production (Table 9.2). Reductions in fallow, conversion of grassland to arable, growing more than one crop per year, increased use of fertilizer, and the colonization of new land are all options, although the environmental consequences of cutting tropi-

Table 9.2 Land not used to its full arable potential, 1988–90 (million hectares)

Region	Presently cultivated	Further potential arable land	Total
Developing countries	721	1816	2537
Sub-Saharan Africa	213	797	1009
Middle East and North Africa	62	16	78
East Asia (excl China)	88	97	184
South Asia	175	38	213
Latin America and Caribbean	185	869	1054
Developed countries	677	200	877

Source: Buringh and Dudal (1987); Alexandratos (1995).

cal rainforest and polluting water courses with chemicals are unacceptable to most observers. For Smil (1994), a 60 per cent gain in food availability is possible without any such environmental damage (Table 9.3) and Penning de Vries *et al.* (1995) estimate that the global food production potential is between four and nine times the present output.

Table 9.3 Possible efficiency gains in food availability by 2050

Changes compared with 1990 practices	Gains equivalent to global 1990 food energy consumption (per cent)
Improved field efficiencies	
• Better agronomic practices (raise average yields 20 per cent)	22
• Higher fertilizer uptake (raise nutrient use efficiency 30 per cent)	7
• Reduced irrigation waste (raise water use efficiency 30 per cent)	7
Reduced waste	
• Post-harvest losses (lower by 20 per cent)	8
• End-use waste (lower by 20 per cent)	8
Healthier diets	
• Limit fat intake to 30 per cent of total energy	10
Total gain	60

Source: Smil (1994).

Much of the literature on macro-scale food production is over-literal in its use of technological and biophysical criteria in making calculations. Little account is taken of uncertain factors such as the boost to productivity that would result if peasant farmers in the less developed countries (LDCs) were given increased financial incentives for their products and were supported by the kind of expensive agricultural research and extension services which are taken for granted in the 'North'. Maybe the answer lies with them because small farms are certainly more productive per hectare than large ones, albeit at a lower level of economic 'efficiency'. Also, their traditional soil and water conservation practices are more sustainable, as is their mixing of crops in each field to reduce the risk of losses from pests and disease.

Nor are physical infrastructure and the organization of marketing often considered. The Former Soviet Union has vast resources of agricultural land which produce only a fraction of their potential because neither the collective farms nor their privatized successors have been able to afford the necessary investment in modern production equipment and inputs, transport, storage and processing facilities, and information and communication networks. Without these, yields are disappointing and much of the harvest rots before it can reach the consumer.

Two further points are important under the heading of agricultural resources. First, national and international politics significantly influence food production (see Chapters 11–13). Traditionally, the rural sector has been used as a taxation milch cow in the early stages of development and has later been seen as a lesser priority for investment than urban-based industrialization. Both types of policy militate against securing a well-nourished rural population. The advanced countries, on the other hand, are frequently embarrassed by surpluses of food production (produced at subsidized prices) and have to formulate policies to control their farming sector (Chapter 11). Achieving the full biological potential of the soil is therefore an aim very rarely encouraged by politicians today in developed countries.

A second and related point arises when LDCs are planning their trading strategies in order to maximize their economic development. Most have accepted the economic theory that specialization can help and they have therefore encouraged the export of raw materials to markets in the North, such as peanuts or cotton, or high value products such as cut flowers or luxury vegetables, which are air-freighted at great expense because of perishability. The amount of scarce land, labour, irrigation water and capital devoted to these cash crops is a drain upon the traditional farming economy and self-sufficiency in basic dietary commodities may be sacrificed.

Under our review of the evidence, we must finally think about food production and availability. About 40 per cent of the world's grain output is used as animal feed in the pursuit of meat and milk production, and there are vegans and vegetarians who argue that a widespread adoption

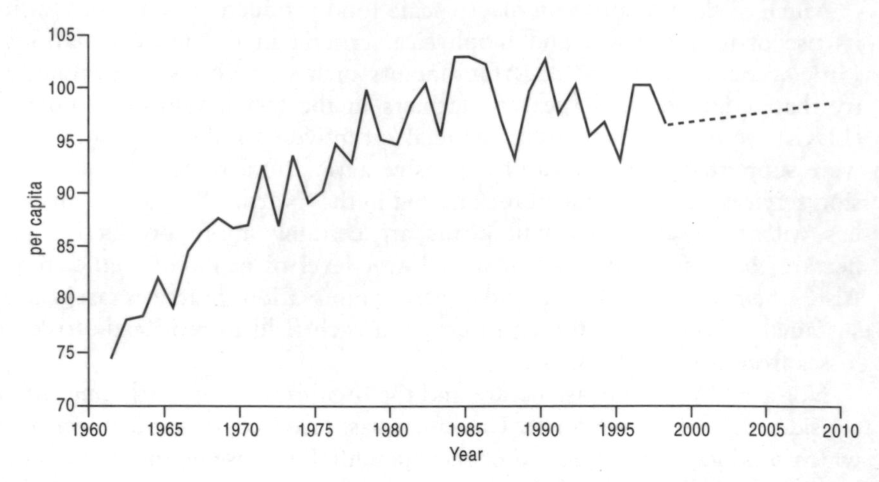

Fig. 9.4 World cereal production per capita, 1961–2010 (1989–91 = 100).
Source: FAOSTAT.

of their dietary practices would reduce the present inefficiency of translating the sun's energy into human nutrition. Apart from the release of grain for human consumption there would probably also be improvements in public health through a reduction in animal fat intake. Realistically, however, such a revolution in eating habits is very unlikely in the short term and we have to look for other sources of consolation on the world's food future.

Figure 9.4 does not look very promising at first sight. It shows very clearly a stagnation in global grain production per head from about 1980, and is perhaps a little surprising when one considers that grain yields have risen steadily throughout the last 30 years, as have the total cultivated area and the irrigated area. In fact this graph demonstrates the dangers of limiting the analysis to the global scale, because a full understanding can only be achieved by disaggregating the trends to a continental and subcontinental level. In the 1980s and 1990s the developed countries in North America and the EU have deliberately limited their cereal output by persuading farmers to set aside marginal land or to grow other crops. At the same time the collapse of the communist governments in Eastern Europe and the former Soviet Union has led to a disastrous period of falling productivity as guaranteed purchases by the state have disappeared. Together, the recent history of cereal farming in these two blocks of countries is sufficient to explain the global slowdown. In contrast, the record of increased food production in Asia has been outstanding (Figure 9.5).

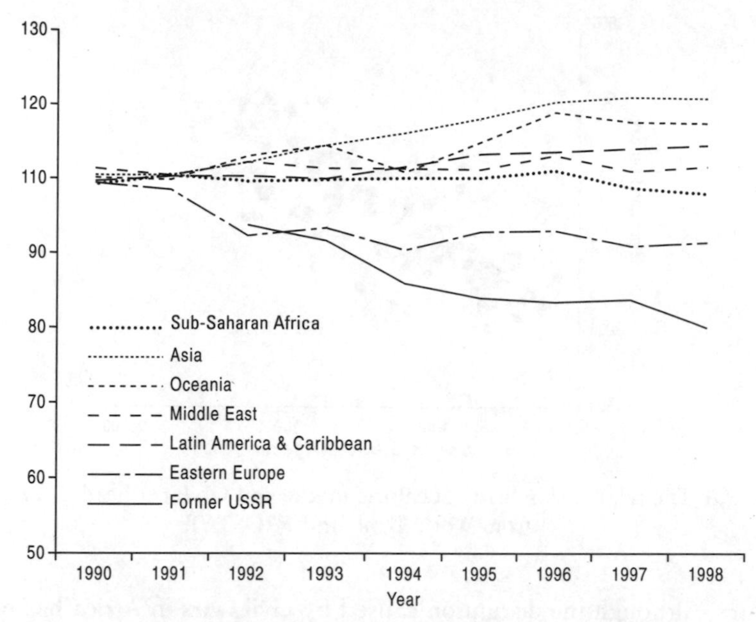

Fig. 9.5 Index numbers of food production in developing countries by region, 1990–98 (1989–91 = 100). *Source*: FAOSTAT.

Table 9.4 Global per capita food supplies for direct human consumption (kg per capita)

	1969–71	1988–90	2010
Cereals	146.3	164.6	167
Roots, tubers, plantains	82.3	65.7	65
Pulses, dry	7.6	6.3	7
Sugar, raw equivalent	22.1	22.7	24
Vegetable oils	6.7	10.1	13
Meat	26.0	31.9	37
Milk	74.6	75.3	72
All food (kcalories per day)	2430	2700	2860

Source: Alexandratos (1995).

Food and population: where concern is justified

Sub-Saharan Africa is usually quoted as the region with most remaining difficulties in its relations between population and resources. Countries such as the Central African Republic, Chad, Angola, Mozambique and Somalia still have high birth rates and their inability to feed their present populations from their own agricultural resources does pose obvious questions about the

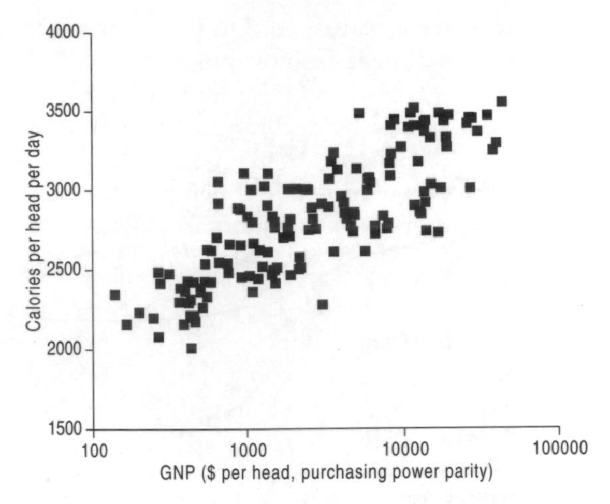

Fig. 9.6 The relationship between calorie intake and GNP per head. *Sources*: data from World Bank and FAOSTAT.

future. Although the disruption caused by civil wars in Africa has been an important factor in food shortages, we return to our original point made in the introduction, that poverty is fundamental. Figure 9.6 provides confirmation in a graph of calorie intake against GNP per capita, where the latter has been adjusted by taking purchasing power parities into account in order to allow for variations in the local cost of living.

Conclusion

Along with Sen (1994), we deplore the artificial and simplistic debate between what he calls 'apocalyptic pessimism, on the one hand, and a dismissive smugness, on the other', and we concur with Dyson (1996) that the reality of the food/population relationship lies somewhere in between these extremes. The latter's impartial analysis of a wide range of the latest official data leads us to the following conclusions:

1. Lester Brown is correct that global food production per capita has been falling since 1984, but this fall is largely due to policies and events in North America, Europe and the Former Soviet Union. Once these are factored out, the trend in the rest of the world is positive. Only Sub-Saharan Africa shows true signs of a food production crisis.
2. The world's grain harvests have become slightly more variable in the recent past, due to extreme weather phenomena associated with El Niño and other macro-environmental changes; but the impact of floods and droughts has been localized and most regions saw a reduction in harvest variability between 1970 and 1990.

3 It is not true that yields have plateaued. On the contrary, they continue to rise more or less in the linear fashion predicted by Malthus (Figure 9.7).
4 The toll of hunger and famine is declining and food security in most countries has improved (*see* Chapter 10).
5 The easing of the world food problem is demonstrated by the secular decline in the real price of grain in the international markets.

Dyson's (1996) regression analysis of food availability found that it was accounted for most satisfactorily by per capita income, and that the addition of population density to the model achieved very little extra explanatory power. This is because the higher income countries have navigated the demographic transition to low levels of fertility and they are able to command ample food for their consumers.

For comparison, the results of Bongaarts' (1996) regression analysis of food supply are shown in Table 9.5. He finds that population density is a significant explanation of the proportion of land cultivated, cropping frequency and crop yield, but not of other supply factors or of calorie intake per capita. The latter are better accounted for by national wealth. We ought to add here that both Dyson and Bongaarts based their work on national-level data and therefore took no account of the variability of access to food within countries.

At times one seriously wonders about the utility of much writing about the global balance between food production and population growth. First, the calculations of future trends are usually based upon heroic assumptions and the margin of possible error is great. To give but one example, everyone seems agreed that global warming will affect agricultural production in the medium and long term; but the forecasts for particular regions are very fragile, because sub-continental-scale models of climatic change are still at an early stage of formulation. At best these predictions are sophisticated 'guestimates'.

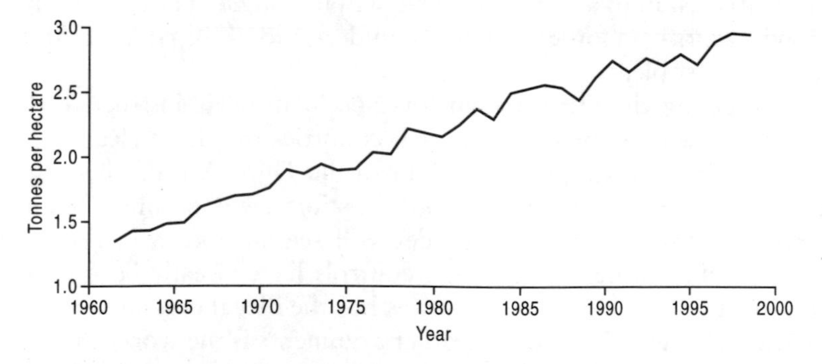

Fig. 9.7 World cereal yields, 1961–98 (tonnes per hectare). *Source*: FAOSTAT.

Table 9.5 Results from regression analyses of the determinants of food supply factors in 90 developing countries, 1989

	Explanatory variable				
			Region		
Supply factor	**Population density**	**GDP per capita**	**Latin America**	**Asia**	**Mid. East N. Africa**
Proportion of land cultivated	+++		–		
Cropping frequency	+++			+	
Proportion of area for food crops			– – –		
Crop yield	++	+++	++	++	
Trade multiplier					+++
Proportion directly consumed		–	– – –	–	– – –
Animal product multiplier		+++	++		
Calories per capita per day		+++			+++

+++ or – – –: $p<0.01$; ++ or – –: $0.01<p<0.05$; + or –: $0.05<p<0.1$.
Source: Bongaarts (1996).

Second, irrespective of the over-flowing grain stores of Green Revolution beneficiaries, such as India, even the most basic of nutritionally balanced diets is still beyond the reach of the very poorest people. Ministers of Agriculture in the Third World may proudly declare their country to be 'self-sufficient' in grain but in reality poverty is still artificially inhibiting demand. It is access to, and economic command over, food that matters to people in the lowest income decile, far more than agricultural or demographic statistics. In the next chapter we will investigate patterns of hunger and food security in order better to understand this paradox of food scarcity amongst plenty.

Deconstructing the literature on 'over-population' is instructive. Most originates and is published in those rich countries that have clear interests in criticizing the demographic behaviour of the Third World. Their biggest underlying concern is that the world's destitute will become 'economic migrants' and that the coming decades will see an upturn in their long-distance mobility. Stiffer immigration controls have already been imposed by virtually all of the advanced countries but the illegal crossing of borders has increased nevertheless. Among other arguments is the worry that larger populations in LDCs will lead to the degradation of 'the commons', such as the acceleration of global warming and ozone depletion by industrialization, and the conviction that a growth in consumption in poor countries will accelerate the demise of so-called non-renewable resources, especially the mineral raw materials so vital for manufacturing industries in the North. It takes no great powers of insight to see that these arguments are essentially selfish.

Further reading and references

Tim Dyson has written an accessible and detailed account of the relationship between food and population. The publications of Lester Brown, the Ehrlichs and the Pimentels are thought provoking but the reader should remember that they are very firmly on one side of the argument. Bongaarts' 1994 paper is a good popular summary of the issues.

Abraham, J. 1991: *Food and development: the political economy of hunger and the modern diet*. London: Kogan Page.

Alexandratos, N. (ed.) 1995: *World agriculture: toward 2010*. Chichester: Wiley.

Bongaarts, J. 1994: Can the growing human population feed itself? *Scientific American* **270** (3), 18–24.

Bongaarts, J. 1995: Global and regional population projections to 2025. In Islam, N. (ed.) *Population and food in the early twenty-first century: meeting future food demand of an increasing population*. Washington, DC: International Food Policy Research Institute.

Bongaarts, J. 1996: Population pressure and the food supply system in the developing world. *Population and Development Review* **22**, 483–503.

Boserup. E. 1965: *The conditions of agricultural growth*. Chicago: Aldine.

Brown, L.R. 1994: *Full house: reassessing the earth's population carrying capacity.* New York: W.W. Norton.

Brown, L.R. 1995: *Who will feed China? Wake-up call for a small planet*. London: Earthscan.

Brown, L.R. 1996: *Tough choices: facing the challenge of food scarcity*. New York: W.W. Norton.

Brown, L.R., Gardner, G. & Halweil, B. 1998: Beyond Malthus: Sixteen dimensions of the population problem. *Worldwatch Paper* **143**. Washington DC: Worldwatch Institute.

Buringh, P. and Dudal, R. 1987: Agricultural land use in space and time. In Wolman, M.G. and Fournier, F.G.A. (eds) *Land transformation in agriculture*. Chichester: Wiley, 9–43.

Dyson, T. 1996: *Population and food: global trends and future prospects*. London: Routledge.

Ehrlich, P.R. and Ehrlich, A.H. 1990: *The population explosion*. New York: Simon and Schuster.

Ehrlich, P.R., Ehrlich, A.H. and Daily, G.C. 1995: *The stork and the plough: the equity answer to the human dilemma*. New York: Grosset/Putnam.

Hardin, G. 1993: *Living within limits: ecology, economics and population taboos*. New York: Oxford University Press.

Penning de Vries, F.W.T., Keulen, H. van, Rabbinge, R. and Luyten, J.C. 1995: Biophysical limits to global food production. *2020 Brief* No. 18. Washington, DC: International Food Policy Institute.

Pimentel, D. and Pimentel, M. 1999: Population growth, environmental resources, and the global availability of food. *Social Research* **66**, 417–28.
Sen, A. 1994: Population: delusion and reality. *New York Review of Books* **41** (15), 62–71.
Simon, J.L. 1981: *The ultimate resource*. Princeton: Princeton University Press.
Smil, V. 1994: How many people can the earth feed? *Population and Development Review* **20**, 255–92.
Tiffen, M., Mortimore, M. and Gichuki, F. 1994: *More people, less erosion: environmental recovery in Kenya*. Chichester: Wiley.

10

Malnutrition, hunger and famine

Introduction

If, as we saw in Chapter 9, access to food varies by income group, class and geographical region, then what are the consequences for the disadvantaged groups and poor countries? This chapter seeks to answer that question by focusing on hunger. After an introductory section on nutrition, we ask questions about the nature of malnutrition and undernutrition. Famine is seen through the eyes of various theorists and examples are given of Bangladesh and Ethiopia.

Perhaps the most helpful general conceptual framework on hunger is that suggested by Watts and Bohle (1993). They have produced a method for the causal analysis of hunger and famine, through what they call the 'space of vulnerability'. This gives some theoretical substance to the investigation of individual famines. They argue that the three dimensions of entitlement, political economy and empowerment are the most important for understanding the various manifestations of vulnerability in social and geographical space (Table 10.1) and they look at a number of case studies in South Asia and Africa in order to visualize the changing nature of this vulnerability through time.

Nutrition and malnutrition

What should we eat in quantity and quality? This simple question has generated a huge industry of information, advice and advertising. By now most of us are aware of official advice to eat less fat, salt and sugar, and we have all seen media adverts and magazine articles encouraging various forms of dieting.

Our bodies have certain basic requirements of food in order to grow and maintain themselves. These will vary with age (Figure 10.2), and also sex,

Table 10.1 Social spaces of vulnerability

	Entitlement relations	Power/institutional relations	Social relations of production/class relations
In social relations	Vulnerability as entitlement problem	Vulnerability as powerlessness	Vulnerability through appropriation and exploitation
Vulnerable groups	The resource-poor	The powerless	The exploited
Critical regions	Marginal regions	Peripheral/dependent regions	Crisis-prone regions

Source: adapted from Watts and Bohle (1993).

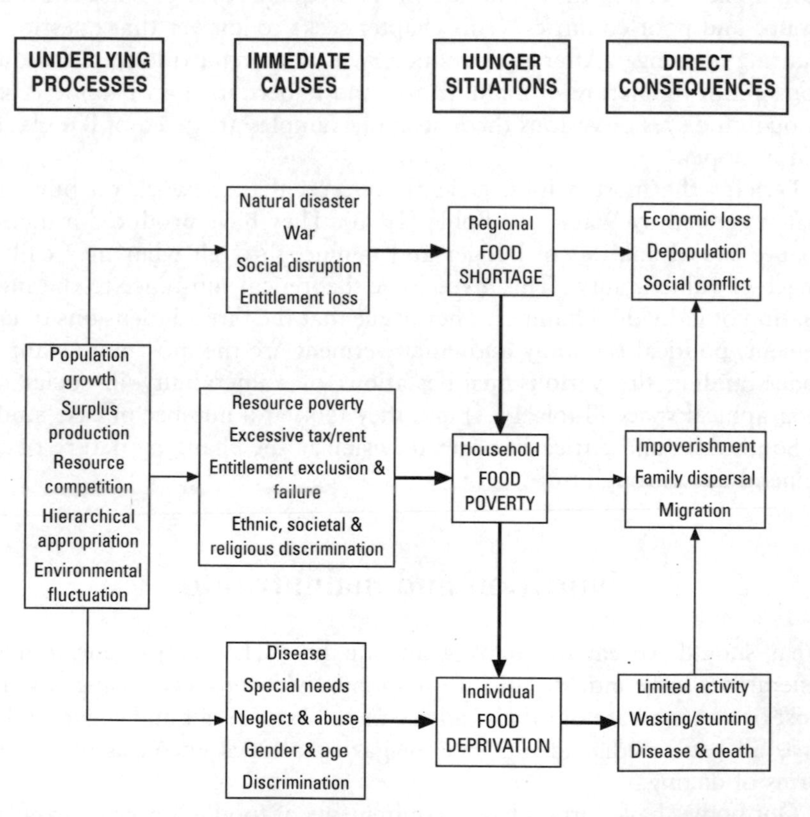

Fig. 10.1 The causal context of vulnerability. *Source*: redrawn from Kates (1992).

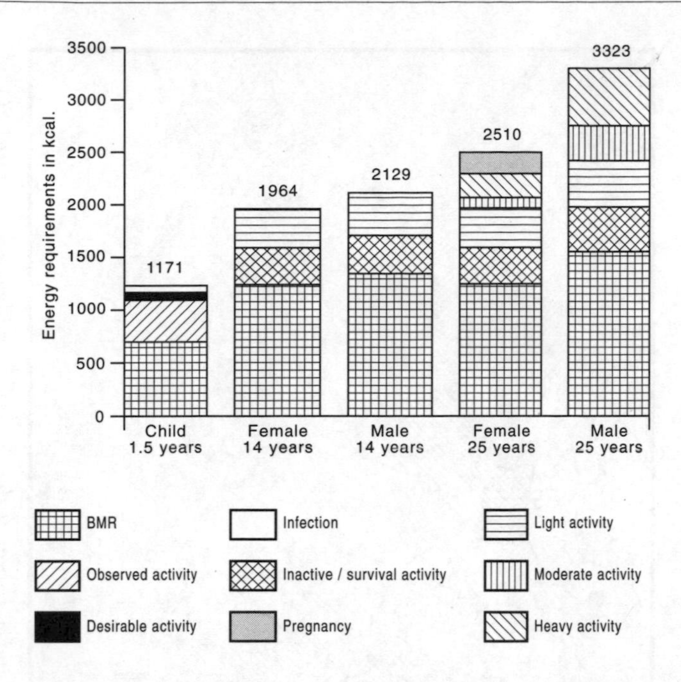

Fig. 10.2 Energy requirements at different ages. *Source*: FAO (1992).

occupation, and other factors. Manual labour requires additional energy, as does pregnancy. In addition to energy, our bodies also need proteins and certain vitamins, minerals and trace elements. These are essential for basic functions such as muscle, bone and tissue growth, and the production of healthy blood.

'Undernutrition' is usually defined as an inadequate intake of calories and 'malnutrition' as an imbalance of nutrient consumption, usually due to a shortage of a key vitamin or mineral (Figure 10.3 and Table 10.2). Occasionally problems may arise where the soil is deficient in certain elements due to geological factors.

It *is* possible to be malnourished even in rich countries, for instance on an unbalanced diet of junk food, or through over-eating, leading to obesity. Thorough historical research (Barker 1992), involving the analysis of individual patients' medical records, has also shown that foetal nutrition is especially important and that any dietary imbalance, including the excessive consumption of alcohol and tobacco by the pregnant mother, has significant consequences throughout the lifespan of her offspring (see Chapter 16). This conclusion has important implications for nutrition programmes in developing countries.

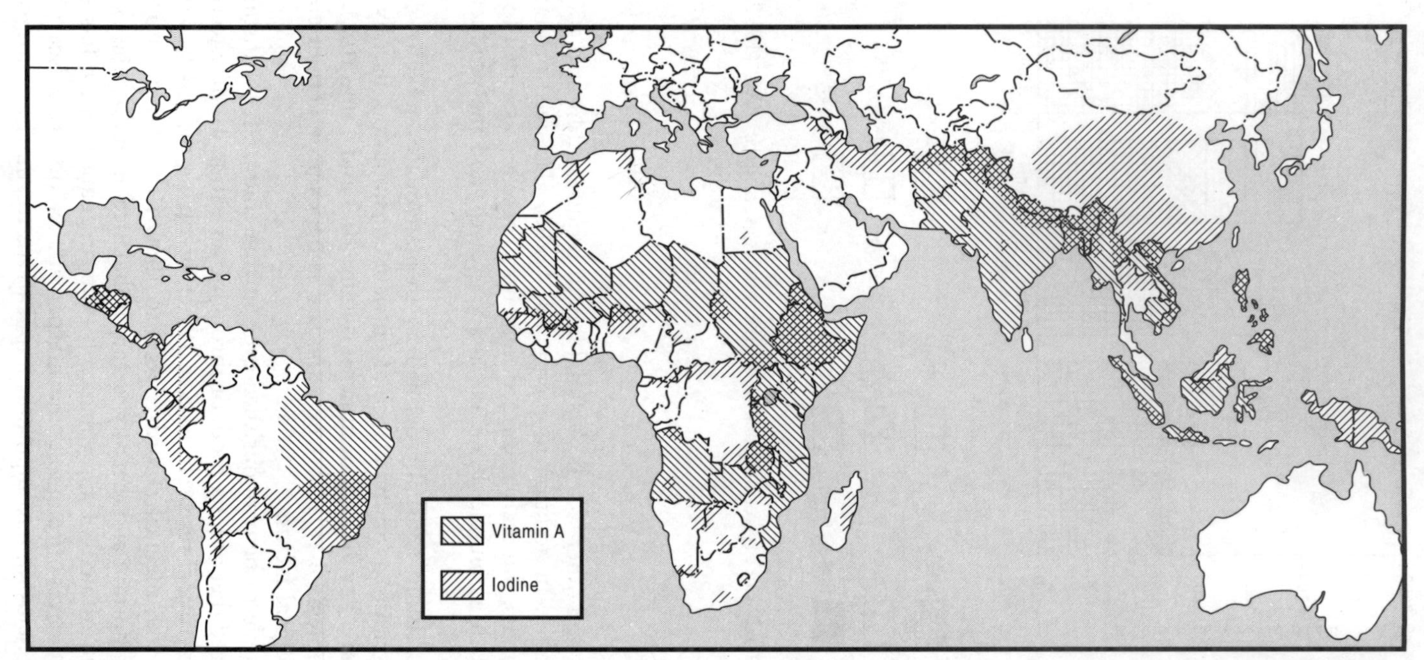

Fig. 10.3 Vitamin A and iodine deficiency. *Source*: FAO 1992: *Food and nutrition at the turn of the millennium*. Rome: FAO.

Table 10.2 Clinical and sub-clinical nutritional deficiencies by region (millions)

	Iodine	Vitamin A	Iron	Obesity
Africa	151	50	438	8
N. and S. America	52	17	266	86
S. and SE Asia	224	126	1435	7
Europe	94	0	150	82
Middle East	152	23	337	10
E. Asia and Pacific	255	42	1085	32
Total	927	259	3709	225

Source: WHO website (accessed 1998).

Most micronutrient deficiencies are readily solved by dietary supplements or by the fortification of foodstuffs while they are being processed and manufactured. Thus in Europe and North America salt is commonly fortified with iodine, bread with vitamins, toothpaste with fluoride, and cornflakes with niacin. Given relatively modest funding and some political will, such policies could solve the malnutrition problem in the developing world. A start has been made, with vitamin A deficiency declining by 40 per cent in the last ten years due to the large-scale distribution of capsules and other forms of delivery. This is a very important achievement because some experts believe that child mortality can be reduced by one third if vitamin A is universally available.

So far we may be excused for thinking that human nutrition is scientific and highly precise, but in reality the science is constantly evolving and the precision is lacking in certain areas where views differ between schools of thought. Two examples will suffice to make this point.

First, analyses of children's growth patterns in the LDCs are often based upon a comparison of actual height, weight, and arm circumference with a standard scale drawn from the experience of western children. This amounts to using a yardstick of maximized growth potential, because that is what results from protein-rich diets. But there is research in the LDCs that suggests that a child can be both slow-growing and healthy, and that small stature does not necessarily indicate undernutrition. A major controversy has arisen between these different viewpoints.

Indeed, it can be argued that a protein-rich western diet is not innately 'superior', and that protein has been a source of confusion among nutritionists. In the 1960s, for example, protein deficiencies were identified as one of the principal causes of malnutrition in the Third World. During the mid-1970s this view was modified, however, and it is now generally agreed that if the supply of energy (calories) is adequate, then an individual will almost certainly also ingest sufficient protein. The main exceptions to this are diets dominated by those small number of starchy staples which are poor in vegetable protein, for instance cassava, yam and plantain in West Africa. Here tiny children may find it difficult to eat enough to satisfy their bodies' needs of both protein and energy.

Second, there has been a heated debate over many years about recommended daily allowances (RDAs) of nutrients. The most fundamental is the dietary energy supply (DES) needed to meet the basal metabolic rate (BMR): the minimum needed for breathing and pumping blood. This varies, according to age, sex and body weight, between 1300 and 1700 kcal per day. To this is added the DES needed for light activity, about 55 per cent of BMR, and any further requirement for heavy work. In the early 1950s the FAO was advocating an overall average of 3200 kcal per day and, as a result, the second World Food Survey (1952) estimated that 60 per cent of the world's population received insufficient calories. This was gradually reduced, following shifts in nutritional orthodoxy, to 2700–2900 kcal in the sixth World Food Survey, including an allowance at the national level for inequalities of access to food (FAO 1996). Lipton (1983) thinks 2100 kcal is a more realistic figure for the definition of undernutrition and ultra-poverty.

Pacey and Payne (1985) and Payne (1992) argue that the RDA approach is flawed. It sets an artificial threshold of undernourishment that often has no meaning in reality. Thus poor urban dwellers in India consume less food than those in rural areas, but they are not necessarily worse off than agricultural labourers, who expend more energy. A better measure is the balance of food input and output of energy expended. It is possible to become fat on 1500 kcal per day by just sitting in an armchair. They suggest that nutritional problems can only be fully understood in their socio-economic and ecological contexts.

Undernutrition and malnutrition are the consequences of the inter-relation of a complex nexus of variables. For instance, women, children and the very old may be at risk when sacrifices are made in order that the active male bread winner might be fed first. They eat 'least and last'. Hidden intra-household maldistribution of food may thus be responsible for undernutrition in villages that might otherwise appear on average to have sufficient food.

Pacey and Payne also point out that the traditional recommended intake approach fails to take into account the following variabilities which may crucially affect people's nutritional status:

- Metabolic variability by genetic inheritance.
- Cultural and religious food taboos and preferences (*see* Chapter 23).
- Seasonal variability. In rural areas of the LDCs, hunger peaks just before the harvest when food stocks are lowest (Figure 10.4).
- Ecological niche (Figure 10.5). Although the monetized economy has reached most of the rural 'South', a degree of self-sufficiency remains important in peasant livelihoods, and the type and range of crops grown locally is therefore still a factor in dietary composition.

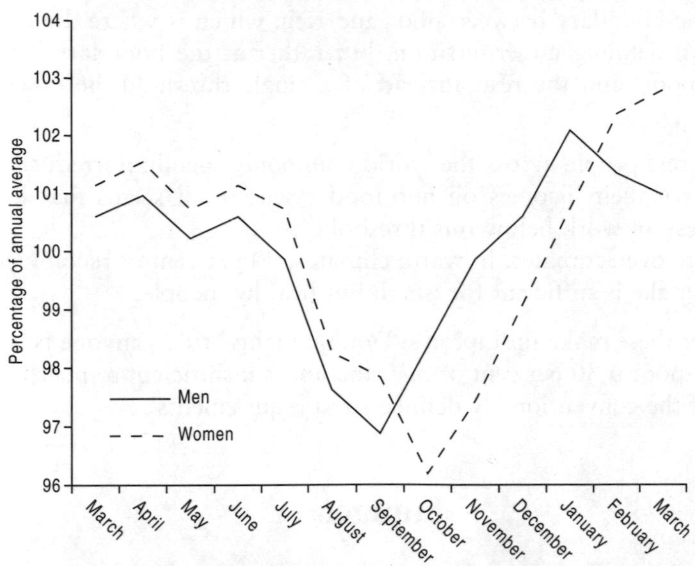

Fig. 10.4 Seasonal adult body weight in a Gambian village (percentage of annual average). *Source*: after Fox, cited by Pacey and Payne (1985).

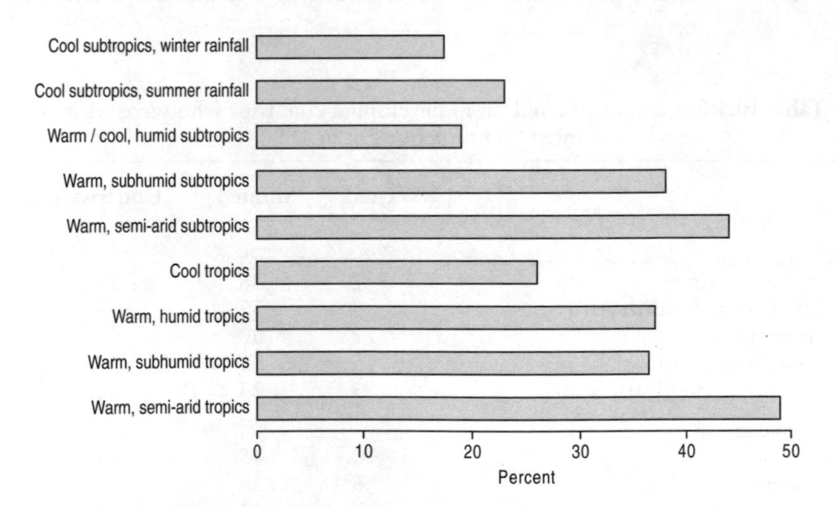

Fig. 10.5 Underweight pre-school children by agro-ecological regions, 1990. *Source*: Sharma *et al.* (1996).

Lipton (1983) argues that the key point of measurement comes not so much at the boundary between poor and rich, which is where the RDA is operative in defining undernutrition, but rather at the boundary between the 'ultra-poor' and the rest. Instead of a single threshold, he takes two together:

- The poorest people across the world commonly spend an irreducible 20 per cent of their incomes on non-food essentials. It seems that Engel's Law does not work below this threshold.
- RDAs are overestimates. In warm climates, 80 per cent of recommended calorie intake is sufficient for 'small but healthy' people.

Together these make up Lipton's 'Double Eighty' rule: 'anyone is judged to be ultra-poor if 80 per cent of their income is insufficient to purchase 80 per cent of the conventionally defined food requirements'.

Hunger

Hunger is more than just a temporary physical discomfort, the most that is ever normally experienced in the nutritionally comfortable developed countries. It is often a chronic (recurring and long-term) and severe condition that may be a precursor to famine and starvation. The physical development of hungry children is impaired to the extent that they may be wasted (poor muscle and body fat), stunted (under normal height) and seriously underweight (Table 10.3). The distribution of undernutrition in Figure 10.6 shows that virtually the whole of Africa south of the Sahara is affected,

Table 10.3 Percentage of children in developing countries who were wasted, stunted and underweight in 1990

	Wasted	**Stunted**	**Underweight**
Region			
Sub-Saharan Africa	7	38	30
Middle East and North Africa	9	32	25
South Asia	17	60	58
East and South East Asia	5	33	24
Latin America and Caribbean	3	23	12
Economic group			
Low income	10	45	38
Middle income	6	29	22
Total	9	41	34

Source: FAO (1996).

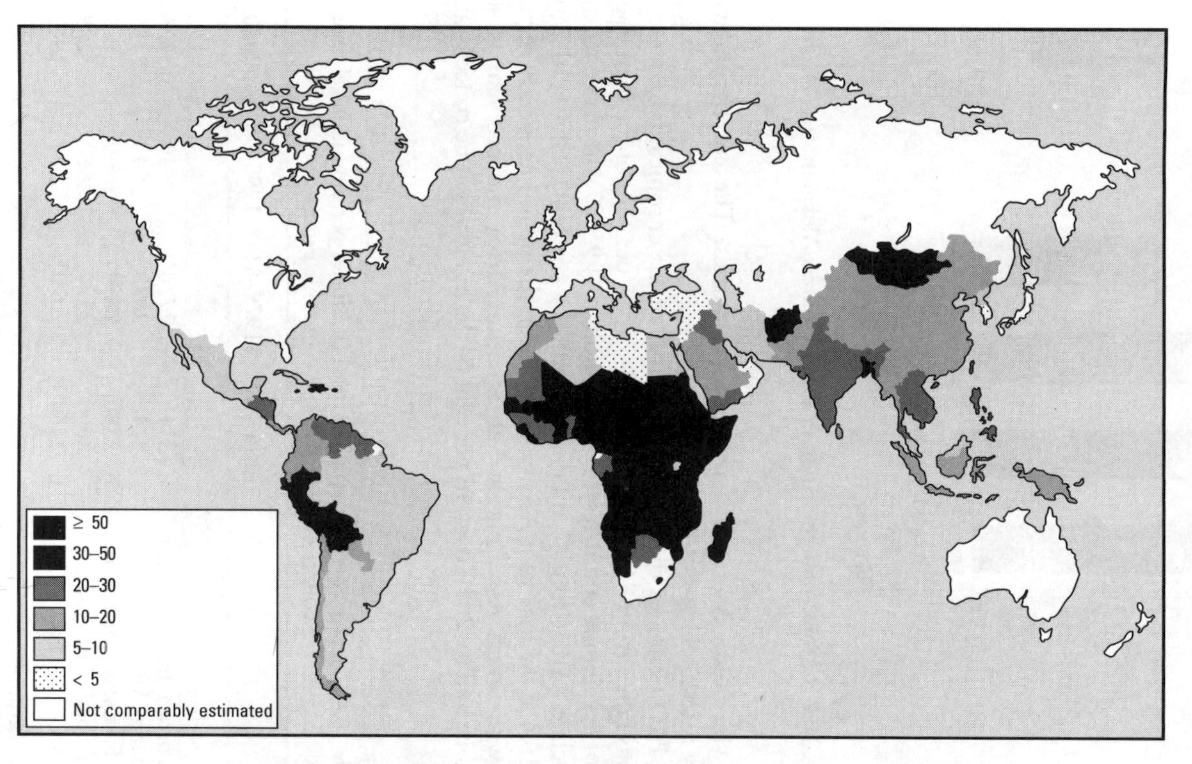

Fig. 10.6 World map of chronic undernutrition. *Source*: FAO (1996).

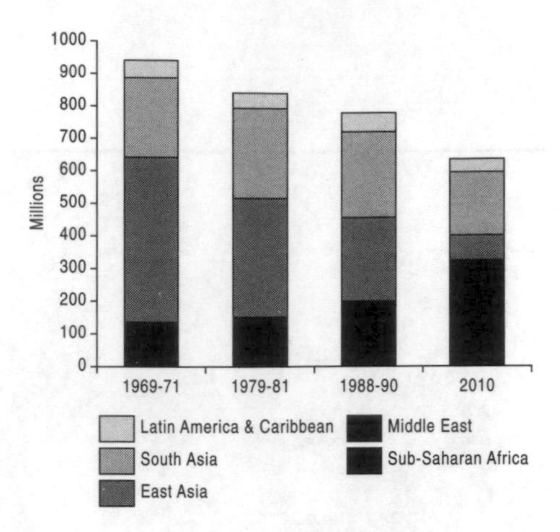

Fig. 10.7 Chronic undernutrition in LDCs. *Source*: Alexandratos (1995).

along with South Asia and a few other countries. The correlation with poverty is compelling but hardly surprising.

The FAO estimates that the total number of the chronically undernourished has fallen from 918 million in 1969–71, 35 per cent of the then population of developing countries, to 790 million in 1995–97 (13 per cent), and will decline further to 637 million in 2010 (11 per cent). Figure 10.7 and Table 10.4 show that East Asia will have been the main beneficiary region, along with South Asia, but in Sub-Saharan Africa the problem is still expanding and by 2010 over 46 per cent of the hungry will be concentrated there.

Table 10.4 Dietary energy supply per head (kcalories per day) in developing countries by region and economic group

	1969–71	1979–81	1990–92	2020
Region				
Latin America and Caribbean	2510	2720	2740	3026
Sub-Saharan Africa	2140	2080	2040	2135
Middle East and North Africa	2380	2850	2960	3114
East and South-east Asia	2060	2370	2680	3239
South Asia	2060	2070	2290	2640
Economic groups				
Least developed	2060	2040	2040	
Low-income food-deficit	2060	2230	2450	
Low-income	2060	2210	2430	
Middle-income	2360	2670	2760	

Source: FAO (1996); Rosegrant *et al.* (1995).

Famine

We all think we know what famine is. *The Oxford English Dictionary* definition reflects the popular understanding of 'extreme and general scarcity of food'. The media, and especially television from the 1960s onwards, has been very powerful in forging our visual image of shrunken, desperate people, robbed of their human dignity as well as of their health. But we should be aware that some of the key elements of this vision do not always apply:

- Famines are not always widespread: they can be geographically very localized and may affect only one group or class of people.
- Famines are not always the result of a general food shortage.
- Starvation is rarely a proximate cause of death. Undernutrition reduces the body's resistance to infection and famished people tend to die from a wide range of diseases.

It is better to describe famine as a socio-economic phenomenon or syndrome. Not all of the symptoms are necessarily present in every famine, but often a combination of these factors is visible months, or even years, before the excess mortality of the famine itself (Table 10.5). The poor and marginalized are familiar with temporary crises in their livelihoods and develop 'coping strategies' to get them through these to better times; but when a famine eventually impacts even these may fail.

Food availability decline

Many explanations of famine have been advanced in the past. By far the most common and influential until recently have been those based upon 'food availability decline' (FAD), which concentrate on problems on the supply side of the food system. Some FAD theorists emphasize short-run environmental disasters, such as droughts and floods, or crop failure due to diseases like the blight which affected the potato crop in Ireland in the 1840s; others prefer medium and longer-term factors, such as over-population (*see* Chapter 9). But both approaches use a blunt instrument. They cannot explain why some groups in society are more vulnerable than others. Moreover they look less and less viable as explanations in a modern world where information flows through high-technology communications equipment reduce the risk of 'hidden famines' in inaccessible locations, and where the world trade in cereals and other basic foodstuffs is flexible enough to react quickly to emergencies. The FAD hypothesis is now best borne out under two sets of conditions. First, food shortages are common in non-monetized peasant subsistence economies afflicted by some natural calamity, such as a crop failure,

Table 10.5 Coping with famine in western Sudan

Source of adjustment	Stages of response		
	Early	**Intermediate**	**Final**
Production	• Change in cropping and planting practices • Increased non-farm home production		
Labour	• Migration in search of employment	• Migration in search of employment (intensified in face of falling expected wage rate) • Separation of family	• Distress migration
Assets/capita	• Sale of small stock (liquid, easily reversible) • Sale of large stock (non-essential)	• Sale of production assets (livestock, tools, land) in depressed market	• Separation of families (possibly permanent
Loans/transfers	• Use of interhouse-hold transfers and loans	• Credit from moneylenders	• Donation (relief assistance)
Consumption	• Switch in expenditure/ dietary composition • Reduction of current consumption level (cut in frequency or size of meals or both) • Adjustments in intra-household allocation	• Reduction of consumption (greater dependence on market)	• Reduction of consumption (survival may be threatened)

Source: Teklu *et al.* (1991).

where the resulting famines are often localized. Second, FAD is possible in broader geographical regions where agriculture is only marginally viable even in good years. The introduction of irrigation in many poor parts of the world has reduced the probability of this sort of famine but there are still vulnerable groups among the nomadic and semi-nomadic pastoralists.

Entitlements

More fashionable, recently, have been the theories of demand-side economists, particularly the Nobel Prize winner Amartya Sen (1981), who argued

Table 10.6 A summary of events in the Bangladesh famine, 1974

- Weak rural economy, with falling real wages, decreasing plot size, and increasing proportion of landless labourers.
- There was civil war in 1971, which led to the splitting of Pakistan and the foundation of Bangladesh. This meant an economic slump and the dislocation of transport.
- The early signs of famine appeared between February and June 1974. People began eating alternative 'famine foods' like plantain saplings and banana leaves. These foods even started to appear on the market. There was a noticeable increase in thefts and begging. People started flocking into the towns, especially Dhaka, looking for relief.
- The peak months of mortality were from July to October 1974. There had been similar timing in the other devastating Bengal famines of 1770, 1866 and 1943. Up to two million people migrated from their homes to find food, some walking 100 miles. There was a disintegration of traditional family bonds, with parents even deserting their children and husbands deserting their wives. People sold or mortgaged their land, sold cattle and agricultural implements, even household utensils: anything that might help them to buy food for the short-term.
- The peak had passed by November/December 1974, by which time about 1.5 million people had died.
- The government did not declare a crisis until September 1974, when they set up 4300 soup kitchens (langarhhanas), but these were closed again in late November. The three hardest hit provinces were also the three to receive the least relief from government stocks.
- Bangladesh was in political difficulties with the USA because it had sold jute to Cuba. This led to the suspension of food aid.
- There were severe floods in 1970 and 1974, and devastating cyclones in 1970 and 1973. In 1974 the floods were worst in July and August when rice is at its seedling and transplanting stage. There was no question of the harvest being washed away and potential food shortage was still several months away at that stage. The crisis was one of under-employment, because there was nothing for labourers to do and they were therefore not being paid.
- Ironically, 1974 turned out to be the best national harvest for 4 years.
- Rice price inflation was rapid in the early months, accelerating from June onwards. Even those still in work found themselves compromised because wages remained static and so purchasing power fell dramatically. Entitlements therefore declined and the famine was widespread because so many people were similarly affected.

Source: Sen (1981).

that famines are the result of a collapse in people's ability to purchase or otherwise acquire food. Such a demand crisis may affect only one social group or class of people and, ironically, it may happen amidst the plenty of a good harvest and plentiful warehouse stocks. Sen's writings are conceptually complex but in essence he advances the idea of what he calls an 'entitlement' to a commodity, such as food, which is based upon assets, power, labour, production, trade, inheritance, and perhaps aid from some organization such as the state or an NGO. Note that income alone is not enough to account for entitlements, because a landless sharecropper and a landless labourer may have the same notional income in money or in kind, but their entitlements during a famine will be different. The sharecropper has at least partial control over a source of food, and may have access to some relief from his/her landlord. The landless labourer, on the other hand, will be on a fixed money wage when work is available, but if the harvest fails s/he will have no work when it is most needed. Wage labour also suffers if wages do not keep up with inflation; and in food-insecure economies the price of food sometimes does accelerate shockingly during an acute shortage or a perceived shortage (Table 10.6).

Sen's work has attracted much praise but also some criticism. In short, in retrospect it now seems to have been somewhat narrowly focused, underplaying especially the background political, cultural and ecological factors that were difficult to bring into his 'conjunctural' analysis of famine events. Also Sen is unable to theorize the process of adjustment and recovery after the crisis itself. In response, Sen has extended his notion of entitlements to a broader one of human 'capabilities' and of gaining access to resources through a cluster of rights, including empowerment and enfranchisement through political agitation (Watts and Bohle 1993).

Conclusion

Looking back to Part II, political economists offer an alternative perspective on the evidence presented in this chapter by explaining famines and other social pathologies in terms of the stress caused by market forces in capitalist economies and also as a consequence of the asymmetrical relations between classes. In effect hunger and famine are seen as outcomes of the increased poverty and vulnerability that are said to accompany commercialization, proletarianization and marginalization. One example might be the so-called 'tragedy of the commons', where the resources that traditionally have been sustainably managed for the good of the whole society are, under the new conditions of a monetized and commoditized economy, selfishly exploited and perhaps degraded by individuals. Desertification, soil erosion and other negative environmental consequences have been explained in this way, with famine at only one remove once the degradation is underway.

In addition, rather than the supply or demand sides being solely implicated, the marketing institutions and mechanisms of balancing the two are seen as materially at fault. The political economy literature identifies peasant societies, newly dependent upon a cash exchange economy, as being especially vulnerable. The marginal livelihoods of poor farmers may force them into 'distress sales' of their harvest at a low price, only for the retail price of grain later in the season to be driven up by the speculation and hoarding of middlemen. Integration with the outside world may be weak and, where balancing supplies from outside are not accessible, the preconditions for famine are laid. The detailed research of Barbara Harriss-White in southern India has helped us to understand the nature of the food marketing process in the developing world (Harriss 1981, 1984; Harriss-White 1995).

Looking forward to Part IV on the political ecology of food, inappropriate government policies may also exacerbate deaths from famine. In some cases they may even cause them, as was the case in the Soviet Union from 1932 to 1934. At that time the collectivization of agriculture led to declining productivity, but the extraction of quotas of grain continued and brought about the deaths of 5–7 million people. Above all, international and civil conflict has been a major factor in famines. For example, estimates made for Africa by Messer *et al.* (1998) suggest that, since 1980, wars have reduced food production per head per year by 2–5 per cent overall and by about 12.4 per cent in those countries directly affected. In 1996 there were 23 million refugees who had migrated across borders, 27 million displaced within their own countries, and a further 30 million trapped in combat zones. These 80 million people were at direct risk of hunger (Messer *et al.* 1998).

We can conclude, therefore, that hunger and famine, in turn the insidious and dramatic manifestations of undernutrition, are really just subsets of poverty and the failure of economic development. Over millennia society has evolved the ability to cope informally with regular seasonal hunger and with the occasional shocks caused by extreme environmental events; but the malign influence of inappropriate macro-economic policies, and of military strife, has in recent decades meant the occasional disastrous collapse of ordinary people's ability to feed themselves. We do not interpret acute food problems of this sort as 'acts of God' or as 'natural disasters', rather we place them in their wider economic and political context. Emergency famine relief is, of course, essential but such humanitarian responses are no substitute for the longer-term development of prosperity.

Further reading and references

Both Pacey and Payne (1985) and Foster and Leathers (1999) are good on the technical background to malnutrition. Amartya Sen's work is outstanding in its penetrating insight into the problem of famine, coupled with a style that is not overladen with jargon. Gordon Conway's book is even more accessible and is very much up-to-date on current issues about food production. Michael Watts is a fine theorist on food geography.

Abraham, J. 1991: *Food and development: the political economy of hunger and the modern diet.* London: Kogan Page.

Alexandratos, N. 1995: The outlook for world food and agriculture to year 2010. In Islam, N. (ed.) *Population and food in the early twenty-first century: meeting future food demand of an increasing population.* Washington, DC: International Food Policy Research Institute.

Barker, D.J.P. (ed.) 1992: *Fetal and infant origins of adult disease: papers written by the Medical Research Council Environmental Epidemiology Unit, University of Southampton*. London: British Medical Journal.
Conway, G. 1997: *The doubly green revolution: food for all in the 21st century*. London: Penguin Books.
Devereux, S. 1993: *Theories of famine*. New York: Harvester Wheatsheaf.
Food and Agriculture Organization of the United Nations 1992: *Food and nutrition at the turn of the millennium*. Rome: FAO.
Food and Agriculture Organization of the United Nations 1996: *The sixth world food survey*. Rome: FAO.
Foster, P. and Leathers, H.D. 1999: *The world food problem: tackling causes of undernutrition in the Third World*. Boulder: Rienner.
Harriss, B. 1981: *Transitional trade and rural development: the nature and role of agricultural trade in a south Indian district*. New Delhi: Vikas.
Harriss, B. 1984: *Exchange relations and poverty in dryland agriculture: studies of south India*. New Delhi: Concept.
Harriss-White, B. 1995: *A political economy of agricultural markets in South India*. New Delhi: Sage.
Kates, R.W. 1992: Times of hunger. In Wong, S.T. (ed.) *Person, place and thing: interpretative and empirical essays in cultural geography*. Baton Rouge, LA: Geoscience Publications, Department of Geography and Anthropology, Louisiana State University, 275–99.
Lipton, M. 1983: Poverty, undernutrition, and hunger, *World Bank Staff Working Paper* No. 597. Washington, DC: World Bank.
Messer, E., Cohen, M.J. and D'Costa, J. 1998: Food from peace: breaking the links between conflict and hunger. *2020 Brief* No. 50. Washington, DC: International Food Policy Institute.
Pacey, A. and Payne, P. (eds) 1985: *Agricultural development and nutrition*. London: Hutchinson.
Payne, P. 1992: Assessing undernutrition: the need for a reconceptualization. In Osmani, S.R. (ed.) *Nutrition and poverty*. Oxford: Clarendon Press, 49–96.
Rosegrant, M.W., Agcaoili-Sombilla, M. and Perez, N.D. 1995: Global food projections to 2020: implications for investment. *Food, Agriculture, and the Environment Discussion Paper* No. 5. Washington, DC: International Food Policy Research Institute.
Sen, A.K 1981: *Poverty and famines: an essay on entitlement and deprivation*. Oxford: Clarendon Press.
Sharma, M., Garcia, M., Qureshi, A. and Brown, L. 1996: Overcoming malnutrition: is there an ecoregional dimension? *Food, Agriculture and the Environment Discussion Paper* No. 10. Washington, DC: International Food Policy Research Institute.
Teklu, T., Braun, J. von and Zaki, E. 1991: Drought and famine relationships in Sudan: policy implications. *Research Report* 88. Washington, DC: International Food Policy Research Institute.
Watts, M.J. and Bohle, H.G. 1993: The space of vulnerability: the causal structure of hunger and famine. *Progress in Human Geography* 17, 43–67.
Webb, P. and Braun, J. von 1994: *Famine and food security in Ethiopia: lessons for Africa*. Chichester: Wiley.

11
Food surpluses

Introduction

In a world plagued by hunger and famine, the existence of food surpluses could be interpreted as, at best, paradoxical or, at worst, morally indefensible. Surely, some argue, the food surpluses of the developed world can be used to eliminate the food deficits of the developing world, as revealed by Chapter 10. But circumstances are never as simple as they might first appear, and the purpose of this chapter is to throw some light upon the causes and characteristics of food surpluses, so as to lay the foundations for an understanding of their place in food aid (Chapter 12) and food trade (Chapter 13).

Looking in brief at the meaning of 'food surplus', in one sense achieving a level of food production surplus to domestic demand is a laudable achievement: such a surplus can form the basis of both international trade, with its economic advantages to participants, and food aid, with its humanitarian benefits. In addition, in circumstances where food production can vary from year to year according to the vagaries of climate and economics, a surplus in one year can help offset the deficits of another. But the term 'food surplus' is not generally used in this positive sense; rather it is applied pejoratively to production in excess of domestic demand that occurs, over a number of years, at production costs and prices above those prevailing on world markets.

Such 'structural' or persistent surpluses can only be sustained by an economic environment subsidized by the state and protected from international competition and its world market prices. Otherwise prices would fall on an over-supplied market, farmers would respond by reducing their production, and a new equilibrium would be established in the market between supply and demand. So as to enter lower-priced international trade, traders of high-cost food surpluses have to be further subsidized by export refunds. Not unreasonably, producer nations operating under world market prices resent this 'unfair' competition and the instability it can bring to international trading relations. Just what 'free market' price levels might be for any agricultural product is uncertain, however, not least because of the distorting effects on trade caused by subsidized food surpluses (i.e. internal market protection and external subsidized exports), and the varying levels of protection given to the farm sector by most developed countries. Rather, 'shadow prices', usually higher than existing international market prices, have to be estimated as the likely global prices of each agricultural product in the absence of aggregate state assistance to the agricultural sector (Ritson and Harvey 1991, 250).

The state and food surpluses

The previous paragraph shows how any consideration of food surpluses, as defined, is drawn inevitably into an analysis of the logic of state intervention in agriculture. Indeed this is a convenient place to consider a topic which was left implicit rather than explicit in our earlier analysis of the political economy of food in Part II. At that time we made only passing reference to the importance of the state in regulating the relations between actors in the food supply chain, choosing instead to leave a more detailed consideration until now. We can begin by placing state intervention into the context of 'real regulation', with its basis in state theory and legal interpretism (Moran *et al.* 1996). Real regulation is the 'administrative manner, style and logic by which the state regulates society in general and the economic landscape in particular' (Clark 1992). For example, under the second food régime, UK agriculture experienced two phases of 'real' regulation, each with long-term consequences for the development of the industry. Prior to 1973 agriculture was subject to regulation by UK agricultural policy; between 1973 and the early 1980s productionist regulations were applied under the CAP of the EU. Since the early 1980s a post-productionist reregulation of agriculture under the CAP has become increasingly influential (Ilbery 1999, 57).

Clark's (1992) account privileges bureaucrats and judges in the writing and interpretation of law, but he also emphasizes regulation as a social practice, having elements of spatial, contextual and historical uniqueness. Rules and legal procedures, therefore, are set within national, sociocultural milieus, so that we need to examine the wider social conditions under which the formulation of regulations takes place. Social action can lead to the reformulation of legislation: real regulations are contested, not fixed, and the process of regulatory change is an arena of contest and challenge. In other words, the regulatory state is interpreted as pluralistic in nature. Thus Moran *et al.* (1996) extend Clark's real regulation by drawing attention to: (1) social groups and social action in the formulation and interpretation of regulations; (2) the inertia effect of existing regulations;

and (3) the importance of ideology. They propose a formulation–enactment–interpretation–reformulation model, similar in concept to Bowler's (1979, 2) agricultural policy process model (i.e. policy goals–policy measures–policy impacts–revised policy goals), and draw attention to the significance of political bodies at local and national levels, such as pressure groups, which may not have full legal force or status. In addition, both Moran *et al.* and Bowler show that present legislation is predicated on pre-existing legislation, with legislative lags acting as a powerful force in the uneven impact of real regulation in different parts of the world.

For the present discussion, three main arenas of state regulation can be identified: the agricultural market, agri-environmental relations and food quality/safety. Here we deal with regulation of the market, with agri-environmental and food quality regulation. We argue that the uneven development of food systems at a global level can be traced, in part, to national differences in the framework of real regulation as developed in each nation-state. Each state has its own structure, process and outcome in real regulation, with consequences for the creation and persistence of food surpluses. In this chapter we use the CAP to illustrate our argument.

Let us turn first to the logic of state intervention in agriculture and its relationship with agricultural surpluses. That logic begins with the reasons for intervention by the state in agriculture and Table 11.1 shows the objectives of the CAP. A useful distinction can be drawn between utility and equity objectives. In the first group lie objectives such as providing security of national food supply, contributing to national economic growth, helping the balance of trade in the economy, and providing an efficient sector in the exploitation of national resources. In the second group can be found equity objectives, such as providing a reasonable return to the capital and labour of those employed in the farm sector, providing stability in the incomes of the farm population, maintaining rural society, and reducing variations in income levels between different types, sizes and locations of farms. A reading of these objectives, therefore, reveals a mixture of economic and social objectives, some of which are in conflict, with a varying balance being struck between them according to national economic context and political ideology. For example, on the one hand the state acts to facilitate capital accumulation by actors in the farm, food and retail sectors; but on the other hand, the state also acts to protect farm businesses from the consequences of capitalist market forces. Indeed a common characteristic amongst developed countries has been the weighting given to social equity objectives, especially as regard farm income levels, in the development of agricultural policy.

Following Bowler (1992, 251), policy measures (mechanisms) are created to meet the objectives sought within each country, and a useful categorization can be made between measures that:

- increase the demand for farm products
- reduce the supply of farm products

Table 11.1 The objectives of the CAP (Articles 38, 39 and 110 of the Treaty of Rome)

- To incorporate agriculture within the common market.
- To increase agricultural productivity by promoting technical progress and ensuring the national development of agricultural production and the optimum utilization of all factors of production, particularly labour.
- To ensure a fair standard of living for the agricultural population, particularly by increasing the individual earnings of persons engaged in agriculture.
- To stabilize agricultural markets.
- To guarantee regular food supplies.
- To ensure reasonable prices in food supplies to consumers.
- To support the harmonious development of world trade.

Source: compiled from Fennell (1997).

- increase farm incomes by either reducing production costs or increasing product prices.

In the first group, in addition to domestic food subsidies and food aid, can be placed intervention buying (Table 11.2) (Gardner 1996, 35; Fennell 1997, 136). This policy measure involves the creation of intervention agencies charged with providing a 'floor' to the market. Agencies act as buyers of last resort, in particular purchasing agricultural produce that is surplus to domestic demand. Using funds provided by the state, they offer a politically negotiated intervention price for surplus food produce and are required to accept all eligible produce that meets specified standards of quality and quantity. The agencies store the purchased products and resell them onto the national or international market as economic conditions allow. In the latter case, state-funded export refunds are available to traders to cover the difference between the (higher) national buying-in price and the (lower) world market selling price. Intervention buying, which provides a guaranteed market for producers of commodities taken into intervention (e.g. processed milk, meat carcasses), lies at the centre of the production and persistence of food surpluses, because food processors, and farmers in their turn, are sheltered from the consequences of their over-production for the domestic market. In addition, the system can only work if the internal market is protected against cheaper food imports. Consequently measures such as variable import levies and customs tariffs are commonly deployed to insulate domestic markets against lower-priced imports (Fennell 1997, 134).

In an effort to reduce structural food surpluses, the second group of measures includes mechanisms to reduce agricultural output, including set-aside, budgetary stabilizers, co-responsibility levies and production quotas (Table 11.2). Under set-aside, for example, farmers are compensated financially for taking a politically negotiated percentage of their cropland out of

Table 11.2 Principal measures of the CAP

- Internal commodity price guarantees, but at levels above world market prices, through target prices and intervention agency buying for each of the main farm products.
- Variable import levy protection of the internal market against imports, based on threshold prices.
- Export subsidies for the main farm products in surplus.
- Direct income supplements for farms in defined areas (e.g. hill livestock compensatory allowances in the Less Favoured Areas).
- Financial incentives for farm modernization, farmer retirement and farmer training.
- Limits on production (e.g. co-responsibility levies, budgetary stabilizers, production quotas, set-aside of arable land).
- A centrally-financed farm budget (e.g. the European Agricultural Guidance and Guarantee Fund).

Source: compiled from Bowler (1985).

production from year to year and in the EU, since 1993, participation in set-aside has become a pre-requisite for receiving financial subsidies on crop production (i.e. cross-compliance). On production quotas, individual farmers are provided with a licence to produce a maximum quantity of a product, for instance milk, poultry meat or eggs, in exchange for a price subsidy. So as to maintain farm incomes, the price subsidy on each unit of the controlled product tends to rise. Thus output-reducing measures have been applied in both the USA and the EU for many decades; but ultimately such measures tend to be defeated by mounting financial costs and increasingly complex bureaucratic procedures.

The third group of policy measures attempts to address farm income objectives more directly by providing direct income supplements, for instance on each cow or sheep on designated farms, or on each hectare of crop planted (Table 11.2). Increasingly the attempt is made to separate, or decouple, the relationship between farm production and the income payment, for otherwise farmers tend to increase their production so as to maximize their payments, as possible under intervention buying. Thus income supplements can be limited by applying maximum stocking densities, or maximum hectares, to eligible production on individual farms, or by placing an upper limit on the total income supplement that can be received by each farm business.

Supply and demand trends for food

We now have to address two critical questions. First, how have structural surpluses arisen in developed countries, such as the USA and those comprising the EU? Second, given the evident problems associated with those surpluses, why has intervention by the state failed to resolve them, at least until recent years?

For the first question, we need to return to Chapter 5 and be reminded of the significant gains in agricultural productivity that have been achieved in developed countries since the World War II. Sustained productivity gains of over two per cent a year have not been unusual, with yields per crop hectare and per head of livestock both on a rising trend. The state has been deeply implicated in these trends, by funding agricultural research and education, providing financial aid for farm modernization, and subsidizing the prices received for farm production. Set against this sustained increase in farm production can be placed far more modest increases in food consumption. Rates of population increase have lagged behind the gains in farm production; at best there has been a modest growth in consumption of some foods partly offset by declines in others (Chapter 20).

In an ideal economy of mobile resources, land, labour and capital would be reallocated from the farm sector into other productive sectors, so as to maintain a market equilibrium between the supply and demand for food. But in reality 'market failure' has occurred because farm families have proved resistant to moving their productive resources, including their human capital, into the manufacturing and service sectors of the economy. Fennell (1997, 154) describes the situation, from a policy-making viewpoint, as 'the underestimation of producer response to price signals'. More deeply implicated, however, have been features such as the lack of transferable skills in farm families, farming as a way of life, the low salvage value of agricultural assets, unemployment in the non-farm sectors of the economy, and the development of pluriactivity (i.e. multiple job holding). The resulting imbalance between supply and demand for food has tended to drive down farm product prices and farm incomes, provoking the state to intervene in the pursuit of equity objectives. Of course, in providing farm subsidies, the needed transference of productive resources has been slowed even further. This, in summary, comprises 'the farm problem', with which successive governments of different political persuasions have been struggling unsuccessfully for decades. As politicians in countries such as New Zealand and Australia found in the 1980s, the only way of resolving the impasse posed by 'the farm problem' is to abandon, both quickly and substantially, equity objectives for the farm sector and permit free market conditions to prevail. Experience in Australasia shows that, after an initial impact of severe dislocation and social cost for individual farm families, including a further reduction in the number of full-time farms, a new, more financially viable, market equilibrium can be achieved in agriculture.

Turning now to the second question posed earlier, if there is a 'market solution' to the problem of food surpluses, why have countries, such as those comprising the EU, not moved earlier and more resolutely to implement it? Here we can introduce the concept of 'non-market failure' (Bowler

1987). Recall the earlier strictures of Moran *et al.* (1996) on the role of social groups and social action in the formulation and interpretation of regulations, the inertia effect of existing regulations, and the importance of ideology. Politicians in the EU, including in other developed countries such as Japan, Canada and Switzerland, have proved unwilling to grasp the political and social consequences of needed reductions in levels of farm subsidies to and protection of their domestic agricultures. The resulting 'non-market failure' to reduce farm subsidies perpetuated food surpluses from the early 1960s to the early 1990s and was based on: (1) the politically powerful lobbying of farmer unions at national and international levels; (2) the power of political ideology based on agricultural fundamentalism; (3) the perceived political importance of a cohesive 'farm vote' at elections; and (4) the successive rounds of 'reforms' needed to overcome inertia in farm policy and move it towards open market principles.

Food surpluses under the Common Agricultural Policy

The most notorious and persistent food surpluses have occurred in the EU, although similar problems were evident in the USA in the late 1950s and 1960s, and periodically during the 1980s. A number of writers have reviewed the policy mechanisms of the CAP that have produced structural surpluses, for example Bowler (1985), Ritson and Harvey (1991), Gardner (1996) and Fennell (1997) in the case of the EU. These reviews, largely covering the 'productivist' or second food régime of agricultural development, are not repeated here. Rather attention is turned to the characteristics of structural food surpluses under the CAP.

Table 11.3 shows the level of self-sufficiency for a selected number of EU member countries and farm products between 1968 and 1995. Figures over 100 indicate a level of production surplus to domestic demand and three features are evident. First, structural surpluses in the EU were present in some countries (e.g. France) and products (e.g. cereals and milk) as early as the 1960s and were a result of national farm policies inherited by the CAP. Second, surpluses in some products continued into the early 1990s despite attempts to reduce them. Third, levels of surplus have varied between countries and products and over time. For example, the surpluses of cereals were reduced in countries such as the UK by the mid-1990s, in part because of EU policies such as set-aside and reductions in support prices. On the other hand, products such as meat and sugar in France, have remained in surplus.

Of course surplus production can be traded between members of the EU and with non-member countries, the latter attracting export refunds. The latter, shown in Table 11.4 for a selection of products, reveal a decreasing total value during the 1990s and a falling percentage of total market

Table 11.3 Self-sufficiency levels in selected agricultural products and countries within the EU, 1968–95 (per cent)

Product		UK	France	Spain	Italy	EU
Cereals	1968*	63	147	–	69	86
	1975*	65	153	–	71	88
	1985†	125	205	89	81	114
	1995†	111	180	59	78	105
Sugar	1968*	34	120	–	94	82
	1975*	26	158	–	82	98
	1985†	63	226	103	90	129
	1995†	56	220	87	102	130
SMP	1968*	91	183	–	0	140
	1975*	160	129	–	0	126
	1985†	172	133	0	0	118
	1995†	95	117	96	0	118
All meat	1968*	62	97	–	77	93
	1975*	74	98	–	74	97
	1985†	81	100	97	74	102
	1995†	91	113	102	76	105

† European Community of 9 members. *: EU of 12 members. SMP: Skimmed-milk powder.
Source: Compiled from Commission of the European Communities (various years) The agricultural situation in the Community: Report. Office for the Official Publications of the European Communities, Luxembourg.

Table 11.4 Export refunds on selected agricultural products in the EU, 1973–1998 (%E: percentage of all export refunds; %TMS: export refunds as a percentage of total market supports)

Product	1973		1983		1993		1998	
	%E	%TMS	%E	%TMS	%E	%TMS	%E	%TMS
Arable crops	6	52	29	62	19	12	9	3
Sugar	4	41	15	58	17	67	19	69
Olive oil	–	–	–	1	–	3	–	2
Fruit and vegetables	2	75	1	5	3	14	1	5
Milk products	50	51	25	30	24	44	30	56
Beef and veal	–	19	16	48	0	0	25	23
Sheep meat	–	–	0	0	3	62	0	0
Wine	–	5	–	3	1	7	1	6
Total (mECU)	1,542*	3,410*	5,221	15,431	8,075	32,205	5,884	40,423

* Million units of account. ECU: European currency unit. – Less than 1.
Source: Compiled from Commission of the European Communities (various years) *The agricultural situation in the Community: Report.* Office for the Official Publications of the European Communities, Luxembourg.

supports, from 45 per cent in 1973 to 14 per cent in 1998. While the total cost of supporting farm prices has continued to increase, including price inflation and the extension of membership of the EU, the reliance on exporting the food surplus problems of the EU to other countries has been substantially reduced. The allocation of funding between products has also changed: from the dominance of milk products in the 1970s (50 per cent of all export refunds), to include sugar and beef in 1998 (19 and 25 per cent of refunds respectively). However, for these three products, export refunds absorb a significant proportion of all their market support funding. For instance, 69 per cent of funds spent in supporting the price of sugar in 1998 was expended on export refunds; the figure for milk products was 56 per cent.

Not all farm produce can be found an immediate market and, as explained earlier, a proportion is taken into intervention. Table 11.5 shows the quantity and value of products in public storage (i.e. intervention) between 1973 and 1998. Four features are evident. First, the total value, and by implication the total volume, of surplus production held in store increased during the 1970s and 1980s, but then, despite increases in the membership of the EU and the consequences of price inflation, was brought under control in the 1980s and 1990s, by methods to be discussed in the next section. Second, there is only a partial relationship between the value and volume of surpluses on a product by product basis. For instance, until the 1990s, cereals comprised the largest surpluses by volume but not necessarily by value. By comparison, the value of beef and milk products in

Table 11.5 Quantity and value of selected agricultural products in public storage in the European Union, 1973–1998

Product	**1973**		**1983**		**1993**		**1998**	
	'000t	**mUA**	**'000t**	**mECU**	**'000t**	**mECU**	**'000t**	**mECU**
Common wheat	1051	184	6806	1279	8903	555	451	47
Barley	182	30	1673	323	7138	404	797	71
Durum wheat	151	30	737	187	2330	128	0	0
Maize	–	–	–	–	2783	229	22	2
Rice	–	–	–	–	75	12	151	36
Olive oil	105	126	121	175	243	217	11	7
SMP	722	709	957	1458	37	28	142	113
Butter	258	581	686	2475	160	129	28	26
Beef carcasses	136	236	301	762	156	91	342	205
Boned beef	79	189	89	280	563	334	281	164
Alcohol ('000 hl)	–	–	–	–	3032	22	579	2
Total	–	2310	–	7035	–	2344	–	749

Source: Compiled from Commission of the European Communities (various years) *The agricultural situation in the Community: Report*. Office for the Official Publications of the European Communities, Luxembourg.

public storage has been higher relative to their volume. Third, significant food surpluses in public storage are restricted to a relatively narrow range of farm products, including cereals, milk products, olive oil and beef. Fourth, there is considerable variation through time in the products in public storage. For instance, butter and skimmed milk powder were major contributors to surpluses in public storage in the 1980s (from surplus milk) but, following the introduction of milk quotas in 1984, their size was reduced to more manageable proportions. By the late 1990s, the main products in public storage by value were beef, milk (as SMP and whole-milk powder) and barley, rather than wheat.

From this brief case study we can see that the problem of structural food surpluses in the EU has been persistent but confined to a relatively few products. Moreover it is a problem of diminishing scale. To understand these recent changes in food surpluses we must now summarize developments in reforming the CAP over the last decade.

Reducing food surpluses through reform of the CAP

One of the features defining 'post-productivism', or the third food régime, is the declining significance of state subsidies to agriculture (Chapter 3). By the 1980s the financial costs of state intervention in agriculture, and the distorting effect on trade relationships of subsidized food exports from the EU and other developed countries, had reached economically, some would argue politically, unacceptable levels. Table 11.6 shows the situation in the EU, where by 1990 the total annual cost of the CAP had reached 28,402 m ECU (equivalent to approximately £18,462 m or 64 per cent of the EU's total budget), with subsidies on product prices (the Guarantee section) comprising 93 per cent of the European Agricultural Guidance and Guarantee Fund (EAGGF).

Table 11.6 The allocation of expenditure under the European Agricultural Guidance and Guarantee Fund

Item	1990	1994	1998*
Budget of the EU (m ECU)	44,379	59,909	81,434
Budget of EAGGF (% EU budget)	64	60	56
Guarantee element (% CAP budget)	93	92	90
Guidance element (% CAP budget)	7	8	10

* Official estimate.

Source: Complied from Commission of the European Communities (various years) *The agricultural situation in the Community: Report*. Office for the Official Publications of the European Communities, Luxembourg.

Incoming governments in developed countries began to remove state supports to farming, dramatically so in New Zealand in 1984; this has exposed the agricultural sectors of national economies to increased world market competition. Significant change in the EU was delayed until the early 1990s, at which time three political events combined to bring about both a reduction in the total cost of the CAP and, although less significantly, the market intervention (guarantee) component of EAGGF (Table 11.6).

The first political event was the United Nations Conference on Environment and Development, held at Rio de Janeiro in 1992, leading to widespread political acceptance of Agenda 21 and its objective of sustainable development (Ilbery *et al.* 1997). Chapter 14 of Agenda 21 legitimized the reduction of public subsidies to the industrial model of agricultural development: a model increasingly recognized as economically and environmentally unsustainable. The beneficiaries have been farming systems that apply principles of balanced agri-ecology to integrated farming systems, are less dependent on agri-chemicals and inorganic fertilizers, and produce environmental goods such as wetlands, heathlands, moorlands and nature conservation. This approach was evident as early as 1985 and 1987 in the use of EAGGF – Guidance (Regulations 797/85 and 1760/87) to enable member states to designate areas (e.g. Environmentally Sensitive Areas in the UK) within which farmers could be offered financial inducements to 'farm the environment'.

The second political event was the introduction of a package of reforms to the CAP in 1992, called the MacSharry reforms, named after the then Commissioner for Agriculture (Directorate General VI). The main features of the Reforms are set out in Table 11.7. Direct crop compensation payments were added to the principal measures of the CAP as a new form of farm income supplement (Gardner 1996, 111), linked in cross-compliance to farm-based set-aside (10 per cent of a farm's 1989/91 base area);

Table 11.7 Selected MacSharry Reforms agreed by the Council of Ministers, 1992

- 29 per cent reduction in cereal support prices over 3 years.
- Compensation for rotational set-aside (a percentage of a base area).
- Cross-compliance between set-aside and area-based income payments on arable production.
- Removal of co-responsibility levies.
- New non-rotational set-aside.
- 'Accompanying measures' providing aid for environmental protection, forestry and early farmer retirement.
- Quotas in the beef, sheep and milk sectors.

Source: adapted from Kay (1998).

Table 11.8 Agri-environmental schemes implemented in the EU under Regulation 2078/92

Element	Number of countries with the element
Reduction of chemical inputs and pollution control	15
Organic farming	15
Extensification	15
Environmental practices	15
Landscape and countryside management	14
Twenty-year set-aside	12
Training and demonstration projects	12
Rearing rare livestock breeds	11
Convert arable into grassland	10
Reduction of livestock density	9
Upkeep of abandoned land	7
Public access and leisure	6

Source: adapted from Ilbery (1999, 95).

livestock headage payments (beef cattle and sheep) were limited to fixed stocking densities; agri-environmental measures (Regulation 2078/92) were introduced through national agri-environmental programmes as 'accompanying measures', including payments for the development of organic farming (Table 11.8) (Wynne 1994). In broad terms, as discussed by Robinson and Ilbery (1993), the attempt was made to decouple farm income support from the volume of agricultural production.

The third political event was the conclusion, in 1993–94, of the Uruguay Round of trade negotiations under the GATT (Fennell 1997, 378). The relevant details, including the setting up of the WTO from 1995 to oversee the greater liberalization of international trade, are considered in the following chapter. It is sufficient to note here that the EU agreed to a significant reduction in levels of price protection for farm products (Regulation 3290/94), phased over a number of years, and an opening of the domestic market to more food imports from other trading nations, such as the Cairns Group of countries (e.g. New Zealand, Australia), the USA and developing countries. As described by Gardner (1996, 119), the negotiations leading up to the conclusion of GATT, particularly the politically acrimonious relationship between the EU and USA over agricultural trade policy, had a significant bearing on the final details of the MacSharry Reforms.

When taken together, and with hindsight, these three political developments may be viewed as a watershed in post-World War II agricultural regulation. Evidence presented in this chapter indicates a down-turn in structural food surpluses as a political and economic problem, with farm support prices in developed countries edging closer to prices on world markets. The question remaining for the longer term is whether or not there is sufficient political will in developed countries to obtain, and then maintain, an even smaller margin between remaining farm support prices and world market prices.

Conclusion

Within the next two decades, the structural food surpluses of developed countries, between the 1960s and the 1990s, could well be viewed as a transitory feature of productivist agriculture and its associated intervention by the state. On the one hand, such surpluses reflected the achievement of national food security for consumers but, on the other, they were achieved at considerable economic, political and environmental costs (Brouwer *et al.* 1991; Kronert *et al.* 1999). At present, food surpluses have been reinterpreted as reflecting the existence of a 'land surplus' (Swinbank 1992) and the search is underway for appropriate alternative rural land uses, such as biomass crops for fuel (Bolsius *et al.* 1993; Bryden 1994; Mangan 1995).

The prospects for the elimination of all food surpluses in the longer term, through the decoupling of farm supports from agricultural production, look reasonable (Brouwer and Lowe 1999), although Fennell (1997, 401) is less optimistic. For the EU, first, the political power of the farm unions appears to have waned and international farm organizations, such as the Comité des Organisations Professionnelles Agricoles (COPA), no longer have such an influence on agricultural policy making (Clark and Jones 1999). Second, under the WTO, the EU is committed to the further liberalization of agricultural trade, not least as Pouliquen (1998) shows, because of the financial implications of maintaining the existing mechanisms of the CAP under the further enlargement of the EU. However, as made clear by Clarke (1992), this development will not be without its environmental conflicts. Third, Agenda 2000 policy in the EU includes a commitment to further movement from price supports to direct income payments and the transferring of more funds to the Guidance section of the EAGGF, including the development of rural policy. This last aspect reflects the integration of the CAP into broader rural development policy for the EU. It includes the redesignation of Objective 1 and 2 regions for receiving funding, through which agriculture becomes integrated into, rather than separated from, the development of regional rural economies.

However, data for the Guidance section of EAGGF (Table 11.6) suggest that this will be a gradual process; indeed the history of the CAP is one of resistance to rapid change. Thus policy inertia may well see structural food surpluses remaining for a limited number of products over the short-term.

Further reading and references

The work of one of the present authors (Bowler 1979, 1985, 1987, 1992) may be consulted for further material on food surpluses and state policies designed to deal with them. Ritson and Harvey (1991), Gardner (1996) and Fennell (1997) are other helpful sources.

Arden Clarke, C. 1992: Agriculture and environment in the GATT: integration or collision? *Ecos* **13**, 9–14.

Bolsius, E., Clark, G. and Groendijk, J. (eds) 1993: The retreat: rural land use and European agriculture. *Nederlandse Geografische Studies* **172**. Amsterdam: University of Amsterdam.

Bowler, I. 1979: *Government and agriculture: a spatial perspective*. London: Longman.

Bowler, I. 1985: *Agriculture under the Common Agricultural Policy: a geography*. Manchester: Manchester University Press.

Bowler, I. 1987: Nonmarket failure in agricultural policy – a review of the literature for the European Community. *Agricultural Administration and Extension* **26**, 1–15.

Bowler, I. 1992: *The geography of agriculture in developed market economies*. London: Longman.

Brouwer, F. and Lowe, P. (ed.) 1999: *CAP régimes and the European countryside*. Wallingford: CAB International.

Brouwer, F., Thomas, A. and Chadwick, M. (eds) 1991: *Land use changes in Europe*. London: Kluwer.

Bryden, J. 1994: Prospects for rural areas in an enlarged Europe. *Journal of Rural Studies* **10**, 387–94.

Clark, G.L. 1992: 'Real' regulation: the administrative state. *Environment and Planning A* **24**, 615–27.

Clark, J. and Jones, A. 1999: From policy insider to policy outcast? Comité des Organisations Professionnelles Agricoles, EU policymaking, and the EU's 'agri-environment' regulation. *Environment and Planning C* **17**, 637–53.

Fennell, R. 1997: *The Common Agricultural Policy: continuity and change*. Oxford: Clarendon Press.

Gardner, B. 1996: *European agriculture: policies, production and trade*. London: Routledge.

Ilbery, B. 1999: *The geography of rural change*. London: Longman.

Ilbery, B., Chiotti, Q. and Rickard, T. (eds) 1997: *Agricultural restructuring and sustainability*. Wallingford: CAB International.

Kay, A. 1998: *The reform of the Common Agricultural Policy: the case of the MacSharry reforms*. Wallingford: CAB International.

Kronert, R., Baudry, J., Bowler, I. and Reenberg, A. 1999: Land-use changes and their environmental impact in rural areas in Europe. *UNESCO Man and Biosphere Series* **24**. New York: Parthenon Publishing.

Mangan, C. 1995: European Commission research policy towards non-food crops and processing for energy and industrial products. *Outlook on Agriculture* **24**, 77–83.

Moran, W., Blunden, G., Workman, M. and Brady, A. 1996: Family farmers, real regulation and the experience of food régimes. *Journal of Rural Studies* **12**, 245–58.

Pouliquen, A. 1998: Agricultural enlargement of the EU under Agenda 2000. *Economics of Transition* **6**, 505–22.

Ritson, C. and Harvey, D. 1991: *The Common Agricultural Policy and the world economy*. Wallingford: CAB International.

Robinson, G. and Ilbery, B. 1993: Reforming the CAP: beyond MacSharry. In Gilg, A. (ed.). *Progress in rural policy and planning*. Volume 3. London: Belhaven Press, 197–207.
Swinbank, A. 1992: A surplus of farm land? *Land Use Policy* **9**, 3–7.
Wynne, P. 1994: Agri-environment schemes: recent events and forthcoming attractions. *Ecos* **15**, 48–52.

12

Food security

Introduction

Food *insecurity* may arise from a number of circumstances. These may include local food availability decline (FAD), perhaps due to a below average harvest, or an entitlement decline due to a collapse in earning capacity (Chapter 9). But there are often other factors, such as distribution bottlenecks in the food system or fluctuations in world prices of grain, which may restrict commercial imports. In this chapter we discuss the types of policies that may be adopted to cope with such food insecurity, including the most obvious one, food aid.

Food security

Food security is defined by the FAO as 'access by all people at all times to the food needed for a healthy and active life', while academic and political interest in food security has increased since the 1970s. Indeed hunger and famine have become such major issues in the public's mind in the television age that political action has become inevitable. Analyses of food problems and policy recommendations have varied a good deal, however, and achieving a consensus has not been possible. What follows is a brief selection from a vast literature and it certainly does not represent the last word on food security activities. Figures 12.1 and 12.2 give two ways (among many) of conceptualizing the issues, in general and in local terms.

Taking Figure 12.1 as a starting point, one very important consideration in the practical assessment of food *availability* is that of measurement and monitoring. Methods of sample survey are discussed in Chapter 20, but these are often time- and resource-intensive. The immediacy of a food crisis may require more rapid action, such as the participatory mapping of vulnerable households or the use of FIVIMS at the regional scale. The latter

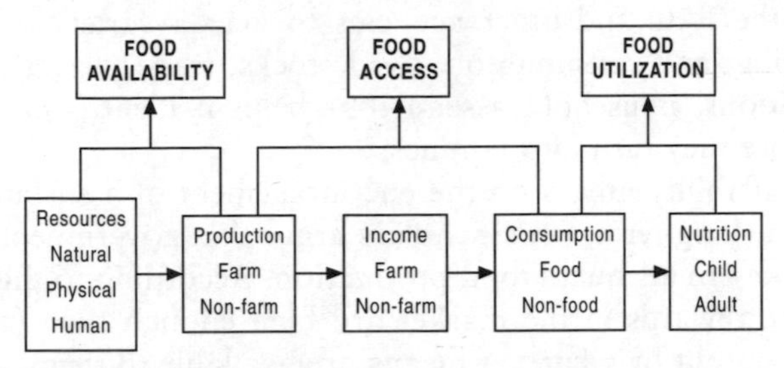

Fig. 12.1 A conceptual framework of food security. *Source*: after Chung *et al.* (1997).

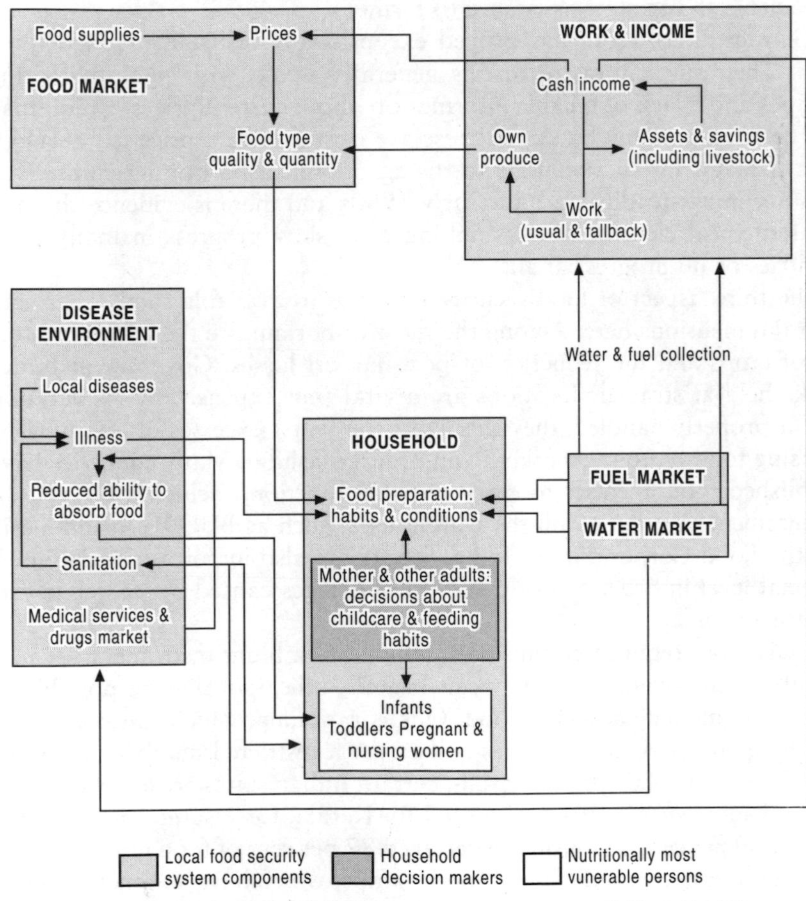

Fig. 12.2 The local food security system. *Source*: Hubbard (1995).

are 'food insecurity and vulnerability information and mapping systems', which allow the FAO and other agencies to map a variety of indicators (food availability and consumption, food stocks, trade, wages and labour market conditions, household assets) that help in identifying local food problems before they turn into famines.

Second, availability hinges on the encouragement of a sustainable agriculture over the long-term. Some scholars argue that government control of pricing is the key to adequate food production. According to this school of thought, if the rewards in the market are high enough then farmers will increase their output by whatever means are available to them, including a greater investment of labour and possibly the adoption of innovations such as fertilizers, irrigation, and HYVs of staple crops. This view was popular in the 1980s but it waned in the 1990s as the IMF and World Bank put pressure on governments to reduce price subsidies on these kinds of agricultural input, and as consumers have pressed for the curtailment of inflation. A further dampener has come recently from the GATT/WTO agreements to cap state assistance to farmers.

'Getting prices right' has proved exceptionally difficult in practice for LICs. Their rural infrastructure is generally weak, with poor marketing facilities and a lack of reliable information about current prices. Credit may also be hard to come by. All of these are obstacles to a price-led development strategy. International aid to the agronomic aspect of agriculture has been declining steadily since the early 1980s and there is evidence that the infrastructural element is also making only slow progress in many poor countries or no progress at all.

The third aspect of food security is *access* to available food. There are several dimensions here. Among the most important are the adequate storage of crops and the reduction of post-harvest losses. Government buffer stocks held at strategic locations are a vital (but expensive) consideration and, if properly handled, they can avert the worst excesses of a famine by releasing food on to the market in an orderly fashion. Many countries have established food agencies to carry out this function, including the role of procurement in order to fill the warehouses, such as BULOG in Indonesia and the Food Corporation of India. Stocks are also important at the international level in order to overcome the shortages caused by poor harvests (Figure 12.3).

Apart from retail price subsidies, which are a blunt instrument because potentially all groups in society can benefit, other policies are possible to improve immediate access to food. One is rationing, which can both regulate the quantity of commodities and target a restricted number of eligible households. The 'fair price shops' in certain Indian states are an example of this and some success has been noted. In Kerala, for instance, in 1977 the poorest 60 per cent of consumers received 87 per cent of food grains distributed. An alternative is the use of food stamps, which have a monetary value rather than being linked to a physical weight of rationed goods. There is

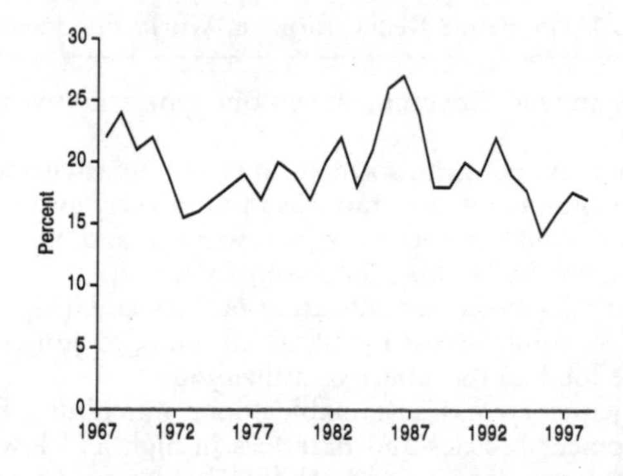

Fig. 12.3 World grain stocks, 1967–98 (percentage of annual consumption). *Source*: World Food Programme.

evidence that rationing and food stamps are most effective in advancing food security in urban areas, mainly because of the bureaucratic complexity of coping with scattered rural populations.

Supplementary feeding schemes are a final approach. Here individuals can be targeted and the intake of a balanced meal actually observed. One of the best developed and most impressive of such schemes is the 'Noon Meals' programme organized by the state of Tamil Nadu in India. This started in 1980 to provide meals for school children but has since been extended to assist other needy groups, such as the aged.

In 1994 the FAO launched its Special Programme for Food Security, with the objective of helping low income, food-deficit countries. It began with a pilot phase and by the end of 1998 was operational in 38 countries, mainly in Africa and Asia. The approach is to identify constraints to production, and then find solutions that involve simple technologies, a participatory approach and a regard for the role of women.

In 1996 the World Food Summit produced a very helpful series of *Technical Background Documents*, which give insights into the thinking of the UN organizations on the nature of hunger and food security. Unfortunately, the Summit itself was at the centre of a fierce controversy about the attitude it should take to a number of sensitive economic and political issues, and this somewhat devalued the authority of its plan of action, the so-called Rome Declaration on World Food Security (Table 12.1).

Famine early warning systems

In the 1970s the international community became increasingly concerned that famine should not only be dealt with as an acute emergency but that precautionary measures should be put in place. A number of attempts were

Table 12.1 The Rome Declaration on World Food Security

'We, the Heads of State and Government, or our representatives, gathered at the World Food Summit:

1 Will ensure an enabling political, social, and economic environment designed to create the best conditions for the eradication of poverty and for durable peace, based on full and equal participation of women and men, which is most conducive to achieving sustainable food security for all.
2 Will implement policies aimed at eradicating poverty and inequality and improving physical and economic access by all, at all times, to sufficient, nutritionally adequate and safe food and its effective utilization.
3 Will pursue participatory and sustainable food, agriculture, fisheries, forestry and rural development policies and practices in high and low potential areas, which are essential to adequate and reliable food supplies at the household, national, regional and global levels, and combat pests, drought and desertification, considering the multifunctional character of agriculture.
4 Will strive to ensure that food, agricultural trade and overall trade policies are conducive to fostering food security for all through a fair and market oriented world trade system.
5 Will endeavour to prevent and be prepared for natural disasters and man-made emergencies and to meet transitory and emergency food requirements in ways that encourage recovery, rehabilitation, development and a capacity to satisfy future needs.
6 Will promote optimal allocation and use of public and private investments to foster human resources, sustainable food and agriculture systems, and rural development, in high and low potential areas.
7 Will implement, monitor, and follow-up this Plan of Action at all levels in cooperation with the international community'.

Source: World Food Summit website.

made to devise 'early warning systems' (EWS) to alert national authorities and to mobilize global food stocks so that they could be deployed in time to prevent deaths.

EWSs rely upon timely, accurate and appropriate predictive information, and may include a wide variety of data and means of analysis (Table 12.2). They cannot be regarded in any way as 'scientific', however, because there

Table 12.2 Selected EWS information sources

- Nutritional surveillance at the community level.
- Supply-side data, especially on harvest, storage and marketing.
- Socio-economic indicators, such as monitoring of coping strategies.
- Satellite remote sensing for environmental information.
- Automated data analysis and cartography using Geographical Information Systems.

Source: based on Walker (1989).

is always a vital component of subjective interpretation that determines the nature of the response. Also EWSs are very much imbedded into the politics of international emergencies and whether any pre-emptive aid is triggered will depend upon a number of factors, including the receptiveness of the host government and the cost to the donors.

Examples of international EWSs include the USAID's Famine Early Warning System (FEWS), based in Arlington, and the Global Information and Early Warning System (GIEWS) on food and agriculture established by the FAO in Rome in 1973. In the latter case information is stored on food production, trade, stocks, food consumption, food aid requirements, pledges and actual deliveries. These include qualitative data on the significance of pest attacks or the failure of rains (Figure 12.4) and they publicize the progress of any emergency and the response. In addition, the GIEWS publishes regular series of reports that include *Food Outlook*, *Foodcrops and Shortages*, *Sahel Weather and Crop Situation*, and *Food Supply Situation and Crop Prospects in Sub-Saharan Africa*. They also maintain three databases on the immediate situation in Africa. All of these are accessible via the GIEWS website.

There are national EWS in some Low Income Countries (for instance in Mozambique, Peru, Senegal, and Zambia), but they are all still at an early stage of development. Access to satellite image-processing and GIS is restricted by the expense of the technology and by shortages of local expertise but their utility is obvious, as demonstrated by the vegetation map of the Sudan in Figure 12.5. Because of skills shortages, international agencies and Northern NGOs are often responsible for at least part and sometimes all of the EWS, and this may cause some friction with local governments. The Sudanese government, for instance, has several times been reluctant to accept EWS signals of hunger/famine and as a result has been slow to call for emergency relief.

The international relief system also takes time to respond to EWS warnings. In 1989–90 the Sudanese harvest was poor, especially in the province of Darfur. Grain reserves were low and warnings of the vulnerability of farming families were issued by several agencies, including GIEWS in April 1990 and USAID in June of that year. In December the FAO and WFP separately calculated a serious shortfall of grain and only then did donors make pledges of aid. The crisis was acute throughout 1991 but deliveries of food were slow to arrive (Figure 12.6).

There are other problems with EWS. The data gathered and means of analysis used differ from one agency to another and this may not only lead to divergence of estimates of at-risk populations, but also to public disagreement about what needs to be done. International donors may delay their intervention until such disputes are settled, but in the case of the Sudanese emergency, and several others in the recent past, it has been the media that have forced an immediate response. Television pictures are especially powerful but newspaper column inches also have an impact

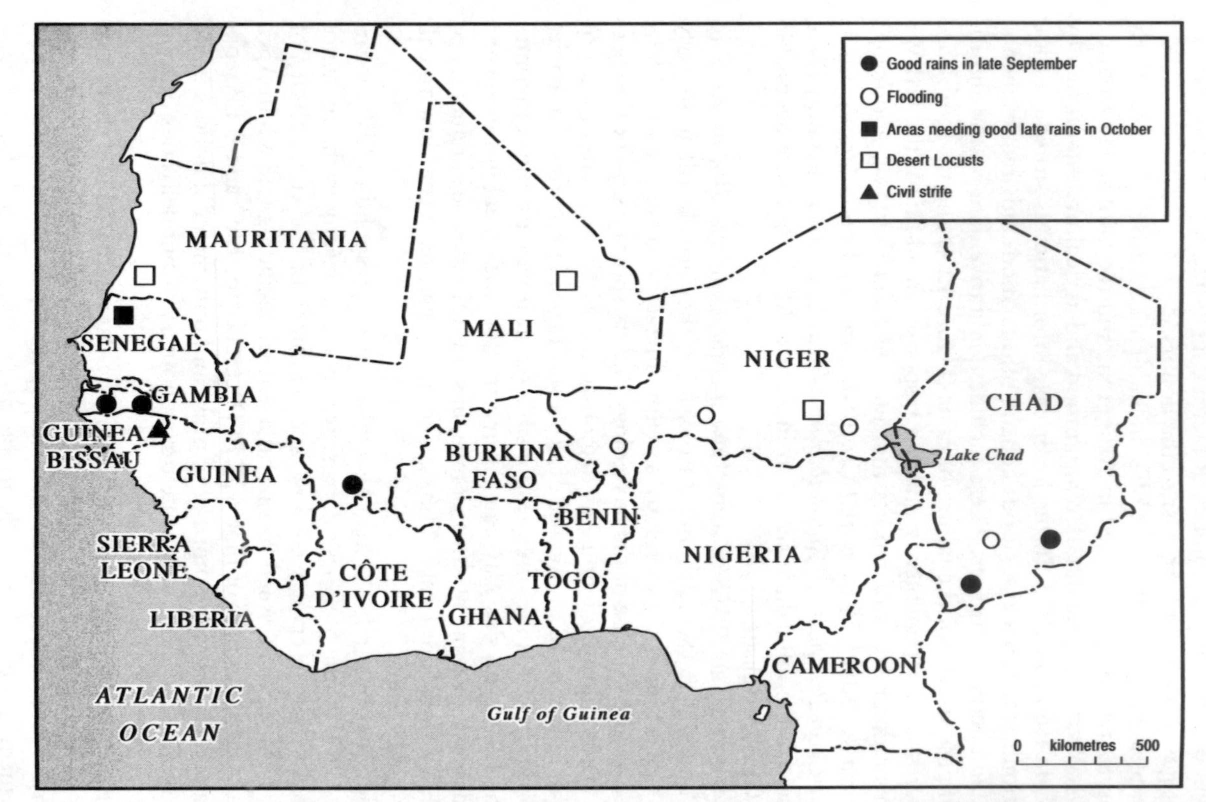

Fig. 12.4 The GIEWS summary of harvest problems in West Africa on 10th October, 1998. *Source*: FAO/GIEWS.

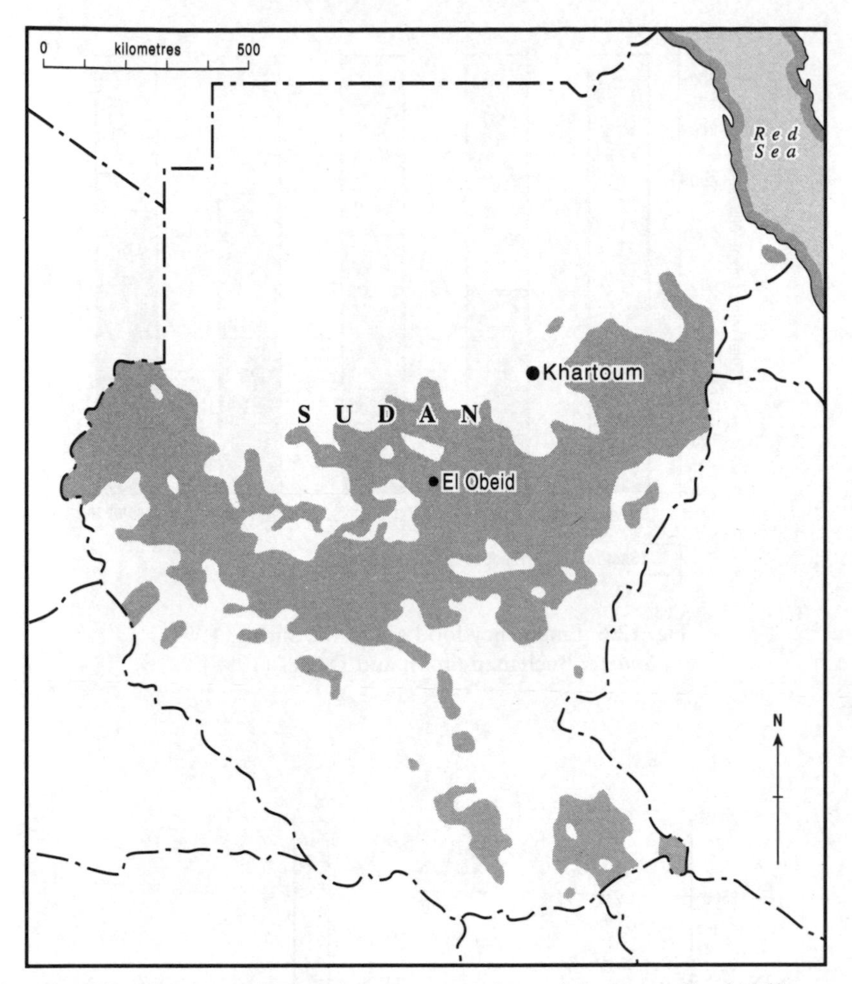

Fig. 12.5 Distribution of higher than average values of the Normalized Difference Vegetation Index in the Sudan in 1988. *Note*: Based on data from the advanced very high resolution radiometer carried on the satellites of the National Oceanic and Atmospheric Administration. The shading shows areas with higher than normal biomass, indicating good rainfall and growing conditions, yet this was a year of famine because of civil war and flooding. *Source*: Hutchinson (1991).

(Figure 12.7).

Food aid

Between 1960 and 1994 approximately $1,400,000,000,000 worth (in constant 1988 dollars) of aid has been made available to poor countries for a variety of purposes (*Economist*, May 7th 1994, p. 21). In view of the

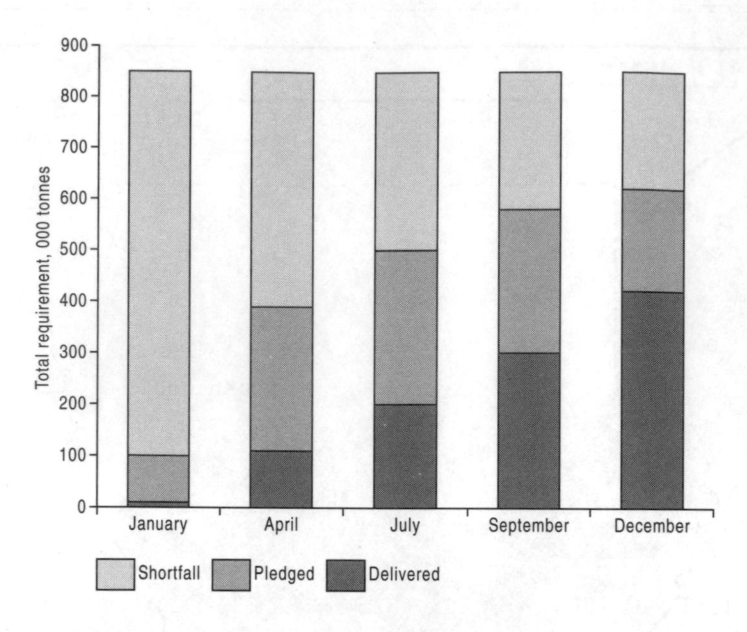

Fig. 12.6 Emergency food aid in the Sudan, 1991.
Source: Buchanan-Smith and Davies (1995).

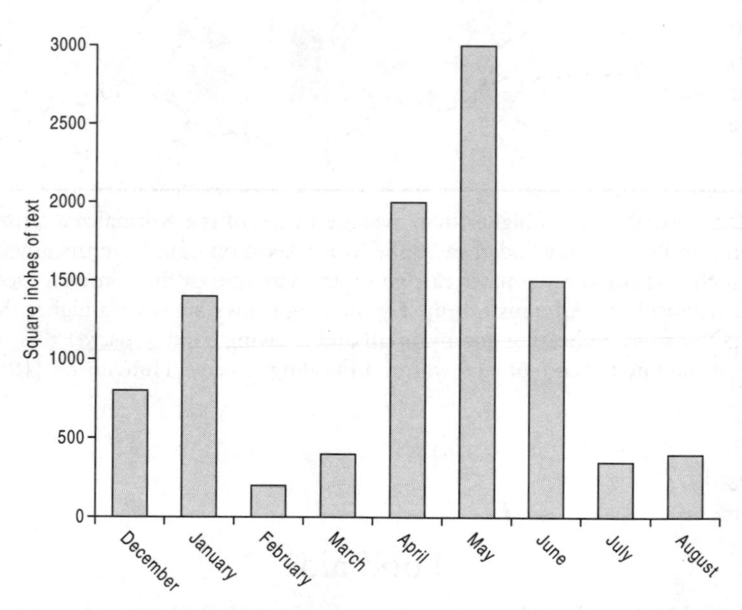

Fig. 12.7 UK newspaper coverage of famine in Africa, 1990–91.
Source: Buchanan-Smith and Davies (1995).

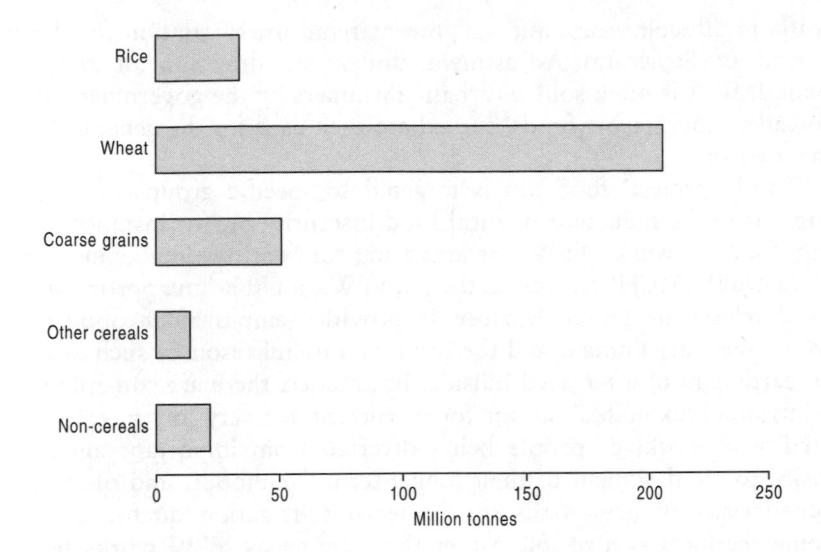

Fig. 12.8 Food aid by commodity, 1971–96 (million tonnes). *Source*: FAOSTAT.

continuing and widespread problems of poverty and food shortages (Chapter 10), one might very reasonably ask whether this vast sum has been well spent. Surprisingly perhaps, in view of its high media profile, food aid constitutes under four per cent by value of total Overseas Development Aid, and that proportion is slipping as its role has been reassessed in the 1990s. Wheat remains by far the most important commodity (Figure 12.8).

There are three main types of food aid. First, there is emergency aid for disaster relief provided by individual governments, NGOs, and by international organizations such as the WFP (a UN agency established in 1961). Some 'emergency' aid is provided to refugees from military conflicts and this commitment may last for years. It may seem churlish to criticize this type of aid but there is evidence that it is often badly organized and sometimes even counter-productive. It is impossible to begrudge starving people the temporary respite that food aid may bring; but we could argue that real help should take the form of structural economic and political changes that treat the causes of poverty rather than the acute symptoms of a food disaster. The media occasionally run stories about incompetence in emergency food aid organization, such as the alleged parachuting of pork and bean rations into the Muslim part of Bosnia during the recent war, or the airlift of dried foods to a Rwandan Hutu refugee camp where there was a severe water shortage. Unfortunately such stories colour the public's view of food aid in general.

Second, 'programme' food aid is aimed at the macro level of the whole economy of a recipient country, especially to meet a shortfall of basic foodstuffs in difficult years, and so prevent retail price inflation, or during a period of Structural Adjustment under the direction of the World Bank/IMF. It is often sold to urban consumers by the government and the so-called 'counterpart funds' earned are then used for the general development effort.

Third, 'project' food aid is targeted to specific groups in society in support of the reduction of rural food insecurity by, for instance, providing 'food for work' (FFW), or arranging for 'wet' feeding at mother and child health (MCH) centres. In theory, FFW should be an appropriate form of development policy because it provides employment opportunities where these are limited, and the result is a useful resource such as a road, an earth dam or a terraced hillside. In practice, there are sometimes criticisms of an 'exploited' labour force working for very low pay; of a poor quality of work; of people being diverted from important agricultural tasks, to the detriment of their longer-term livelihoods; and of the major beneficiaries of new roads or cleaned-out irrigation ditches ultimately being the local rich people rather than the needy. FFW works better in some countries than others. In South Asia it seems to be effective, but in Africa it means an extra burden for women, who do much of the physical labour. Most observers now seem to agree that FFW is better partly or wholly paid for in cash, or 'monetized', and that it should be conceptualized more broadly as Rural Public Works Programmes, a means of entitlement protection.

MCH and school meals programmes have a more positive image because they are targeted at vulnerable sections of society, but even here there have been problems. The most serious is the issue of 'displacement', where poor people may understandably consider MCH meals as a substitute for feeding at home and therefore the potential for better nutrition may not be accomplished. And school meals are no use to the very poorest children who cannot attend classes because they have to work in order to support their family's income.

Why give food aid?

Many members of the general public think that the answer to this question is obvious. One often hears the opinion that the surplus commodities stored in Europe and North America as a result of an over-productive farming sector should be transferred to food-deficit countries. The harrowing pictures of ghost-like starving people would then disappear from our television screens. The motive here is humanitarian, coupled with a sense of morality. To the cynical this popular view may seem naïve but it has support in international law. The 'right to food' was mentioned in the Universal Declaration of Human Rights (1948) and was formalized in the Interna-

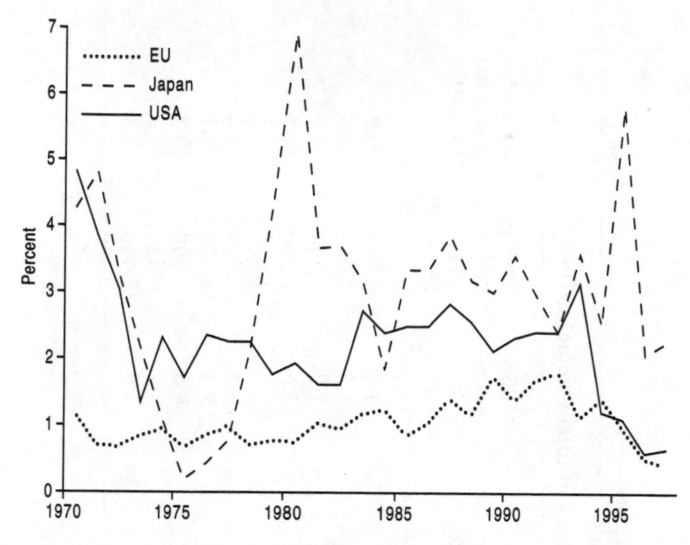

Fig. 12.9 Cereal aid as a percentage of national production, 1970–97. *Source*: FAOSTAT.

tional Covenant on Economic, Social and Cultural Rights (1966). It was reaffirmed at the WFS in 1996, when the Rome Declaration on World Food Security stated that it is 'the right of everyone to have access to safe and nutritious food, consistent with the right to adequate food and the fundamental right of everyone to be free from hunger'. In the real world of hard politics and free market economics, such idealism is unfortunately rarely deployed in its fullest and purest sense. There are emergency responses to famines, of course, but the general run of food aid works to a highly complex agenda and there are many different views about its degree of success.

North America and the European Union are, and have for decades been, the world's main sources of food aid. Their apparent generosity is tempered somewhat, however, when this food aid is set against their overall output of food and its monetary value as a proportion of their national wealth (Figure 12.9).

The roll-call of the top recipients (Table 12.3) will be very surprising to those who think that food aid should be driven by humanitarian ideals. It shows several countries that might indeed be expected according to their poverty and history of food insecurity, such as Bangladesh, Ethiopia, Mozambique and the Sudan. But there are others, including Korea, Egypt, Tunisia and the former Soviet Union and its satellites, which are unlikely to be at the top of anyone's list of at-risk nations. It is clear beyond question that the reasons for giving food aid are by no means dominated by genuine short- or long-term need, and that geopolitical considerations are important. Such assistance is frequently used as a means of furthering the

Table 12.3 The main cereal food aid recipients

(a) in kg per capita, 1988–1997				(b) in million tonnes of total volume, 1970–97			
Country	kg per head	GNP per capita (PPP$ 1996-97	Percentage food deficit (1990–92)	Country	Million tonnes	GNP per capita (PPP$ 1996–97)	Percentage food deficit (1990–92)
Jamaica	63.0	3,480	5.6	Egypt	33.5	2,940	0.9
Jordan	50.7	3,430	0.5	Bangladesh	32.7	1,050	8.8
Guyana	48.5	2,490	5.6	India	15.9	1,650	4.9
Liberia	45.9	1,100	23	Ethiopia	13.6	510	28
Mozambique	28.0	520	29.2	Pakistan	12.6	1,590	3.5
Ethiopia	27.3	510	28	Indonesia	10.5	3,450	2.2
Mauritania	25.4	1,870	4.4	Korea Rep	9.1	13,500	0.1
Suriname	24.3	3,150	4.9	Sudan	8.6	860	10.9
Malawi	23.0	700	16.4	Mozambique	8.1	520	29.2
Bolivia	22.6	3,000	11.9	Sri Lanka	7.9	2,460	6.4

Sources: FAOSTAT; FAO (1996); World Bank (1998).

foreign policy aims of the donor nations.

The development literature has, over the last 20 years or so, echoed to a chorus of dissent about food aid. Apart from the points made above concerning programme and project aid, two general criticisms have been particularly damning. The first is the accusation that imported food, especially grain from Europe and North America, has come to dominate the market in some countries. This has acted as a disincentive to local producers, who may find it difficult to compete in terms of quality and price. As a result, food aid may achieve the exact opposite of what is needed: it may destroy the local ability to be self-sufficient. This is not inevitable, however, first because the food can be targeted at those who would not otherwise buy, thus bringing marginal groups into the market and, second, counterpart funds can also be used to support local producer prices. Rather than exporting their own destabilizing surpluses to LICs, the donor countries might be better advised to purchase any food to be used for aid in the region close to the target location. As long as this does not lead to further price distortions in the source area, this would be a more appropriate strategy. But in 1997, local purchase represented only 9.5 per cent of cereal food aid and 4.0 per cent of non-cereal aid arranged by the WFP.

Second, the 'post-development' school has questioned the whole basis of food aid. According to Arturo Escobar (1995), for instance:

> the body of the malnourished – the starving 'African' portrayed on so many covers of Western magazines, or the lethargic South American child to be 'adopted' for $16 a month portrayed in the advertisements of the same magazines – is the most striking symbol of the power of the First World over the Third. A whole economy of discourse and unequal power relations is encoded in that body.

In this reading at least, food aid is an instrument of power through which the North perpetuates a feeling of superiority and reproduces its political hold over the 'poor' and the 'hungry'. In a sense, food aid can be thought of as an important corner stone of a development 'industry' that seems actually to have exacerbated the problems it set out some 50 years ago to solve.

Partly as a result of these doubts about food aid, and partly because of improved food security in countries such as India and Bangladesh, international donations have fallen in recent years, by about two-thirds in volume since 1993, and by about 80 per cent by value since the high point of the mid 1980s (Figure 12.10). The WFP has devoted more of its energy to emergencies (Figure 12.11) and in the early 1990s the USA diverted a portion of its food aid to Eastern Europe in an attempt to support the transitional stage of the newly democratic régimes.

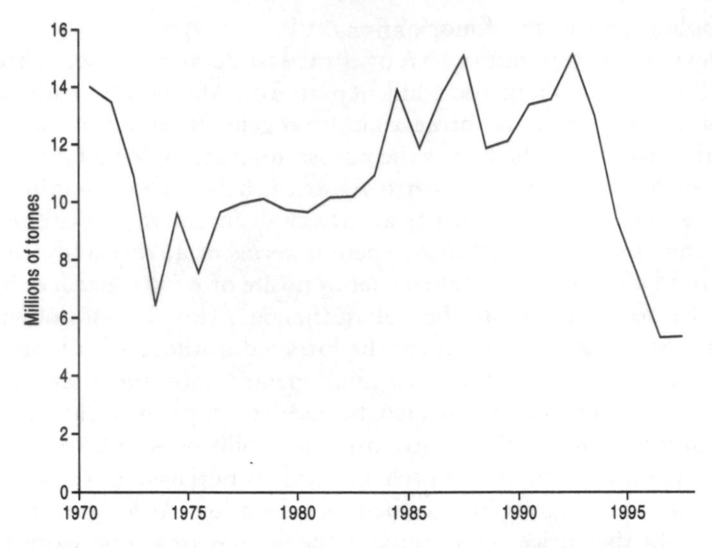

Fig. 12.10 World food aid donations, 1970–97 (million tonnes per annum). *Source*: FAOSTAT.

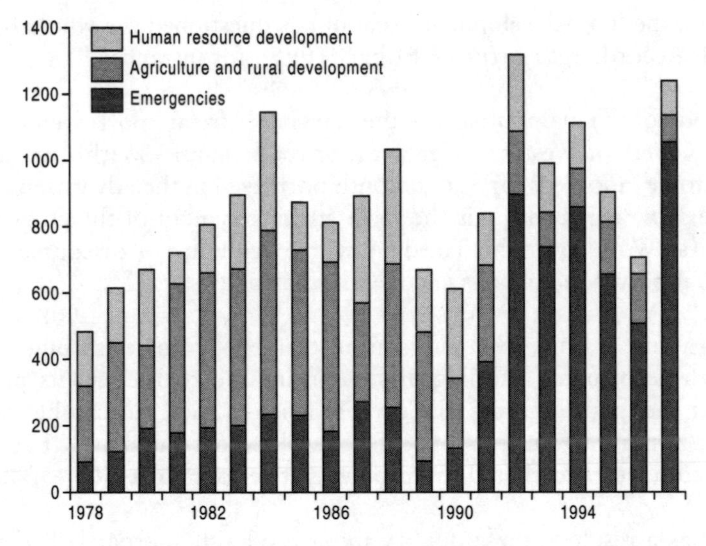

Fig. 12.11 WFP commitments, 1978–97 (million $). *Source*: WFP.

Food wars

Food trade has frequently become embroiled in politics. An excellent example is the 1971 decision by the USA administration to allow the Soviet Union to import surplus American grain. This was an attempt to mitigate the tensions caused by the Cold War but it had unintended consequences. The following year the USSR entered secret negotiations with several trading companies and arranged for the purchase of 19 million tons of cereals, such a large amount that it subsequently distorted the world price. American politicians were embarrassed because the grain was exported with government subsidies and the Soviet consumer was therefore benefiting from the largesse of the US taxpayer.

Such collisions between the interests of the superpowers were the very stuff of the Cold War. During the 1990s the scale of geopolitical action has shifted from the global to regional theatres of conflict: either tensions between neighbouring countries or internecine strife between established governments and rebel groups. Again food has been at the centre of this action, either as a weapon used by the protagonists or as an unintended consequence of shortages cause by the disruption of war. The WFP somewhat disingenuously attributes such food crises to 'man-made disasters', but they really mean war and civil disturbance. Figure 12.12 charts the number of occasions that the WFP has disbursed aid for this reason between 1975 and 1997. The continent of Africa is depressingly prominent in this record of what Messer (1994) calls 'food wars'.

Macrae and Zwi (1994), De Waal (1992, 1997) and Cohen and Pinstrup-Andersen (1999) have argued that war and insurrection have played a much larger role in hunger and famine than allowed for in the models of Sen and most other writers. Three types of outcome might be cited. The first is a hot war where bombing, shelling and ground fighting lead to a disruption of communications and the destruction of productive assets such as fields, irrigation canals and storage facilities. In Cambodia

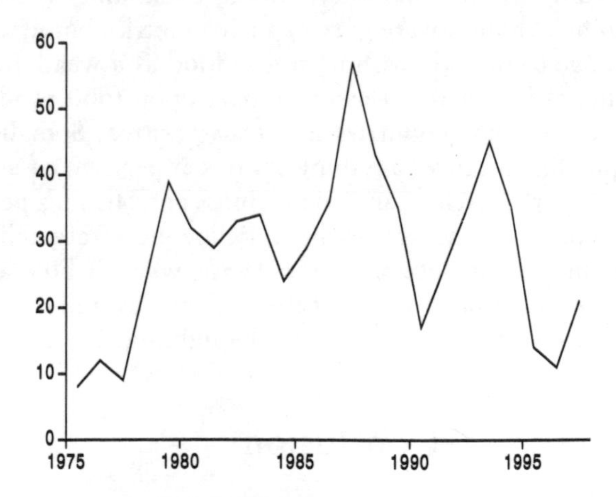

Fig. 12.12 Number of WFP emergency operations approved annually to combat 'man-made' disasters, 1975–97. *Source*: WFP.

between 1969 and 1973 rice production fell by 73 per cent nationally, and by up to 98 per cent in one province. American B52 bombing and the activities of the Khmer Rouge were responsible, leading in 1975 to the collapse of civil society, a famine, and the bloody take-over by Pol Pot. In Mozambique observers estimate that between 500,000 and one million people died from the effects of insurrection and famine between 1977 and 1992. A rebel group, RENAMO, deliberately destroyed health and education facilities, and disrupted agricultural production by the forced villagization of scattered communities, but both sides used scorched earth tactics when invading each other's area of control. Again, soldiers from both armies attacked food aid convoys and disrupted the humanitarian efforts of international relief agencies, diverting the food to their own war effort. A RENAMO official articulated their policy:

> Food is a tool of war, we use it to make strategic gains . . . The key to our success is that our forces have full stomachs. FRELIMO is the word for hunger – the people know this and join us (De Waal 1992).

Second, sieges have occasionally been used as a tactic to cut off populations and so starve them into submission. Bosnia was a very high profile example in the 1990s, with the largely Muslim capital Sarajevo in such a condition of hunger that supplies were airlifted in by the international community at great risk of aircraft being shot down. Examples from other continents include Afghanistan and Angola. More insidious, and certainly less reported in the media, has been the tactical sealing of southern Sudan in the 1980s and 1990s by the northern government in the hope of denying the rebel army its subsistence. There has been disruption of food production, killing of livestock, burning of crops and grain stores, and prevention in certain localities of the importation of humanitarian aid across borders or its use for feeding troops. The international community is understandably reluctant to breach the sovereignty of such countries but inaction gives a tactical advantage to the side wishing to use food as a weapon.

Third, land mines have had a serious impact upon food production on several continents. In some countries in Africa (Eritrea, Somalia, Angola and Mozambique, for instance) anything up to ten per cent of agricultural land has been rendered useless and many innocent farming people have been killed or injured. Some armies have deliberately targeted civilians, resorting to banditry, as in Liberia in the 1990s where whole areas were cleared and their populations robbed of their assets, including food, before they were allowed safe passage away from the fighting.

Conclusion

There has been a substantial restructuring of the world food economy in the last 50 years. Food security, if measured in the crude terms of the availability of energy and protein, has improved during this period (Chapter 10), but there are countries and classes for whom this is tenuous. A range of policies has been tried, some with a considerable measure of success; but it is worrying that the most obvious policy, food aid, has been found wanting in several respects. It should be remembered that food aid was originally used in the furtherance of foreign policy by the USA and other donors, and it has only been since the 1980s that humanitarian motives have come to the fore. Ironically, this has coincided with a general decline in political enthusiasm for this type of assistance.

International food trade, as a means of food security, seems destined to increase, at least in part due to the globalization of capital. But one cannot without qualification use trade as an indicator of potential prosperity for poor farmers in the future. Indeed, the conclusion of the agriculture portion of the GATT Uruguay Round was heavily influenced by the United States and EU, and was structured mainly for their benefit and in the interests of the food-based MNCs. It is not yet clear whether poor farmers will also profit.

By far the most effective means of increasing food security appears to be an effective programme of rural development and poverty reduction. This might include policies on credit provision, employment guarantee schemes of rural infrastructure works, land titling and redistribution, improved marketing arrangements, and the empowerment of ordinary people in local decision-making about resource use (Table 12.4). A discussion of these points, vital though it is for the Third World, would, however, draw us away from our central concern with food and we will leave it for another forum.

Table 12.4 Food security priorities drawn up by the NGO Forum to the World Food Summit, 1996

- Strengthening of family farming.
- The concentration of wealth and power must be reversed and action taken to prevent further concentration.
- Agriculture and food production systems that rely on non-renewable resources, which negatively affect the environment, must be changed toward a model based on agri-ecological (sustainable) principles.
- Responsibility of governments to improve food security, and also to implement policies of poverty reduction and democratization.
- IMF/World Bank structural adjustment programmes should be suspended.
- The participation of people's organizations and NGOs at all levels must be strengthened and deepened.
- International law must guarantee the right to food, ensuring that food sovereignty takes precedence over macro-economic policies and trade liberalization.
- Hunger and malnutrition are fundamentally a question of justice. Unless we agree that the right of every human being to the sustenance of life comes before the quest for profit, the scourge of hunger and malnutrition will continue.

Source: WFS website (accessed 1998).

Further reading and references

There is a very large literature on food security, paralleling the interest among the general population and the action taken by politicians. General discussion may be found in Kracht and Schulz and there are many articles and books on specific case studies. It is worth noting a recent trend towards questioning the effectiveness of the food security measures taken by the international community and by individual donor countries in terms of food aid. Arturo Escobar (1995) and Alex de Waal (1997) are particularly effective iconoclasts in this regard.

Buchanan-Smith, M. and Davies, S. 1995: *Famine early warning and response: the missing link*. London: Intermediate Technology Publications.

Chung, K., Haddad, L., Ramakrishna, J. and Riely, F. 1997: *Identifying the food insecure: the application of mixed-method approaches in India*. Washington, DC: International Food Policy Research Institute.

Cohen, M.J. and Pinstrup-Andersen, P. 1999: Food security and conflict. *Social Research* **66**, 375–416.

De Waal, A. (ed.) 1992: *Conspicuous destruction: war, famine and the reform process in Mozambique*. New York: Human Rights Watch.

De Waal, A. 1997: *Famine crimes: politics and the disaster relief industry in Africa*. London: Africa Rights.

Escobar, A. 1995: *Encountering development: the making and unmaking of the third world*. Princeton, NJ: Princeton University Press.

Food and Agriculture Organization of the United Nations 1996: *The sixth world food survey*. Rome: FAO.

Hubbard, M. 1995: *Improving food security: a guide for rural development managers*. London: Intermediate Technology Publications.

Hutchinson, C.F. 1991: Uses of satellite data for famine early warning in sub-Saharan Africa. *International Journal of Remote Sensing* **12**, 1405–21.

Kracht, U. and Schulz, M. 1999: *Food security and nutrition: the global challenge*. New York: St Martin's Press.

Macrae, J. and Zwi, A. (1994): Famine, complex emergencies and international policy in Africa: an overview. In Macrae, J. and Zwi, A. (eds) *War and hunger: rethinking international responses to complex emergencies*. London: Zed, 6–36.

Messer, E. 1994: Food wars: hunger as a weapon of war in 1993. In Uvin, P. (ed.) *The hunger report: 1993*. Langhorne, Pennsylvania: Gordon and Breach.

Walker, P. 1989: *Famine early warning systems: victims and destitution*. London: Earthscan.

World Bank 1998: *World development indicators*. New York: World Bank.

13

Food, world trade and geopolitics

Introduction

In Chapter 12 we saw how wars, particularly civil wars and guerrilla insurgencies, have had a negative impact on food security. We also demonstrated the political calculations and motivations behind food aid. This chapter continues the political theme, first by looking at food riots in the context of Structural Adjustment and, second, by suggesting that international food trade has undertones of geopolitical manoeuvring. In short, we will investigate global regulatory frameworks and their effect upon food.

Geopolitics and food riots

Walton and Seddon (1994) trace the origins of modern austerity protests to the mid-1970s. The classical food riot, which has been common throughout history, usually in response to a popular perception of profiteering or anger about shortages, was here transformed. A worldwide economic slump, triggered by a rapid rise in oil prices, had led to much political dissatisfaction and civil disturbance, but the major new element of austerity was introduced in 1976 when the IMF began imposing 'conditionality' on its loans to developing nations. First in Peru and then in Egypt the IMF insisted on changes to government economic policies, and in the early 1980s this became a more formal programme of 'structural adjustment'. The IMF felt that it had diagnosed an acute sickness in those Third World countries which were becoming over-indebted and which had over-reached themselves in the establishment of costly social improvement programmes and various forms of state enterprise. The symptoms of this sickness were hyper-inflation and general economic instability.

The IMF/World Bank structural adjustment programmes were tailored to deliver a shock to the economic structure of these countries. Between

1980 and 1990 the IMF signed agreements with 85 developing countries and the World Bank gave 220 adjustment loans to 63 countries. Currencies were devalued and exports encouraged; public spending was reduced and state-owned enterprises privatized; protectionism against TNCs was reduced; wages and prices were stabilized; and, above all, government subsidies on goods and services were reduced or abolished. Very few poor countries have been in a position to resist this new framework, which has helped to spread a free-market liberalism to a new stratum of the world economy. Without state subsidies on basic foodstuffs, prices of staples increased significantly, sometimes overnight, and throughout the 1980s the experience of protests and riots was repeated across Africa and Latin America (Figure 13.1).

One IMF/World Bank policy has been to encourage crops for which there is thought to be a comparative advantage, such as luxury vegetables or flowers. Unfortunately there is little evidence of low income, food deficit countries benefiting significantly from this wisdom. In fact Africa's share of the world market in its basic commodities has fallen. In reality comparative advantage is never a 'given': it is socially constructed or mediated through state policies and must therefore be recognized as such and nurtured.

Walton and Seddon claim to identify a pattern in the 'IMF riots'. These have all been city-based and occur mainly in countries which are 'overurbanized' in the sense of having large populations of poor people unable to find regular work in the upper circuit of the economy. In such environ-

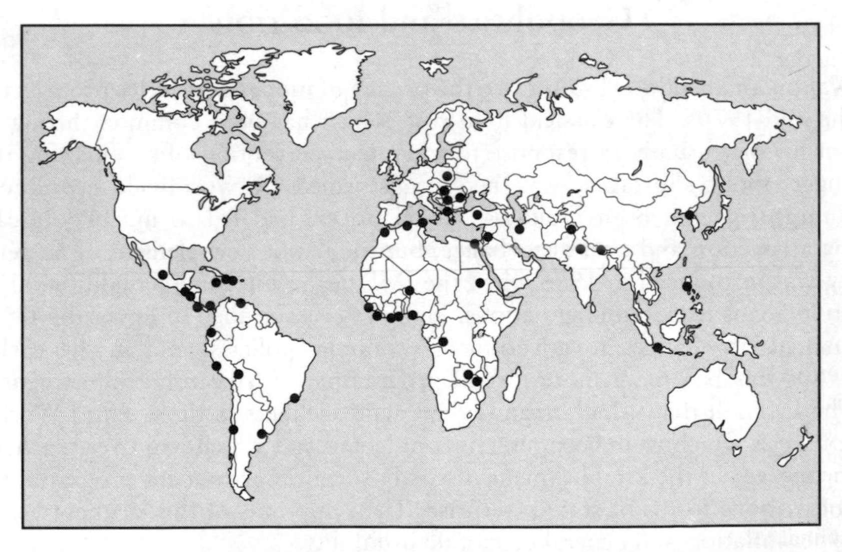

Fig. 13.1 Food riots, 1975–92. *Sources*: Walton and Seddon (1994); various newspapers.

ments working-class political organization often flourishes and provides a base for activism. Governments have recognized the tinder box nature of their big city slums and shanty towns and some have sought to buy a social pact with poor urban people through subsidies on their diet. The sundering of this arrangement by the IMF's adjustment has released pent-up political energy. Civil society was giving judgement and several governments came under threat as a result.

International food trade

Food constituted 8.5 per cent of international trade by value in 1995. This is more than oil and fuels at 7.7 per cent. It is well to remember, however, that the vast bulk of food is consumed in the country of origin and that trade across borders is still a relatively small part (6.8 per cent for crops and 5.7 per cent for livestock products in 1998) of final consumption.

Fruit and vegetables and grain are the most important food commodities traded (Table 13.1), although animal and fish products are mounting a challenge as rising incomes encourage a more protein-centred diet. When discussing grain, it is important to remember that food aid in the form of grain actually represents only 4.2 per cent of overall international flows by weight, and that a significant proportion of cereals are shipped as animal feed. Fruit and vegetables are increasingly popular exports for LDCs but this is built upon a solid foundation of a century or more of plantation-based fruit cultivation in the tropics.

The developing world is at an increasing disadvantage vis-à-vis the HICs. The export of high-value, luxury vegetables is very little compensation for the decline in real value of their other commodities, and in the last 30 years many have become net-importers of basic foodstuffs. Moreover, much of their trade is handled by TNCs, whose power is immense (Chapter 4), and

Table 13.1 Principal food commodities in world trade, 1995

Commodity	$ billion	Commodity	$ billion
Fruit and vegetables	69.8	Sugar, honey, sweets	18.0
Cereals	56.1	Tea, coffee	17.0
Fish	46.0	Other edible products	15.4
Meat	45.1	Cocoa, chocolate	12.8
Dairy products	29.6	Live animals	9.9
Alcohol	26.9	Non-alcoholic beverages	4.4
Tobacco	23.5	Spices	1.8
Animal feed	20.2	Margarine, shortening	1.8

Source: UN (1996).

Table 13.2 Exporters and importers of food, 1995

	Exports ($ billion)	Imports ($ billion)
EEC	156.3	158.3
USA	42.2	31.0
E. Europe and former USSR	10.7	16.2
Other developed	31.3	49.2
Africa developing	10.2	10.9
Americas developing	39.8	17.6
Middle East developing	6.3	12.9
Asia developing	49.6	39.1
Other developing	1.8	2.7

Source: UN (1996).

regulated by international bodies dominated by the USA and EU. Even the traditional LDC monopoly of tropical products is threatened by competition from highly capitalized sub-tropical producers in the HICs, such as rice in the southern states of the USA.

Trade and politics have become intertwined, each affecting the other. Governments have traditionally had difficulty in balancing the interests of their producers with those of their consumers and in developed countries the historical trajectory of policy has very much depended upon the lobbying power of the various interest groups. The EU's CAP, for instance, has favoured farmers through subsidies and by tariff and other barriers to imports from beyond Europe (Chapter 11).

Finally, there is instability in world food markets due to occasional periods of rapid price movements as a result of shortages or over-supply. One example was the so-called 'world food crisis' of 1973–74, which was triggered by a series of poor harvests and by the Soviet bulk purchase of grain from the USA. International wheat prices increased from about $60 per tonne in mid 1972 to $220 in February 1974, but this proved to be a temporary blip in the inexorable downward pressure on food commodity prices, which is predicted to continue (*see* Table 13.3).

Bananas: a study of export dependence

Western supermarkets are nowadays host to a wide variety of fruits and vegetables (Chapter 7). Many previously seasonal lines are now available throughout the year, being sourced globally in order to keep up a constant flow of product, and the largest stores are now also selling exotic species that go under the collective name of 'queer gear'. Before the 1980s these

Table 13.3 Projected food trade and commodity prices

	World trade (million tonnes)			Commodity prices (1990 $ per tonne)		
	1990	2020	Percentage change	1990	2020	Percentage change
Wheat	86.4	138.9	61	156	132	–15
Rice	10.1	24.2	139	231	181	–22
Maize	55.7	76.4	37	109	84	–23
Other coarse grains	24.7	40.0	62	89	67	–25
Soyabeans	25.1	55.4	121	247	219	–11
Roots and tubers	27.0	34.9	29	148	122	–18
Beef	3.0	8.7	187	2062	1947	–6
Pigmeat	1.5	2.4	64	1664	1500	–10
Sheepmeat	1.0	1.8	85	1907	1825	–4
Poultry	1.6	5.2	227	739	662	–10

Source: Rosegrant *et al.* (1995).

same shelves were stocked mainly with the much smaller range of fruits, such as bananas and oranges, which had dominated the international fruit trade throughout the century. They had done so because under the right conditions, they could be transported by sea over long distances without spoiling, whereas the counter-seasonal fruits and vegetables require a sophisticated 'cool chain'.

So important have bananas been to tropical producers that the 'banana republics' of Central America and the Caribbean have used fruit exports as the foundation of their development strategies. However, from the outset these specialist economies have been dominated by large corporations. Today, 80 per cent of the world's trade in bananas is controlled by an oligopoly of just five companies: Del Monte, Dole, Chiquita (formerly United Brands), Fyffes (Irish) and Noboa (Ecuador) (Table 13.5). In the past they have been the subject of complex mergers and acquisitions, in common with many other agri-food corporations. Dole merged with Castle & Cooke and Standard Fruit; Del Monte has been owned by US, Mexican, Chilean and now Middle Eastern interests; Fyffes was part of Chiquita, and was nearly sold to Dole in 1993; and Geest (UK) has been taken over by Fyffes/WIBDECO (a consortium of Windward Island governments).

These countries have been encouraged to devote much of their infrastructure to the efficient production and export of this fruit and their economies have become dangerously over-dependent on one commodity. Honduras, for instance, derived 45 per cent of its export income from bananas in 1970–72, although this had fallen to 18 per cent in 1995. About 62 per cent of this came from the plantations of United Brands and 38 per

Table 13.4 The large transnational banana corporations

Company	Founded	Products	Turnover ($ billion)	Main source of bananas
Dole	1851	Fruit and vegetables, flowers	4.3	Costa Rica, Honduras, Nicaragua, Philippines
Chiquita	1870	Bananas	2.8	Panama, Costa Rica, Colombia, Honduras, Philippines
Fyffes	1888	Fruit and vegetables	2.4	Small-holders in Caribbean
Del Monte	1886	Fruit and vegetables	1.5	Costa Rica, Honduras, Philippines

Source: Company reports.

cent from Standard Fruit. In Panama, Chiquita alone produces nearly half of the country's agricultural value and one-third of total goods exports. It employs 40 per cent of the permanent agricultural workforce.

In the past the banana companies were powerful enough in Central America to make or break governments. They dominated the regional economy and sought special treatment for their operations. This is well illustrated by the past activities of the United Fruit Company (now Chiquita) in Honduras. In the 1920s it received land from the government in exchange

Table 13.5 Main banana exporters by value

Country	$ million	Percentage of all commodity exports	Country	$ million	Percentage of all commodity exports
Ecuador	692.2	18.0	Jamaica	46.1	3.8
Costa Rica	680.4	24.6	Cameroon	45.0	1.6
Colombia	431.2	4.2	Guadeloupe	40.9	25.2
Philippines	223.7	0.8	China	40.1	–
Panama	197.2	16.2	Dominica	30.5	56.2
Guatemala	138.6	7.2	St Vincent	25.7	44.4
Honduras	120.9	18.4	Dominican Republic	17.1	0.9
Martinique	97.3	40.2	Belize	14.9	24.6
Cote d'Ivoire	80.6	22.0	Nicaragua	14.2	2.8
St Lucia	58.0	48.4	Indonesia	8.6	–
Mexico	48.1	0.1	Suriname	8.1	2.2

Source: UN (1996).

for building a railway, but the line that was laid only served the banana-growing regions and was never extended to the capital, Tegucigalpa. United Fruit's chairman at this time summed up the corporate attitude when he instructed his lawyer to:

> obtain rigid contracts of such a nature that no-one can compete against us, not even in the distant future . . . We must obtain concessions, privileges, franchises, repeal of custom duties, freedom from all . . . taxes and obligations which restrict our profits and those of our associates. We must erect a privileged situation in order to impose our commercial philosophy and our economic base (Thomson 1987).

The genetic and environmental implications of banana cultivation are far-reaching. Consumers like a large fruit without blemishes, of a quality and standardization that is reminiscent of a factory production line. To satisfy this market, most producers use varieties derived from the 'Cavendish', which is partly resistant to Sigatoka disease. They ignore the smaller, sweeter varieties such as burro, ninos, manzanos and others, some of which have a different flavour and coloured flesh. This is a risky strategy because it minimizes genetic variation within the plantation and maximizes the chance of attack by pests and diseases. Fields are treated with heavy doses of chemicals (sixteen times more per hectare than intensive agriculture in the HICs), and there are inevitable environmental and human consequences. Banana plantations are vulnerable to any winds over 25 mph and the frequent hurricanes in the Caribbean can be very destructive, as witnessed by the devastation caused in Honduras and Nicaragua in November 1998 by Hurricane Mitch.

The 'dollar bananas' of Central American plantations are in direct competition in world markets with the product of 100,000 small holders of the ACP group of countries (Windward Islands, Jamaica, Belize, Suriname, Cameroon, Ivory Coast, Martinique and Guadeloupe). Under the Lomé Convention (renegotiated 1990) the latter have been given preference in the EU at the request of their former colonial powers, mainly Britain and France (Coote 1996). But lobbying by the fruit TNCs made this a contentious issue with the United States administration, which has insisted on the abolition of such preferences. The Caribbean producers have an innate disadvantage because their small farms are often on difficult, hilly terrain and cannot compete on price with dollar bananas (costs are $500 a tonne as against $162). Without some protection whole economies would be devastated. The EU's banana régime is currently based upon the Framework Agreement of December 1994 (Figure 13.2), which went some way to meeting US objections but in 1999 the latter threatened a trade war unless this was further relaxed.

The transnational banana corporations are vertically integrated. They own banana plantations and ship the fruit to distant markets. They also control the ripening and wholesale parts of the chain. It is only at the retail

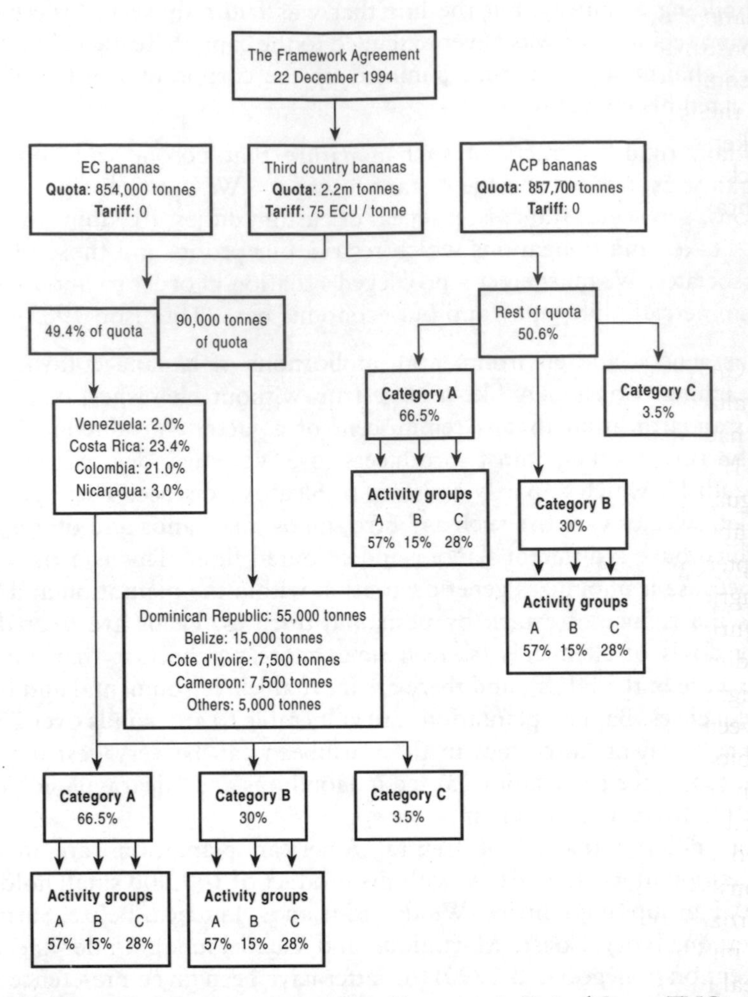

Fig. 13.2 The EU's banana régime, 1994. Category A: United States TNCs; Category B: EU operators; Category C: new EU companies; Activity Group (a): primary importers; Activity Group (b): secondary importers; Activity Group (c): ripeners. *Source*: Thagesen and Matthews 1997.

stage that the bananas finally change hands. Wresting part of this chain from corporate control has proved problematical for the producing countries. Some have nationalized plantations but the companies are flexible enough then to strike agreements with 'associated producers', who were really under their close control. Other countries have tried export levies but the 'banana war' of 1974–76 between the Panamanian government and United Fruit over a $1 per box tax showed that these would be very problematic to impose.

Labour has historically also been in a weak position on the banana plantations. Although they often live in company housing and earn wages that are comparable to or better than those of the non-banana sector, nevertheless the workers' employment depends upon the vagaries of the world market for fruit and upon the avoidance of hurricanes and pest/disease attack. There have been various exposés of the risks run by those employed to spray the plants with a cocktail of chemicals.

The GATT, the WTO and other bodies

The GATT was an international means by which trade was to be reformed and liberalized. It originated in 1947 as an attempt to counter the protectionism that had brought the world to its knees in the Great Depression of the 1920s and 1930s. However, agriculture was effectively excluded from this until the Uruguay Round of negotiations and, as a result, the European Community was able to build into its CAP features that were blatantly anti-free-trade (*see* Chapter 11). In 1995 Western developed nations spent $182 billion subsidizing agriculture, equivalent to about 40 per cent of the value of output.

During the long history of the GATT there have been many misunderstandings and disagreements between trading rivals that have led to outright hostility. 'Trade wars' in food products have been especially bitter between the USA and the EEC/EU, no doubt because both are major trading blocs that are striving for hegemony in this very lucrative and power-laden sector. Threats of retaliation have been frequent, in commodities as varied as oilseeds and bananas, but we will choose one example, the so-called 'chicken war' of 1961–64. Here the USA was objecting to EEC limitations and levies on imported poultry, which were threatening her market for frozen birds in Germany. In 1962 the duty reached 40 per cent by value and imports inevitably declined. Bilateral negotiations failed, as did an appeal to the GATT Secretariat, and in 1963 the US government therefore introduced tariffs on imported European goods, such as trucks and brandy.

The Uruguay Round was acrimonious and drawn-out but there was eventually agreement to cut subsidized food exports over 6 years by 36 per cent in value and 21 per cent in volume, domestic farm support by 20 per cent, and tariffs on imports by an average of 36 per cent. Developing country signatories also have to adhere to these new non-trade-distorting or 'green box' policies, although with a concession of somewhat lower thresholds (24, 14, 13 and 24 per cent respectively). In theory, the agri-industries in these countries should benefit from their comparative advantage of lower labour costs, but liberalization also opens their vulnerable economies to competition from TNCs. Estimates suggest an overall benefit of $510 billion by 2005 for all trade, of which about one fifth will accrue to developing and transitional economies, and the FAO predicts a rise in world trade of agricultural commodities of $53 billion between 1987–89 and the year 2000. Economists disagree on the long-term regional impacts that this will have upon agriculture and many think that the winners will not be in poor countries. Interestingly, the simple and effective subsidies on farm-gate and export prices are being phased out but direct payments to farmers, such as for set-aside land, are untouched. These are standard practice in the USA and EU but would be exceptionally difficult and costly to implement in the South. In effect they are an indirect form of subsidy, which will simply mean a change in the centre of gravity of farm support in the North.

In theory, the reduction of subsidized agriculture and consequent increase in trade on the basis of comparative advantage should have a number of positive effects other than the purely economic. Government support has in the past encouraged production on marginal land, much of which has been environmentally degraded as a result. Externalities such as soil erosion, pollution and deforestation should therefore be reduced. Whether this will happen in practice remains to be seen.

The WTO was born out of the Uruguay Round and started operations in 1995. It has become the official body to settle disputes and will initiate a new round of negotiations in 1999. The process of liberalizing world trade in agricultural products is far from over, both in terms of reducing tariff and other economic barriers to trade, but also concerning food quality and safety, and environmental considerations.

The International Standards Organization is a global regulatory body that deserves more notice. Based in Geneva, it sets the technical standards that apply to a wide range of food products, including the environmental conditions under which they are produced. For instance, the standard ISO 14001, concerns environmental management systems and was adopted in 1996. Those companies which claim to be environment-friendly will have to adopt it, along with future standards such as ISO 14020 on eco-labelling (Egger 1998).

Also important is the Codex Alimentarius Commission (CAC) set up by the UN in the 1960s to agree international standards for healthy food, with the goal of facilitating international trade. It has recently been accredited by the WTO as the scientific arbiter of what are called 'sanitary and phytosanitary measures', along with the International Office of Epizootics and the International Plant Protection Convention (Engels 1998). The CAC has so far proposed 237 standards for foods; set safe residue levels for 3274 pesticides and acceptable daily intake levels for 780 chemical additives; and evaluated the impact of 25 industrial food contaminants and 54 veterinary drugs (FAO 1998).

Conclusion

Although this chapter has dealt with a range of issues associated with trade and globalization, there is one important point that has stood in the background. This is the role of the nation state. Fully involved in the sense of attempting to regulate the flows of capital and commodities, the state nevertheless lost a slice of its power at the end of the twentieth century. The conclusion of the GATT Uruguay Round saw governmental subsidies on agricultural production and export eroded and scope for tariff barriers on imports reduced. Poor countries in particular have lost sovereignty as a result and will find themselves relying on the WTO to look after their interests in future. Even wealthy countries can feel the hot breath of food TNCs affecting their policy-making process because of the huge sums of tax revenue and the number of jobs to be lost if a company decides to disinvest.

These TNCs have 'a more global vision of food-system coordination than any given nation state and are the active stakeholders, whereas the nation states are much more passive "receivers" of commodities produced through global production systems' (Heffernan and Constance 1994). They have good information systems and this allows them to manipulate governments and if necessary play one off against another in the wider corporate interest. We have very clearly crossed the line into a food régime in which new power structures are being forged but predicting the outcome is difficult. The future of the global food trade will depend to a certain extent upon the nature of the resistance that is mounted by governments and consumers to the corporate domination of our food system and our diet.

Further reading and references

Walton and Seddon have written the best account of food riots. Dicken's book is an outstanding account of the global economy and in particular transnational corporations, although he does not deal in depth with food companies. See Moberg or Dosal for some historical background to the banana trade. The book by Josling *et al.* is probably the best introduction to the GATT and its role in regulating the international food trade.

Coote, B. 1996: *The trade trap: poverty and the global commodity markets.* Oxford: Oxfam.

Dicken, P. 1998: *Global shift: transforming the world economy*. 3rd edn. London: Paul Chapman.

Dosal, P. 1993: *Doing business with the dictators: a political history of United Fruit in Guatemala 1899–1944*. Wilmington, DE: Scholarly Resources Books.

Egger, M. 1998: *Are ISO standards a suitable instrument for supporting a sustainable banana economy?* Background paper for the International Banana Conference.

Engels, R. 1998: Possible consequences of international standards set by the Codex Alimentarius Commission on sustainable banana production. Background paper for the International Banana Conference.

Food and Agriculture Organization 1998: *The state of food and agriculture, 1997*. Rome: FAO.

Friedland, W.H. 1994: The new globalizaton: the case of fresh produce. In Bonnano, A., Busch, L., Friedland, W.H., Gouveia, L. and Mingione, E. (eds)

From Columbus to ConAgra: the globalization of agriculture and food. Lawrence, Kansas: University Press of Kansas, 210–31.

Grant, R. and Agnew, J. 1996: Representing Africa: the geography of Africa in world trade, 1960–92. *Annals of the Association of American Geographers* **86**, 729–44.

Heffernan, W.D. and Constance, D.H. 1994: Transnational corporations and the globalization of the food system. In Bonnano, A., Busch, L., Friedland, W.H., Gouveia, L. and Mingione, E. (eds) *From Columbus to ConAgra: the globalization of agriculture and food.* Lawrence, Kansas: University Press of Kansas, 29–51.

Josling, T.E., Tangermann, S. and Warley, T.K. 1996: *Agriculture in the GATT.* Basingstoke: Macmillan.

Moberg, M. 1996: Crown colony as banana republic: the United Fruit Company in British Honduras, 1900–1920. *Journal of Latin American Studies* **28**, 357–81.

Rosegrant, M.W., Agcaoili-Sombilla, M. and Perez, N.D. 1995: Global food projections to 2020: implications for investment. *Food, Agriculture, and the Environment Discussion Paper* No. 5 Washington, DC: International Food Policy Research Institute.

Thagesen, R. and Matthews, A. 1997: The EU's common banana régime: an initial evaluation. *Journal of Common Market Studies* **35**, 615–27.

Thomson, R. 1987: *Green gold: bananas and dependency in the eastern Caribbean.* London: Latin America Bureau.

United Nations 1996: *International trade statistics yearbook 1995.* New York: UN.

Walton, J. and Seddon, D. (eds) 1994: *Free markets and food riots: the politics of global adjustment.* Oxford: Blackwell.

Wiley, J. 1998: The banana industries of Costa Rica and Dominica in a time of change. *Tijdschrift voor Economische en Sociale Geografie* **89**, 66–81.

PART IV

A POLITICAL ECOLOGY OF FOOD

14
Introduction

In Part IV we intend to select a number of food-related issues and discuss them under the umbrella of 'political ecology'. Rather than a coherent theory, this is a broad conceptual framework that encourages studies of the interface between nature and society, where political economy meets ecology. Its first strand is a concern for the environmental change wrought by food producing activities such as agriculture and food processing. A number of the major themes of environmental management, such as agro-ecosystem modification, soil erosion, deforestation, desertification, and pollution, are relevant here but we will not follow them up because there are already a number of excellent existing texts (for instance Blaikie 1985; Blaikie and Brookfield 1987; Mannion 1995; Bryant and Bailey 1997).

A second strand is represented by Escobar's (1996) paper. His post-structural political ecology seeks to shed light on the various discursively constructed understandings of the environment and, in so doing, to reveal the social, ideological and linguistic bases of knowledge about the world. For him, a scientifically objective 'material reality' has no place in this schema because all 'facts' are derivatives of the articulation of knowledge and power. Blaikie (1996) adds the observation that scientific knowledge about the environment is both unstable and negotiable. He shows how views of degradation have changed and how institutional structures and policy prescriptions based upon previously accepted environmental 'facts' and 'truths' have in recent years been undermined. This has led to 'dissent, confusion and political machination' (Blaikie 1995, 211).

Third, there is the notion of risk, as discussed by scholars such as Beck (1992, 1995) and Giddens (1991). Beck identifies a 'modern' form of risk that is of a general nature. It is not necessarily traceable to individuals but is a collective responsibility of societies that have embraced modernity. Such risks may not be spatially limited and can be experienced widely, for instance the consequences of global warming. Beck goes on to argue that modernity has an element of reflexivity, because it creates both an awareness of risk and a desire to combat it. There is no shortage of recent examples in the realm of food consumption, for instance Mad Cow Disease. As the risk society replaces the industrial society ecological politics become normalized. This is a theme picked up by Latour (1999), who argues that green movements are not modifying the modernist project but giving it, on the contrary, a new lease on life. Now we have more reason for the monitoring, regulation and standardization of foods than ever before.

There are many other specific interpretations of political ecology, but our version in this book is very broad. We take food itself as an environmental issue for humans because its consumption is one of our most intimate interactions with nature. Chapters 15 and 16 see it as a fundamental part of the environmental health context of modern life and therefore susceptible to the political ecology approach. The argument is that our diet is one of the most hazardous aspects of nature that impinges directly on our daily lives, yet the food system is controllable. The organization of food production and processing is a matter of investment and regulation, and priorities for the reduction of risk are therefore issues of commercial and political judgement.

Chapter 17 investigates the claims of the sciences of plant breeding and biotechnology – that they have the answers to many of the food problems faced by humanity. This is appropriate because of the vast literature generated by the 'green revolution' and because of the worldwide interest caused by the current agitation against the release of GMOs. In Britain disquiet about GMOs has been generated from civil society rather than from the dogma of conventional politics. In the final chapter in this section, Chapter 18, an effort will be made to place food within the context of such popular debates, for instance the ethical ones about vegetarianism and animal welfare. The notion of food ethics and food justice will then be enlarged to include the human rights discourse on the 'right to food' of all human beings.

Many of these issues are controversial and are open to interpretation from different viewpoints. They are 'political' in the sense of arguments about priorities for action and especially about priorities for government spending. Some would argue that the relationship between society and environment is the principal problem of the age and that in future debates should centre on 'green politics'. The term political ecology has been used by some writers in that way (Lipietz 1995; Latour 1998), although this is not our intention in the chapters that follow.

Further reading and references

Blaikie's (1995) paper is a provocative starting point for thinking about environmental facts and knowledge claims. Beck (1992) has written one of the most influential books of the late twentieth century.

Beck, U. 1992: *Risk society: towards a new modernity.* London: Sage.
Beck, U. 1995: *Ecological politics in an age of risk.* Cambridge: Polity Press.
Blaikie, P. 1985: *The political economy of soil erosion.* London: Longman.
Blaikie, P. 1995: Changing environments or changing views?: a political ecology for developing countries. *Geography* **80**, 203–14.
Blaikie, P. 1996: Post-modernism and global environmental change. *Global Environmental Change* **6**, 81–5.
Blaikie, P. and Brookfield, H.C. (eds) 1987: *Land degradation and society.* London: Methuen.
Bryant, R.L. and Bailey, S. 1997: *Third World political ecology.* London: Routledge.
Escobar, A. 1996: Constructing nature: elements of a postructural political ecology. In Peet, R. and Watts, M. (eds) *Liberation ecologies: environment, development, social movements.* London: Routledge, 46–68.
Giddens, A. 1991: *Modernity and self-identity: self and society in the late modern age.* Stanford: Stanford University Press.
Latour, B. 1998: To modernise or ecologise? That is the question. In Braun, B. and Castree, N. (eds) *Remaking reality: nature at the millennium.* London: Routledge, 221–41.
Latour, B. 1999: *Politiques de la nature: comment faire entrer les sciences en démocratie.* Paris: La Découverte.
Lipietz, A. 1995: *Green hopes: the future of political ecology.* Oxford: Polity Press.

15

Food quality

Introduction

This chapter is an extension of the argument in Chapter 7 on food retailing; it also prefigures some of the food/health debate in Chapter 16 because food quality and food hazards are two sides of the same coin. State regulation of food, also treated briefly here, appears again in Chapters 11 and 18, which both touch on food policies.

The weight of recent public opinion has brought food quality to the forefront, to the extent that, in Britain for example, the government has felt obliged to initiate a Food Standards Agency. The implications are wide ranging: for health, and also for the likely restructuring of the food retail and farming sectors. As argued in Chapter 6, many farmers, food processors and manufacturers and food retailers see quality as a means of creating new demand or adding value to their existing output.

What is food quality?

As children, the authors remember listening to older relatives reminiscing of a time when 'food had some taste to it'. The point being made here was that modern varieties of fruits and vegetables and intensively produced livestock products are often bland by comparison with the foods produced 50 years ago. We were sceptical of this as just another story about the 'good old days', but the retail grocery industry has taken the idea more seriously. Always alive to an opportunity, they have recently introduced what they claim are more flavoursome products to their supermarket shelves.

Flavour is only one part of food quality, of course. Table 15.1 shows six other criteria that we would identify, while Ilbery and Kneafsey (1998) have classified objective and subjective measures of quality into four dimensions: certification, association, specification and attraction. *Certification* is a

Table 15.1 The dimensions of food quality

- Freedom from germs. The food health scares of the last 10 years have alerted consumers to the presence of micro-organisms causing food poisoning, BSE, tuberculosis and a number of other diseases.
- Low in additives from the food manufacturers and residual contaminants from farming.
- Food that has been sustainably produced, for instance from organic farming without negative environmental side effects.
- From a source that can be trusted. Important here is knowledge about the origins of the food, through certification/labelling or even purchase directly from the producer. Trust may also be put in a brand.
- Knowledge of the constituents (fat, sugar, salt, etc.) and preparation, allowing the food to be consumed as part of a balanced diet.
- Miscellaneous qualitative aspects that imply quality: fresh, exotic, luxurious, expensive, highly refined, traditional.

form of regulated quality achieved by satisfying conditions set by the state or by a professional organization. It is represented by a symbol or quality mark. *Association* is some kind of link with a place of origin, as in Scotch Whisky. *Specification* makes plain the nature of the production process, via a traditional recipe, especially good raw materials, or the particular skills of the production team. *Attraction* comes through the food's physical properties of taste, texture or appearance.

Quality has also become more tangible in recent years with the introduction of quality assurance schemes (QAS) and benchmarking, means of monitoring and setting standards for anticipated outcomes. Central government and the local state have become involved in defining and guaranteeing food quality on behalf of the consumer, while supermarket buyers have also exercised their power by requiring predictable quality from their suppliers.

Quality regulation

Consumer constructions of food quality are well known to the food industry and it has been common for manufacturers to label their products in ways that they know will appeal to customers. Words such as 'natural', 'home-made', 'fresh' and 'wholesome' would be open to wide interpretation and therefore to abuse if they were not regulated. The modern process of regulation began in the late nineteenth century as a result of a growing disquiet about the then common adulteration of foodstuffs, for instance by adding water to milk or skimming off the cream (Chapter 18). Governments were forced to define the legally acceptable composition of some foods and local authorities were empowered to analyse samples to detect fraud and to check that labels were not misleading.

In addition to state intervention of this kind, there has also been a long history of self-certification. The French *appellations d'origine contrôlée* are an example, particularly for wine (Chapter 21), where the good name of certain producing regions is preserved by a well-organized and carefully circumscribed scheme of certification, with official approval. Wine producers in Australia and other countries have proved that they can also market high quality products, often using grapes very similar to those grown in France, but their wines do not yet have the prestige guaranteed by the controlled labels on French bottles.

Ilbery and Kneafsey's (2000) study of speciality food producers in south west England suggests that relatively few have been quality-certified or even see the need for certification. They stress instead, first, the specification of production methods, raw materials used and the close involvement of the owner, and, second, the attractiveness of their products in terms of design, texture, flavour, freshness, and appearance. For them quality was meaningful only in terms of market satisfaction and reputation.

Morris and Young (2000) acknowledge the significance of QAS established by contemporary British food supply chains. Since the 1990s these QAS have become unavoidable and producers and processors are learning to live with the extra demands placed upon them. Three especially problematic aspects of QAS for the food industry and for the consumer are the plethora of standards agencies (such as the British Standards Institute, and the International Standards Organization); the hugely complex domestic and international food safety legislation; and the many quality marks and codes of practice issued by organizations such as the Soil Association, the Royal Society for the Prevention of Cruelty to Animals, and many more.

Supermarkets now exercise a degree of control over their suppliers that would have seemed impossible only 20 or 30 years ago. Their QAS are not just about the quality of the product as delivered from processing or packaging plants to the supermarket shelf. They also reach back to the conditions of production on the farm, including the variety of potato that can be grown, and the feedstuff supplier who can be used when fattening livestock. Morris and Young (2000, 108) quote a Tesco statement that 'in the long run we must look to be exercising greater control on the entire process, from womb to tomb'.

QAS are one way of producing traceable links in the food chain and are therefore important in its greater integration through time. They are an extension of farming under contract, as discussed in Chapter 5, and as such there are winners and losers; small farmers are again potentially the most vulnerable if they cannot supply in a timely fashion, in sufficient quantities, of a particular grade. Not all want to lose their independence, and there has been resistance by the farming community over many years to imposed notions of quality. This is reported in the farming press as a matter of anxieties over costs, changes in traditional methods, increased paperwork, and regular on-farm inspections (Morris and Young 2000).

Location-specific quality

Ilbery and Kneafsey (2000) investigate the activities of Taste of the West, a constituent group within Food from Britain. Both bodies are charged with assisting small locally-based speciality food producers with their marketing, as a means of increasing quality and value added in the food chain, and ultimately to provide rural employment. Taste of the West members in counties such as Cornwall, Devon, Somerset and Dorset do not, however, seem to feel that they derive any great benefit from the geographical context in which they find themselves because regional association is not an important indicator of quality for most foods in Britain. This suggests that the 'relocalization' process has not yet advanced to the point where either producers or consumers associate food with place.

In recent years the European Union has recognized the commercial potential of what the Italians call 'typical' foods. Following legal cases in which the producers of, for instance, Champagne and Feta cheese, have sought to prevent the use of their name on the labels of unrelated or imitation products, the concept of effectively copyrighting a food-place association has arisen. In 1993, under Regulations 2081/92 and 2082/92, the EU introduced the Protected Designation of Origin (PDO) and the Protected Geographical Indication (PGI) (Table 15.2). A PDO must have its place link at all stages of production, whereas a PGI has a link at only one stage. By the end of 1998, 18 PDO designations had been granted to UK products and 16 PGIs. Ilbery and Kneafsey (1999) suggest that the PDO/PGI recognition process has only just begun. Their research shows that so far it is regarded by manufacturers, mostly of cheese and drink, as an opportunity

Table 15.2 PDO and PGI designations created by EU Regulations 2081/92 and 2082/92

- *Protected Designation of Origin (PDO)*
  Open to products that are produced, processed and prepared within a particular geographical area, and with features and characteristics that must be due to the geographical area. The methods used to produce the product must be unique in that area.
- *Protected Geographical Indication (PGI)*
  Open to products that must be produced or processed or prepared within a geographical area and have a reputation, features or certain qualities attributable to that area.
- *Certificate of Specific Character (CSC)*
  Open to products that are traditional or have customary names and have a set of features which distinguish them from other similar products. These features must not be due to the geographical area the product is produced in nor entirely based on technical advances in the method of production.

Source: MAFF.

to stifle competition for their place-based labelling, and they do not yet think that consumers will be aware of this type of quality marking.

Another locational element of quality in the mind of many consumers is that of locally produced food which can be purchased from well-known and trusted individuals, as argued in Chapter 7. There is an element of such direct producer–consumer relations in many pre-industrial agricultural systems, especially where farmers use much of their own output for self-provisioning, and where sales to relatives, friends and neighbours are part of the taken-for-granted world. Money may not even be involved where goods are traded reciprocally. Gilg and Battershill (1998) find that such direct sales are also commercially viable for small farms in France, where the premium charged on high quality products may actually help small farmers to stay in business.

Organic food

Organic food, as introduced in Chapters 6 and 7, forms another dimension of food quality. Sales in Britain, for instance, reached £546 million at the end of 1999 and are set to rise to £1 billion by 2001 (Teeman 1999). The enthusiasm of consumers has not so far been matched by the production of home farmers, who can meet only 30 per cent of this demand. This may change, as government subsidies under EU regulations are being deployed to encourage about 1100 farmers to convert to organic production in the period 1999–2001, but there are others who are simply not able or willing to meet the stringent criteria laid down by the various certifying bodies (Table 15.3). So far only 1.5 per cent of British farmers have an organic enterprise, by comparison with 10 per cent in countries such as Sweden and Austria.

Ilbery *et al.* (1999) have provided the most up to date account of organic production in England and Wales. They recognize that such alternative farming may be more profitable than the intensive production which, as we write (May 2000), is in its worst crisis for decades. On the other hand, they also point out that many organic farmers are 'lifestylers' who are motivated more by a wish to escape modern urban living than by profit. A geographical

Table 15.3 Organic food organizations in the British Isles

- UK Register of Organic Food Standards
- Organic Farmers and Growers Ltd
- Scottish Organic Producers Association
- Soil Association
- Bio-Dynamic Agricultural Association
- Irish Organic Farmers and Growers Association

analysis of the distribution of organic production shows a concentration in the south of England and south Wales. Climate is a factor in this, along with a variety of issues associated with networks of information and support, and the availability of marketing outlets (Cudjoe and Rees 1992). In Canada it seems that organic production is a feature of the urban fringe, where it is associated with innovative farmers who take risks and who have the advantage of proximity to market (Beauchesne and Bryant 1999).

Organic farming and the organic food market have a significance well beyond the bald statistics of sales figures or farming subsidies. There is a deeply held philosophy shared by many producers and consumers that organic food is both kinder to the environment and healthier to eat. The minimization of chemicals is especially appealing, although the quality and health claims have been widely contested by sceptics.

Goodman (1999) reminds us that the organic food movement, and its regulation by a mixture of private and public bodies, represents an appropriation of the 'natural' and the 'ethical' (Chapters 6 and 18). The public now widely believes that foods grown with the aid of factory-produced chemicals are somehow inferior in quality to those produced 'organically'. The latter is more open to public scrutiny and represents a form of quality assurance, but it is nevertheless a form of bio-politics. Goodman (1999, 32–3) argues that the organic movement has 'an alternative world view [that] directly subverts the modernist dichotomy of nature/society'. Its alternative vision is 'a collective that enacts different relational rules of the game, emphasizing the metabolic unity between the cultivation and consumption of food, and encompassing a moral community that transcends the instrumentalist division between people and nature'.

Quality and food safety

Food safety is an increasingly important aspect of quality. Health issues surrounding the food chain have been driven to the top of the political agenda by food scares in the 1990s and the global nature of the concerns in this area is enshrined in the Sanitary and Phytosanitary Agreement of the World Trade Organization and the activities of the Codex Alimentarius Commission (Chapter 13).

Helpfully, Henson and Caswell (1999) have developed a four-part commentary on food safety. The first part addresses the criteria that are used in instituting food safety regulations. The scientific rationale (risk assessment, risk management and risk communication) is now well established, although there are many occasions when the science is either not yet formulated or is uncertain, and on these occasions the precautionary principle is usually employed (Calman and Smith 2000).

The second part of food safety is the relationship between private and public control systems. We have already commented on these under QAS, but we might add here that the UK Food Safety Act of 1990 was a threshold piece of legislation, introducing the notion of 'due diligence', which may be used by food companies as a defence in cases of food poisoning or other hazard (for America *see* Buzby and Frenzen 1999). If they have exercised due diligence then their liability to prosecution is limited, and this has been the driving force behind much of the QAS activity.

Third, there is public food safety regulation. Henson and Caswell identify three approaches: target, performance, and specification standards. Target standards insist that a supplier should not 'knowingly sell a product harmful to health'. Performance standards impose specified levels of safety at the time of sale, such as a low level of pesticide residues, but leave the trade to decide on how to achieve that. Specification standards lay down procedures to be followed, for instance the heating of milk to a certain temperature during pasteurization in order to kill microbes. A subset of specification in recent years has been the adoption of the Hazard Analysis and Critical Control Point System (HACCP), which identifies particular points in the production or processing of food when risks of contamination are greatest. If these points are regularly monitored then safety can be enhanced in a cost-effective fashion.

Compliance with and enforcement of these safety standards and regulations is a very different matter from the 'good idea' behind the legislation. Many small firms in agriculture, food processing, or catering do not seem to know what is required of them. Some cannot bear the cost of making the necessary improvements: they either go out of business or hope that the relatively weak inspection régime will not pick up their deficiencies and they will therefore be able to continue in business. Table 15.4 shows that two-thirds of the businesses questioned by Henson and Heasman (1998) found it either difficult or very difficult to identify the costs of meeting specific food standards.

Enforcement at the local level is by Environmental Health Officers (EHOs) and Trading Standards Officers. The role of EHOs in particular was clarified and strengthened by the Food Safety Act (1990) but their

Table 15.4 Ease with which costs of compliance with food regulations are identified

Degree of ease	Percentage response
Very easy	0.0
Easy	31.3
Difficult	52.2
Very difficult	13.4
Don't know	3.0

Source: Henson and Heasman (1998).

actual activities have been determined by several phases of policy. Marsden *et al.* (2000) describe a more insistent approach between 1990 and 1992, with a greater readiness to prosecute. By 1992 food retailers were complaining of unnecessarily harsh treatment, and a regulatory retreat followed in that year. After that there was a rapprochement between the retailers and the EHOs but the swings of policy were damaging to the credibility of local regulation.

The fourth dimension is that of the strategic response to food safety regulation (Loader and Hobbs 1999). In the private sector this has often been the capture and use of food safety by large manufacturing and retail companies as a means of extending their control upstream in the system. For a government's strategic viewpoint, a cost-benefit analysis shows that food safety measures are well worth careful implementation. In the USA the annual cost has been estimated at $100 million, set against a public loss of $5–10 billion from illness associated with food-borne disease (Antle 1999).

Theorizing quality

We can now return to an earlier argument in Chapter 4, on food networks, as a means of theorizing food quality issues. As we discussed earlier, Marsden and Arce (1995) view food networks as an integral element in the processes of globalization. Quality is tied up with this because it has become a means of competition between producers and a lever used by food companies in the construction of a circuit of compliant suppliers. The economic implications for the participants in such global networks are potentially profound but there are also important social consequences for the labour forces, particularly in developing countries where conditions of work are often poor and wages low.

The interconnexions between actors in global, or even local, food networks are becoming a focus of scholarship. The embeddedness of marketing networks in social relations allows us to understand the forging of the trust necessary to hold links together (Granovetter 1973, 1985; Lyon 2000). Actor Network Theory (*see* Chapter 5) has provided further insights into the nature of inter-relationships (Kneafsey and Ilbery 2000) and has proved to be an especially helpful basis for understanding the distribution of power. ANT is also wedded to the idea of a symmetrical treatment of the social and the material. It yields a degree of priority to objects and to nature that is not to be found in other social theory (FitzSimmons and Goodman 1998; Goodman 1999). This makes it ideal for studies of food networks at the micro-scale, with detailed ethnographic observation as its main tool.

Conventions theory (Storper and Salais 1997; Wilkinson 1997) has also been employed to look at food quality. It has advantages in that it allows comparison between networks, where ANT's somewhat idiosyncratic inductive epistemology is more introspective (Murdoch *et al.* 1998;

Murdoch and Miele 1999). Conventions theory is also more suited to a spatial perspective.

Conventions are the agreed bases for the characteristics and qualities of products. They may be taken-for-granted, informally agreed, or enforced through a set of rules, practices and contracts. Market demand and technological factors are important but it is the interpersonal world of contacts that is vital for establishing and maintaining the trade in speciality and high quality foods. Conventions theory concentrates on these contacts and gives a greater emphasis than hitherto to the constellation of activities within which the product forms. In a study of Wales, Murdoch *et al.* (1998) highlight the varied conventions responsible for organic products such as cheese, yoghurt and lamb, and argue that their approach could be employed more generally 'to gain a greater insight into the complicated geography of agro-food production'. Murdoch and Miele (1999) attempt a similar analysis for organic eggs and meat in Italy, where they find an increasing complexity in the food system, where companies can exist simultaneously in several worlds of production.

Conclusion

This chapter has shown that quality is a fundamental consideration in food studies. It maps into a bundle of cognate issues, such as health, food policy, and the economic restructuring of the food system. Our conclusion is that the 1990s were a crucial decade in making food quality a pivotal area, but the early years of the new century will also be important in determining the extent to which quality will be driven by market forces or by the democratic will of the consumer.

Further reading and references

The reader will note that most of the references below have publication dates in the late 1990s and 2000. This indicates that the topic of food quality is relatively new and rapidly expanding. Marsden *et al.* (2000) tackle many of the issues raised in this chapter but no one reference is comprehensive.

Antle, J.M. 1999: Benefits and costs of food safety regulation. *Food Policy* **24**, 605–23.

Beauchesne, A. and Bryant, C. 1999: Agriculture and innovation in the urban fringe: the case of organic farming in Quebec, Canada. *Tijdschrift voor Economische en Sociale Geografie* **90**, 320–28.

Buzby, J.C. and Frenzen, P.D. 1999: Food safety and product liability. *Food Policy* **24**, 637–51.

Calman, K. and Smith, D. 2000: Works in theory but not in practice? Some notes on the precautionary principle. *Public Administration* (forthcoming).

Cudjoe, F. and Rees, P. 1992: How important is organic farming in Great Britain? *Tijdschrift voor Economische en Sociale Geografie* **83**, 13–23.

FitzSimmons, M. and Goodman, D. 1998: Incorporating nature: environmental narratives and the reproduction of food. In Braun, B. and Castree, N. (eds) *Remaking reality: nature at the millennium*. London: Routledge, 194–220.

Flynn, A., Marsden, T. and Ward, N. 1994: Retailing, the food system and the regulatory state. In Lowe, P., Marsden, T. and Whatmore, S. (eds) *Regulating agriculture*. London: Fulton.

Gilg, A.W. and Battershill, M. 1998: Quality farm food in Europe: a possible alternative to the industrialised food market and to current agri-environmental policies: lessons from France. *Food Policy* **23**, 25–40.

Goodman, D. 1999: Agro-food studies in the 'age of ecology': nature, corporeality, bio-politics. *Sociologia Ruralis* **39**, 17–38.

Granovetter, M.S. 1973: The strength of weak ties. *American Journal of Sociology* **78**, 1360–80.

Granovetter, M.S. 1985: Economic action and social structure: the problem of embeddedness. *American Journal of Sociology* **91**, 481–510.

Harrison, M., Marsden, T. and Flynn, A. 1997: Contested regulatory practice, the local state and the implementation of food policy. *Transactions of the Institution of British Geographers* (NS) **22**, 473–87.

Henson, S. and Caswell, J. 1999: Food safety regulation: an overview of contemporary issues. *Food Policy* **24**, 589–603.

Henson, S. and Heasman, M. 1998: Food safety regulation and the firm: understanding the compliance process. *Food Policy* **23**, 9–23.

Ilbery, B., Holloway, L. and Arber, R. 1999: The geography of organic farming in England and Wales in the 1990s. *Tijdschrift voor Economische en Sociale Geografie* **90**, 285–95.

Ilbery, B. and Kneafsey, M. 1998: Product and place: promoting quality products and services in the lagging rural regions of the European Union. *European Urban and Regional Studies* **5**, 329–41.

Ilbery, B. and Kneafsey, M. 1999: Niche markets and regional speciality food products in Europe: towards a research agenda. *Environment and Planning A* **31**, 2207–22.

Ilbery, B. and Kneafsey, M. 2000: Producer constructions of quality in regional speciality food production: a case study from south west England. *Journal of Rural Studies* **16**, 217–30.

Loader, R. and Hobbs, J.E. 1999: Strategic responses to food safety legislation. *Food Policy* **24**, 685–706.

Lyon, F. 2000: Trust and power in farmer–trader relations: a study of small scale vegetable production and marketing systems in Ghana. Unpublished PhD thesis, University of Durham.

Marsden, T.K. and Arce, A. 1995: Constructing quality: emerging food networks in the rural transition. *Environment and Planning A* **27**, 1261–79.

Marsden, T., Flynn, A. and Harrison, M. 1997: Retailing, regulation, and food consumption: the public interest in a privatized world? *Agribusiness* **13**, 211–26.

Marsden, T., Flynn, A. and Harrison, M. 2000: *Consuming interests: the social provision of food*. London: UCL Press.

Morris, C. and Young, C. 2000: 'Seed to shelf', 'teat to table', and 'womb to tomb': discourses of food quality and quality assurance schemes in the UK. *Journal of Rural Studies* **16**, 103–15.

Murdoch, J., Banks, J. and Marsden, T. 1998: An economy of conventions? Some thoughts on conventions theory and its application to the agri-food sector. Paper for Economic Geography Study Group Conference on Geographies of Commodities, University of Manchester, 1–2 September.

Murdoch, J. and Miele, M. 1999: 'Back to nature': changing 'worlds of production' in the food sector. *Sociologia Ruralis* 39, 465–83.
Storper, M. and Salais, R. 1997: *Worlds of production: the action frameworks of the economy*. Cambridge, Mass.: Harvard University Press.
Teeman, T. 1999: 'The most important issue of our time'. *Guardian* 6th November.
Wilkinson, J. 1997: A new paradigm for economic analysis? Recent convergences in French social science and an exploration of the convention theory approach with a consideration of its application to the analysis of the agrofood system. *Economy and Society* 26, 305–39.
Young, C. and Morris, C. 1998: New geographies of agro-food production: an analysis of UK quality assurance schemes. Paper for Economic Geography Study Group Conference on Geographies of Commodities, University of Manchester, 1–2 September.

16

Food and health

Introduction

This chapter selects a number of themes on food and health. All are examples of issues live at the time of writing and they show how we, the consumers, have become aware of the risks implicit in what we eat. Fischler (1980) has articulated this in what he calls the 'omnivore's paradox' (*see also* Chapter 21). Humans can produce and consume a very wide variety of foodstuffs. An even greater range is available now in the modern era because of the technological advances made by the food processing industry. Yet we are becoming increasingly anxious about the food we eat at the very moment that the choice is seemingly unlimited.

Food scares are becoming common. There have been examples throughout the twentieth century, but the recent history of such scares in Britain was initiated in December 1988 by a statement of Edwina Currie that most eggs were infected with salmonella. At the time she was a Conservative MP and a junior minister in the Department of Health. Although her statement was substantially accurate, she was pressed into resignation soon afterwards because of the disarray that it caused in the egg industry. Sales fell by 50 per cent as a result of high levels of media interest and adverse publicity. This was followed by a number of minor scares in the early 1990s, and then in 1996 by the extraordinary *furor* over mad cow disease, or bovine spongiform encephalopathy (BSE). As we write (June 2000) another story has been running for many months. This is the public's concern about genetically modified crops and their appearance in product form (especially GM soya in a wide range of processed foods) on supermarket shelves, without adequate testing for the environmental and human health implications. Whether the science supports such worries or not seems to be a marginal matter, because a media 'feeding frenzy' has raised the debate to such a pitch that consumers are making purchase decisions on the basis of fear rather than fact. Beck (1992) thinks that he can identify the historical evolution of public awareness about risks towards what today we may call the 'risk society'.

By the late 1990s food safety was very firmly on the political agenda in Britain, other European countries and the USA. The New Labour government elected in 1997 undertook to create an independent Food Standards Agency (FSA) that would allay the public's fears that the close relations between the MAFF, farmers and food manufacturers were detrimental to the interests of consumers. In the words of James (1998), this amounted to a 'widespread crisis of confidence in the UK system for ensuring food safety'. James himself was asked to prepare a plan for the FSA and a modified version of this was implemented when the agency began operations in April 2000.

Intrauterine nutrition, infant health

We turn now to consider a number of food health issues, beginning with intrauterine nutrition in infant health. About 24 per cent of newborn babies in developing countries are estimated to be malnourished, largely as a result of receiving inadequate nutrition in the womb. The vast majority of such cases are found in Asia and they may be attributed to maternal anaemia, gastro-intestinal and respiratory infections, and diseases such as malaria. Intrauterine growth retardation (IUGR) is associated with poor cognitive and neurological development in infancy, and increased risk of high blood pressure, cardiovascular disease, diabetes and renal problems later in life.

Table 16.1 The relationship between weight at 12 months of individual British males and their record of disease in later life

	Men born 1911–30			Men aged 59–70
Weight at 12 months (lb)	**Ischaemic heart disease (standardized mortality ratio)**	**All causes (standardized mortality ratio)**	**Fibrinogen (g/litre)**	**Impaired glucose tolerance**
<19	100	89	3.21	43
19–20	84	89	3.10	32
21–22	92	85	3.13	30
23–24	70	68	2.97	18
25–26	55	73	2.93	19
>26	34	58	2.93	13
All	78	79	3.04	25

Source: Barker (1992).

Table 16.2 Non-infectious diseases with a dietary connexion

Disease	Hypothesized causes
Acute appendicitis	Insufficient dietary fibre
Atheroma and coronary heart disease	Diet high in cholesterol, obesity, hypertension, diabetes mellitus, smoking, lack of exercise
Bladder stone	High cereal consumption
Breast cancer	Some correlation with total fat, animal protein and animal calories in diet; also genetic and other factors
Cancer of endometrium	Possibly due to fat or carbohydrate
Cancer of large bowel and rectum	High calorie intake, meat consumption
Cancer of oesophagus	Fungal toxins in food, tobacco, alcohol
Stomach cancer	Nitrosamines in food
Cancer of pancreas	Probably diet and smoking
Prostate cancer	Possible relation with saturated animal fats
Ulcerative colitis (inflammation of large bowel)	Hypersensitivity to dairy products, additives
Type 2 diabetes mellitus	High in refined carbohydrate, low in fibre, also genetic factors
Diverticular disease (colon)	Insufficient fibre
Gallstones	High in fat, refined carbohydrate
Gout (high serum urate levels)	Possible dietary connexion
Haemorrhoids	Insufficient fibre
Hiatus hernia	Insufficient fibre
Hypertension (high blood pressure)	High salt intake
Kidney stones	High animal protein, refined sugar, low in fibre
Chronic panchreatic disease	High alcohol or cassava consumption
Deep vein thrombosis, stroke, brain haemorrhage	Due to hypertension (qv) and atherosclerosis
Tropical ataxic neuropathy	Cyanide in cassava
Duodenal ulcers	Refined carbohydrates
Varicose veins	Insufficient fibre
Vitaminosis (rare)	Excess vitamin consumption (especially vitamins A, B-6, D)

Note: In many of these diseases diet is only a contributory cause.
Source: Hutt and Burkitt (1986).

Professor Barker and his team at Southampton University have made similar, clear connexions in the British context between foetal/infant health and the incidence of adult diseases. Table 16.1, for instance, indicates that low weight at 12 months is a good predictor in later life of ischaemic heart disease, all deaths, stroke (concentrations of plasma fibrinogen), and diabetes (glucose intolerance) in men. In short, morbidity and mortality are to a certain extent conditioned by the womb experience, and one of the greatest contributions a pregnant mother can make to her offspring's future health is to eat and drink sensibly and not to smoke. It seems that preventative medicine starts before birth.

Lifestyle, healthy eating and government advice

Whereas infectious disease accounted for many deaths in the nineteenth century, targeted public health measures from the Victorians onwards reduced its incidence. Theories of disease, which in the last 150 years have been dominated successively by the role of the environment and by germs, in recent times have tended more and more to stress 'lifestyle'. The neat causal relationship between microbe and mortality has been broken and replaced by the altogether vaguer notion that the key to health is how we live our lives. Diet is an important part of this (Table 16.2), but so are exercise, smoking, stress and body weight.

Certain foods have come under scrutiny (Table 16.3) in the last 40 years or so and many millions of people have altered their diets as a consequence of medical comment and government advice. It is important to note, however, that studies of the links between diet and disease are still at an early stage and it has often proved methodologically difficult to separate out the impact of food from the host of other factors that may contribute.

Coronary heart disease is one example of a likely correlation between morbidity/mortality and diet (COMA 1994). A diet rich in saturated fat seems to be linked to atherosclerosis, which is the formation of deposits of fatty material in the arteries. This in turn raises the risk of the coronary artery being blocked and the decay of heart muscle (myocardial infarction). Smoking is also a factor, doubling or trebling the risk of a heart attack because nicotine accelerates the heart rate and constricts the blood vessels. Figure 16.1 summarizes the latest understanding of the risk factors.

Table 16.3 Foods most often linked with disease

- *Sugar*: a source of energy but not of any other nutrients. It is known to be responsible for tooth decay and there may be links with other forms of ill health.
- *Saturated fats*: linked to disease of the heart and blood circulatory system. Saturated fats are found especially in red meat, butter, cheese, and some vegetable oils.
- *Salt*: a contributory cause of high blood pressure. Eighty per cent of salt consumed in Britain is hidden in processed foods where it is used by manufacturers as a flavour-enhancer, preservative and processsing aid.
- *Refined foods*: high in refined carbohydrate and low in fibre. White bread is an example. Fibre is essential for the efficient operation of the digestive and other systems.

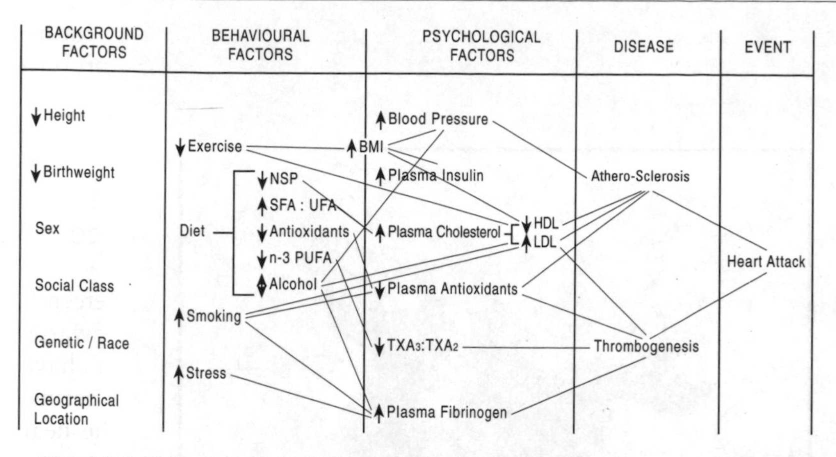

Fig. 16.1 Behavioural and physiological risk factors in coronary heart disease. NSP: non-starch polysaccharides; SFA: saturated fatty acids; UFA: unsaturated fatty acids; PUFA: polyunsaturated fatty acids; BMI: body mass index; TX: thromboxane; HDL: high density lipoprotein; LDL: low density lipoprotein. *Source*: Ashwell (1997).

Not all advanced countries necessarily have a heart disease problem. The Japanese have an industrial economy and consumers who can afford an affluent lifestyle. They use more salt and smoke more than the British, but interestingly their rate of heart disease, until recently, was much lower. This is probably because their diet was traditionally dominated by rice, vegetables, fish and unsaturated vegetable oils and, as a result, their blood cholesterol levels are lower by about 30 per cent. Since World War II, however, Japanese diets have become more westernized and heart disease has increased.

The so-called Mediterranean diet, dominated by cereals (pasta, bread, couscous), fruits and vegetables, and olive oil, is also associated with low rates of heart disease and of certain cancers. This mix of foods has become increasingly popular in northern Europe in the 1990s as a result of a widespread discussion in the media of the health benefits, and also as a result of many consumers' direct experience on package holidays in the Mediterranean. As a result, sales of olive oil and pasta have soared in Britain, although ironically several Mediterranean countries are moving away from their own traditional foods (Gracia and Albisu 1999). France, although it has a Mediterranean coast, already has a diet that is more like that of northern Europe, high in saturated fat and coupled with cigarette smoking, that looks on paper to deliver a high risk of heart disease. Yet the so-called 'French paradox' means that coronary heart disease is lower than might be expected. This is possibly due, in part, to the anti-oxidant properties of the red wine favoured by many consumers in France.

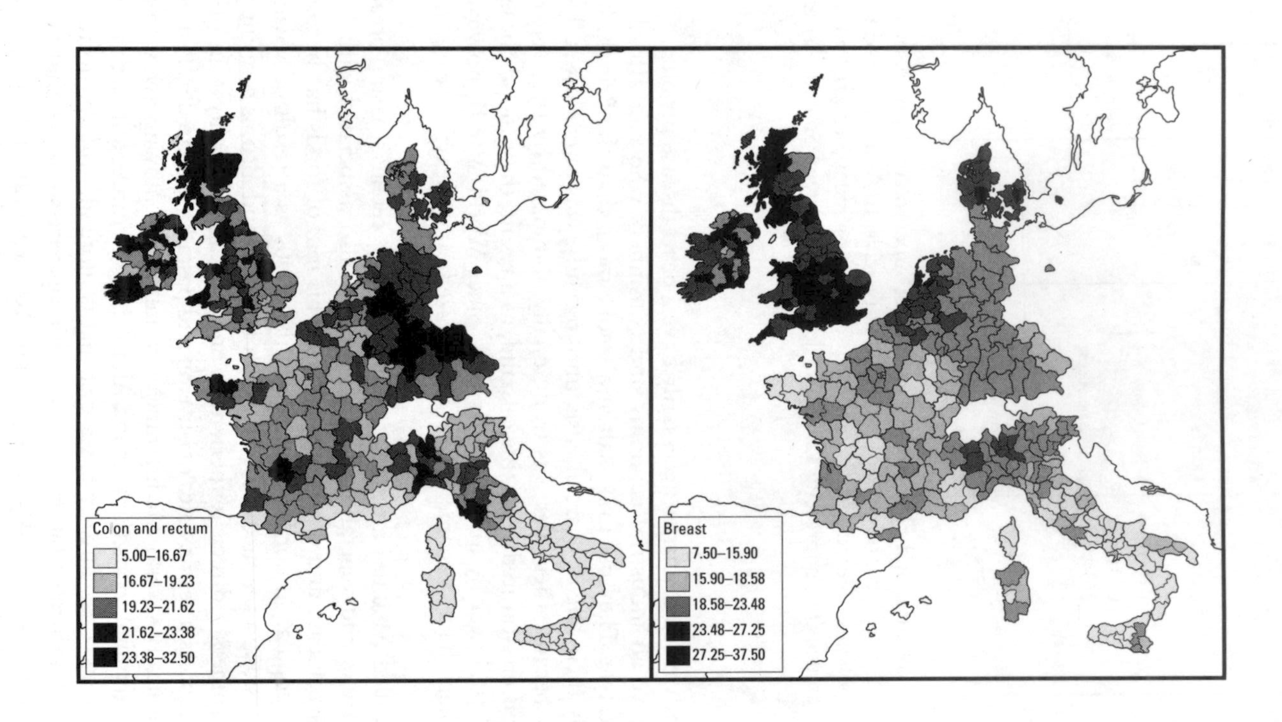

Fig. 16.2 Cancer mortality per 100,000 in the EEC, 1971–80 (age-standardized). (a) Colon and rectum (males). (b) Breast cancer (females). *Source*: Smans *et al.* (1992).

The WHO estimates that one-third of the 10 million annual global cancer cases (including colon and breast cancers: Figure 16.2) could be avoided by feasible and appropriate diets and by physical exercise, and a further third by the avoidance of tobacco. Advice to help minimize this includes reduced alcohol intake and increased consumption of fruits and vegetables (Table 16.4). The last dietary change alone would cut cancer deaths by 20 per cent because it is the anti-oxidant quality of fruits and vegetables in their Vitamins A, C and E that helps to neutralize the degrading effect of free radicals on cellular DNA. This would be a significant saving of life; in the U.K, for example, cancer is second only to heart disease as a cause of death. Recent research has also identified grapes, red wine, green tea, cooked tomatoes, garlic, cruciferous vegetables (broccoli, cauliflower), fish oil, flaxseed, soya, rosemary, turmeric and carrots as having anti-cancer properties, because they contain a range of helpful chemicals such as polyphenols, lycopene, allyl sulphides, sulphoraphane, omega-3 fats, isoflavones, and Cox-2 inhibitors. These are called 'functional foods'.

After these observations about coronary heart disease and cancer, it is hardly surprising that the public has become more health-conscious about diet in recent years than ever before. Most governments now offer dietary advice, for instance in the UK through the Committee on Medical Aspects of Food Policy (COMA). These have produced a stream of literature in the last 20 years, one of the most controversial of which was the National Advisory Committee on Nutrition Education's report on *Proposals for Nutritional Guidelines for Health Education* (1983), which recommended reductions in consumption of 50 per cent for sugar and salt, and 30 per cent for fat. The following year COMA, this time in the specialized context of coronary heart disease, came to similar conclusions (Table 16.5).

The Department of Health's White Paper *The health of the nation* (1992) proposed turning a modified version of these guidelines into dietary targets in order to reduce premature deaths under 65 by at least 40 per cent and deaths in the range 65–74 by 30 per cent. Anecdotal evidence suggests that these messages do have an impact on food habits, although the influence of the state is counterbalanced by the advertising of food companies, who have a clear profit motive in making health claims for their products.

Table 16.4 The relation between diet and cancer

- Up to 80 per cent of bowel and breast cancer preventable by dietary change.
- Diet contributes to lung, prostate, stomach, oesophagus and pancreatic cancers.
- Fruit, vegetables and fibre protect against cancer.
- Red and processed meats increase risk.
- Smoking, alcohol and overweight increase risk.
- Physical exercise protects against cancer.

Source: Cummings and Bingham (1998).

Table 16.5 The COMA recommendations, 1984

- A decrease in the consumption of saturated fatty acids to no more than 15 per cent and of total fats to no more than 35 per cent of total food energy.
- Moderate alcohol consumption of less than 43 ml per day for men and 29 ml for women.
- A decrease in salt consumption.
- Increased eating of bread, cereals, fruit and vegetables.
- Reduction in smoking and in obesity.

Source: COMA (1984).

Social exclusion

Income is by far the most important explanatory variable in the relationship between diet and health (for income and food consumption *see also* Chapter 21). In Britain, as in many other 'advanced' countries, the lower income groups exhibit the characteristics shown in Table 16.6. Such inequalities have a clear spatial expression that has been analysed extensively by medical geographers and sociologists (Townsend and Davidson 1992; Townsend, Phillimore and Beattie 1988).

Corner shops were common in densely populated working-class districts of industrial cities throughout the Western world until the 1960s. They provided a limited range of goods, sold in small portions, but they were conveniently located and were open for long hours. Since that time the restructuring of the retail environment has been led by supermarkets, most recently in out-of-town locations that are inconvenient for people without cars or ready access to public transport (Chapter 7). Young, single mothers on welfare with young children find access particularly difficult, and low wage families generally cannot buy their food in bulk because of cash flow

Table 16.6 Low-income dietary and health links

- Low consumption of fruit and vegetables
- Less food rich in dietary fibre
- Low intake of anti-oxidants, some vitamins and minerals
- High sodium diet, leading to high blood pressure
- Reduced growth rates *in utero*
- High rates of female obesity
- Low incidence of breast-feeding
- Greater dental decay
- Greater maternal smoking during pregnancy
- Higher alcohol-related morbidity

Source: Acheson (1998).

problems and a lack of adequate storage space at home. They need to shop 'little and often', but many of the corner stores, and also village shops, which catered for this need have closed, unable to compete with the sheer market power of the large chains. The dietary health of old people and ethnic minorities is also at risk as a result of accessibility problems.

The supermarkets, which have replaced independent stores, are not necessarily cheaper for food items than smaller shops, and several surveys have indicated that their prices are often higher in poorer areas. Accessibility to a healthy diet is especially compromised by a meagre choice of affordable fresh fruit and vegetables (Cummins and Macintyre 1999) (*see also* Chapter 20). The health implications of this kind of social exclusion are worrying because these foods are protective against some cancers and coronary heart disease and risk levels are therefore raised amongst certain groups in society, through no fault of their own.

Clark *et al.* (1995), McKie *et al.* (1998) and Skerratt (1999) have studied what amounts to social exclusion in the western isles of Scotland. The problems are: low levels of disposable income, high cost of food, limited choice, and the limited availability of fruit and vegetables. The isolated geographical location and economic decline of this peripheral community make the promotion of healthy eating messages exceptionally difficult.

Food poisoning

Food poisoning is among the commonest forms of illness. The WHO estimates that there are 1.5 billion episodes of diarrhoea occurring each year globally, of which 70 per cent are caused by biologically contaminated food. As a result, three million children under the age of five die annually, mainly in developing countries.

Food poisoning is on the increase in developed countries. In Germany the reported incidence of infectious enteritis rose from eleven per 100,000 in 1965 to 193 in 1990. In the USA there are 7000 deaths per year from salmanellosis alone. Since the early 1980s, notifications of food poisoning increased over seven-fold in England and Wales, and more than four-fold in Scotland, to over 160 per 100,000 people (Figure 16.3). As a result, about 200 die each year in England and Wales, mainly the young and very old. About 30,000 cases of *Salmonella*, 40,000 of *Campylobacter* and 700 of *Escherichia coli 0157:H7* (also known as VTEC) are reported annually, but the true numbers are certain to be higher (perhaps by as much as ten times). Other agents include *Listeria monocytogenes, Staphyloccocus aureus* and a number of viruses (Table 16.7).

The reasons for the secular increase are bound up with the changing food system. First, food production has become much more intensive over the last 50 years and there are pressures on farmers and food processors to cut corners in order to reduce costs. Abattoirs and food processing plants have

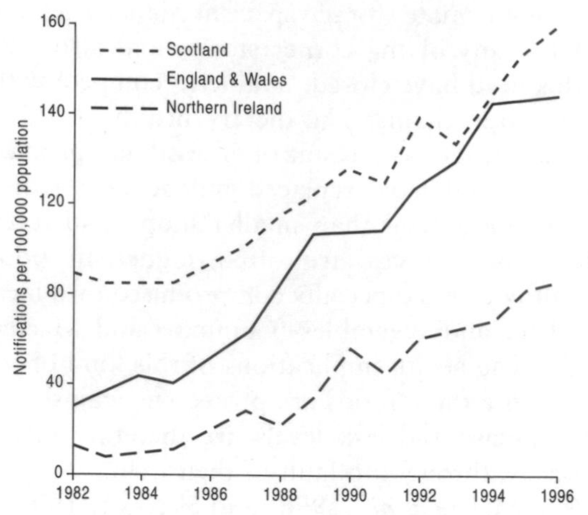

Fig. 16.3 Food poisoning notifications per 100,000 population, 1982–95. *Source: The Guardian*, 14th January 1998.

been shown to be major points for infection and, because they are growing larger and larger, a single incident can have a big impact. In Britain, the Public Health Laboratory Service estimates that up to 30 per cent of frozen chickens have levels of *Salmonella* in their carcases high enough to cause illness if the birds are not thoroughly cooked, and one-sixth contain microbes such as *Listeria*, another agent of serious food poisoning. The equivalent figures in Sweden are very low (under one per cent), suggesting that controls can be enforced, given the political will.

Second, consumers are demanding more convenience foods but these are often not properly stored. Refrigerator temperatures over 5°C are dangerous, as is food kept beyond its use-by date, and also frozen food not fully defrosted before cooking. An example is the so-called cook-chill food that became popular in the 1980s. This is factory-prepared and then cooled before transport to a retail outlet. Unfortunately there are several food poisoning organisms that can multiply at or below freezing point and the refrigerator and cook-chill cabinet are, therefore, both potential sources of infection.

Third, traditional cooking skills have often been substituted by technological innovations such as microwave ovens, which may give a false impression that food has been thoroughly heated. Only the outer couple of centimetres is penetrated by the microwaves and with large food items this might mean the survival of bacteria in the centre despite a superficial appearance of the cooking process being complete. The average cooking time of a main meal in 1934 was 2.5 hours but this has been reduced now to only 15 minutes.

Fourth, there has been insufficient public control over catering establishments. Between 1992 and 1997, 83 per cent of outbreaks of *Campylobacter* and *Salmonella* food poisoning in England and Wales

Table 16.7 Food poisoning

Agent	Source	Symptoms
Campylobacter	Raw or undercooked meat (especially poultry), unpasteurized milk, bird-pecked milk on doorsteps, untreated water, domestic pets	Abdominal pain, profuse diarrhoea, malaise
Salmonella	Red and white meats, raw eggs, milk, and dairy products, following contamination of cooked food by raw food or failing to achieve adequate cooking temperatures	Diarrhoea, vomiting and fever
Clostridium perfringens	Meat, poultry, stuffing, gravies, cooked foods held at inappropriate temperatures	Diarrhoea, abdominal cramps, headache, and chills
Staphylococcus aureus	Any food contaminated by infected person	Vomiting, abdominal pain, diarrhoea
Bacillus cereus	Meat products, soups, vegetables, puddings and sauces, milk and milk products	Abdominal pain, watery diarrhoea, vomiting and nausea
E. coli 0157	Beef products (e.g. undercooked beefburgers), milk, and vegetables	Bloody diarrhoea, kidney damage
Listeria monocytogenes	Wide range of food products, including soft cheeses and meat based patés	Influenza-like illness; spontaneous abortion
Clostridium botulinum	Underheated low-acid tinned foods (e.g. green beans, mushrooms, tuna), vacuum-packaged meats, sausage, fish	Nausea, vomiting, diarrhoea, fatigue, headache, dry mouth, double vision, droopy eyelids, muscle paralysis, trouble speaking and swallowing, breathing difficulty
Small round structured viruses (SRSVs), Norwalk virus	Raw foods, especially shellfish	Nausea, vomiting, diarrhoea
Cyclospora cayetanesis	Any food contaminated by infected person	Explosive, watery diarrhoea, fatigue, nausea, vomiting, muscle aches, low-grade fever, abdominal cramping, loss of appetite
Cryptosporidium parvum	Any food contaminated by infected person	Severe watery diarrhoea, coughing, low-grade fever

Source: Public Health Laboratory Service and USDA websites (accessed 1998).

originated in restaurants, fast food outlets, hotels and other institutions, but inadequate funding of environmental health services means that these are checked frequently in only one-fifth of the local authorities responsible.

Poor meat hygiene often begins in the slaughterhouse. Until better practices were enforced in the mid-1990s, meat in British abattoirs was often smeared with faeces when intestines were accidentally ruptured and because animals arrived from farms or markets in a filthy condition. Butcher knives were frequently not properly sterilized.

The *E. coli 0157* outbreak in November 1996 in North Lanarkshire, Scotland, perhaps most fully symbolizes the crisis of modern food hygiene. A total of 21 people died and about 500 were affected after eating infected meat supplied by a butcher who allegedly had paid only lip-service to the regulations imposed upon him by environmental health officials from the local authority. As a result, an enquiry was established under the chairmanship of Professor Hugh Pennington of Aberdeen University and his committee made a number of recommendations. Among these was the need for raw and cooked meat to be separated, hands washed, equipment cleaned, and butchers who sell cooked meat to be registered and appropriately supervised. Education in food hygiene should be organized for abattoir workers, butchers, and also for school children. There should be the widespread adoption of a method of hygiene called Hazard Analysis and Critical Control Points in order to minimize the risk.

One possible means of reducing the risk of food poisoning is to irradiate food. Many food hygiene experts (including a report in 1998 by a joint Study Group of the FAO, WHO and International Atomic Energy Agency) agree that it is safe and does not reduce nutrients any more than cooking does, but irradiation has nevertheless had a very negative press in recent years. There is, for instance, the suspicion that the food industry will relax its standards and come to rely on irradiation as a means of bringing its products up to the bare minimum of standards required by the state. Irradiation is now permitted in 41 countries, including the UK, which since 1986 has allowed irradiation up to specific limits, in seven categories of food. The process involves passing food through either an electron beam or gamma rays from a cobalt-60 source. The ionizing radiation kills food poisoning micro-organisms in meat, fruit and vegetables, and may assist with storage by inhibiting processes such as sprouting and ripening (potatoes, onions) and killing insects (cereals, fruit). No radioactivity is left in the food.

Food processing

In a sense we have been 'processing' food ever since meat was first cooked and grain ground into flour. Traditional techniques of manipulation to create variations of foods and new tastes, along with the need for preservation, have been common since ancient times, but it was the Industrial Revolution that provided the technology, entrepreneurship, organization of production and investment capital, which, together, accelerated and restructured the food system of advanced nations (*see* Chapter 6). Milling, distilling, refining and fermentation were all industrialized for the efficient mass production of bread, beer, jams, sugar and wine (Abraham 1991).

In the nineteenth century mechanized roller mills, finely sieved flour, and assembly-line baking facilitated the introduction of the standardized, white, wheaten loaf, the sliced version of which is still with us today. White bread replaced the coarser barley and rye breads that had still been common in the eighteenth century, and it represented a decline in nutritional quality because the bran and wheat germ were extracted, removing much of the valuable fibre, protein, vitamins and minerals. In the twentieth century chemicals were used in bread making, such as chlorine to bleach the flour and potassium bromate to give it better baking qualities. Softeners or 'surfactants', such as mono and di-glycerides, retard the appearance of staleness and calcium propionate prevents mouldiness. Many processed foods are similarly dependent on chemicals for their appeal to the customer, and the huge expense of providing the necessary equipment and inputs has encouraged the concentration of production into fewer and fewer hands (Abraham 1991).

Other food technologies, such as the refrigeration of meat and canning of fruits and vegetables, were also important in transforming diets in the nineteenth and early twentieth centuries. Chocolate and sugar confectionery, biscuits, breakfast cereals, carbonated drinks, potato crisps and other snack foods have all provided the basis of major food industries, and increasing numbers of hot take-away meals are rapidly processed *in situ* by chains such as McDonald's (Chapter 7). All of this adds up to a greatly increased range of foods for the consumer to choose from, but it also presents a challenge to her/his health.

Over the last 200 years there has been a declining intake of fibre in modern diets. Wholemeal bread and potatoes are less popular, although there has been some minor compensation in the increased consumption of other vegetables and fruits. Meanwhile the rise of sugar in the diet as a source of energy, has been remarkable, as has the fat in meat and dairy products. The former has been called a 'curse of civilization' by one writer (Yudkin 1988) and the latter is implicated in coronary heart disease and some cancers. We might reasonably conclude that a combination of push factors, emanating from the steadily deeper and broader industrial organization of the food system, and pull factors, following from higher incomes and lifestyle changes, has been responsible for exposing consumers to a range of new risks that are rather different from the pre-industrial hazards of malnutrition and poor sanitation. These risks vary with income, social class and region, emphasizing the point that there are complexities of economic, social and cultural processes that make for varied geographical outcomes of the diet/health relationship.

A good example of the impact of food processing on health is the addition of ingredients that are in effect hidden from the consumer. For instance, in Britain 70–80 per cent of salt intake comes from processed foods, but this has meant an average consumption that is ten times more than the recommended level. Salt is implicated in hypertension (high blood pressure), possibly leading to coronary thrombosis; therefore, for the tenth of the population who are vulnerable, the use of processed foods may actually be dangerous for their health.

Food additives, food intolerance and food allergy

A distinction can be drawn between 'food intolerance' and 'food allergy'. The former rises when the body is unable to digest an ingredient, such as the common inability to assimilate milk sugar (lactose) or a range of chemical additives. Some writers include food poisoning under the heading of intolerance.

Allergic reactions are different because they involve the immune system, which is stimulated and releases the histamines and other substances that cause headaches, stomach-aches and skin rashes. The main problem foods seem to be chocolate, dairy products, nuts, fish and shellfish, soya, wheat, egg, sweets and soft drinks. Twenty per cent of Britons believe that they suffer from a food allergy.

The alteration of natural foods to make them more appetizing or to preserve them has been a feature of the food industry for hundreds of years, for instance amongst the traditional craft processors whose techniques of smoking fish or making cheese have been handed down from generation to generation. We have come to accept such practices as a matter of course and, indeed, processed or manufactured foods constitute about three quarters of our diet and seem quite normal. Normal, that is, until we begin to look at the label (Table 16.8) on the package and find a long list of ingredients, many of which originate in laboratories.

About 3800 additives are used in our daily food, for three basic purposes. First, there are cosmetic chemicals that make products look more

Table 16.8 Legal requirements for food labelling in Britain

- Date, such as 'sell by' for foods with a shelf life of under 6 weeks or 'best before' for foods which can be stored up to 18 months.
- Name and address of manufacturer.
- Nutritional information is provided on some labels.
- Ingredients, in order of weight. It is not compulsory to give a breakdown of weights or percentages of the individual ingredients.
- There is no information on whether food has been irradiated, and whether foods contain residues of pesticides and other chemicals used in the production process.

Table 16.9 E numbers

E number	Additive
E100–180	Colours.
E200–297	Preservatives (extend shelf-life by preventing growth of food decay micro-organisms).
E300–321	Antioxidants, which protect food against deterioration caused by oxidation, such as fat rancidity and flavour or colour changes.
E322–385	Antioxidants, acidifiers and emulsifiers (which aid in the formation and maintenance of the uniform dispersion of two or more substances which would not normally mix, such as oil and water in margarine).
E400–495	Thickeners, stabilizers, gelling agents, gums and emulsifiers (used to maintain physical state of food and to stabilize, retain or intensify colour of food).
E500–578	Acidity regulators and processing aids, such as anti-caking agents (to ensure free flow, as in dried milk or salt), raising and firming agents.
E620–640	Flavour enhancers.
E900–1520	Glazing agents (to produce a protective coating or polish/sheen on surface of confectionery or fruit); waxes, resins, bleach, preservatives, improving agents, packaging gases, propellants, sweeteners, modified starches.

Source: http://www.bryngollie.freeserve.co.uk/Enumbers.htm; http://www.cintamani.net/Cosmetic/E-Numbers%20.htm; http://www.additives.8m.com/

attractive to the senses, especially colouring agents, flavours, sweeteners and texture modifiers, such as emulsifiers and stabilizers. Second, there are preservatives, including antioxidants and sequestrants, which add life to a product. Third, processing aids assist the manufacturing process, for instance by preventing food from sticking to machinery. About 380 of these additives had officially been approved by the EU by 1987 (Table 16.9).

A small but significant group of people are allergic to individual or groups of additives. E102 (tartrazine) has, for instance, been linked to hyperactivity in children, who consume it in fish fingers, cakes, sauces, soft drinks or marzipan. It is banned in Norway and Finland but not in the UK. Such allergic reactions can be sudden and dramatic, but perhaps even more worrying is the unknown and insidious long-term effect that food additives and chemical residues may have. Cannon (1987) and Abraham (1991) discuss the tortured history of government regulation of additives and conclude that it has been more heavily influenced in its policy by the food industry than by the interests of public health.

Illegal tampering with food was commonplace in the nineteenth century (Chapter 15). Milk was especially vulnerable, with a quarter of the average pint being water added by unscrupulous dairymen. Sometimes this added water was not clean and milk became a source of infectious diseases such as typhoid. There were other types of adulteration. Bakers routinely used alum for whitening poor grades of flour, and to make their inferior bread saleable. Brewers put a variety of noxious chemicals into their beer to give it 'body', and tea was adulterated with shrivelled and dyed leaves from hedgerow shrubs.

Such crude and occasionally poisonous practices have long since disappeared under the pressure of food inspection by the authorities. More subtle forms of adulteration persist, however. One example is the use of the phosphate group of chemicals (E450) to increase the water retaining character of foods, especially fish, red meat and poultry.

If our account seems to be loaded against processed food, consider that most additives are not linked in any way with human health risk and that the quality of our food supply generally is a great improvement on that of the nineteenth century. Modern processing technologies produce foods that most of us could not willingly do without, for instance snack and convenience foods that enable us to live a life less tied to the kitchen than previous generations. These fundamentally positive aspects of the food industry may be sufficient to outweigh the negative points referred to above. Each consumer has to decide, and we should not forget that the food industry only supplies what we, the customers, demand or are willing to pay for.

Agri-chemicals and drugs

Insecticide, rodenticide, herbicides, and fungicides are of concern to the consumer because of the residues that appear in food. In the UK, the MAFF's routine testing has found that one-third of food samples are contaminated by these and other agri-chemicals and that one in a hundred have levels above those permitted legally.

Drugs are also routinely used by farmers to increase the growth or yield of their animals and to protect them from disease. Some species kept in intensive housing units, for instance pigs, are especially susceptible to debilitating or life-threatening complaints and therefore must be constantly monitored if the enterprise is to remain profitable. The most common preventative measure is for farmers to add one or more of eight antibiotics to animal feed (Lacey 1991). It is estimated that in Britain alone, £500 million is spent annually on the prophylactic antibiotic products of companies such as Pfizer and Eli Lilly, and that 66–80 per cent of cattle, sheep, pigs and poultry are dosed at one time or another during their lifetime. In December 1998, the EU decided to ban four such growth promoters on the grounds that, over a period of time, bacteria become immune to them and that this immunity might pass through the food chain to make bacterial infections in humans more difficult to treat.

A similar issue is raised by the use of hormones as growth promoters. These were banned in Europe in 1989 but are still used in the USA, Canada and Australia. Friction developed when the EEC restricted the importation of US beef because of the traces of hormones, and the WTO ruled that the ban flouted the rules of free trade. In the EEC itself hormones and harmful drugs such as *clenbuterol* have in the recent past been widely available on a black market organized by criminals in Belgium and other countries, and were used by the veal and beef industries for rapid fattening purposes. Other illegal chemicals include *papain,* which used to be injected into animals before slaughter in order to tenderize their meat.

Bovine Somatotropin (BST) is a naturally produced hormone that controls the cow's metabolism, which in turn determines her milk production. Early experiments showed that the injection of BST from the pituitary glands of slaughtered animals boosted milk production by as much as 20 per cent, and more recently a genetically modified, recombinant version has been successfully tested.

BST was legalized in the USA in 1994 and now more than a quarter of dairy cattle there are in treated herds (Institute of Food Science and Technology website 1998). In Britain, the companies that manufacture BST, Lilly Industries and Monsanto, first applied for licences in 1987, and their products were eventually given the imprimatur of the MAFF's Veterinary Products Committee and the Medicines Commission. However, the EU has imposed a moratorium, most recently extended in 1999, while further research is conducted. In reality the EU has a milk surplus and does not need BST to increase output. The only gainers would be dairy farmers, whose profitability would rise on individual animals.

There are two problems with the use of BST as a yield enhancer. First, there are worries about the increased level of insulin-like growth factor (IGF-1) in the milk of cows treated with BST, since IGF-1 has been linked to breast and prostate cancer in humans. This is probably an unfounded concern because pasteurization reduces the chemical activity of BST by 90 per cent; but scientists do not know what effect it will have over a thirty year period of daily consumption.

Second, there are implications for animal health. BST-injected cows have a 50 per cent increased risk of mastitis, an unpleasant udder disease, and their fertility is reduced by 20 per cent. The increased yield of milk places a very heavy burden on the energy system of a cow and upon its skeletal structure (Chapter 18). Such animals may suffer from anaemia, diabetes and reduced immunity to infection (Lacey 1991).

Conclusion

Much of this chapter has been based on statistics and examples from the UK, but we do not want to give the impression that the issues discussed here are country-specific. With the possible exception of BSE, the problematic links that we have identified between food and health exist in most other countries in Europe and North America. This is because food systems in the wealthy nations are dominated by chemical-based agriculture, by capital-intensive processing, manufacturing and retailing, and by similar patterns of food consumption and general lifestyles. As a result, some health problems, exacerbated by faults in the food system, are on the increase, particularly food poisoning and a wide range of non-infectious complaints caused by chemical additives and residues.

We do not wish to imply that capitalism and the profit motive must shoulder the full blame for food-related ill health, however. Farmers, food manufacturers and retailers all realize that acute illness and mortality caused by their products is bad for business. Along with general improvements in sanitation, this is one reason why infectious diseases spread by food, such as typhoid or tuberculosis, are less common in the food chain than they used to be. Probably more important in the long term are the state-inspired food policies and regulations which are vital if these improvements are to be consolidated and scandals such as BSE banished forever. The USA has its Food and Drug Administration and the British government its Food Standards Agency for this purpose, setting standards and monitoring food quality. It is in the consumers' interest that such bodies should be given strong powers and be made to work.

Further reading and references

Abraham and Cannon are good, critical writers on health and the food system generally, and Ulrich Beck is insightful on the nature of risk in modern life. For the BSE story, the best source of information is the BSE Official Enquiry website.

Abraham, J. 1991: *Food and development: the political economy of hunger and the modern diet*. London: Kogan Page.

Acheson, Sir Donald (Chairman) 1998: *Independent inquiry into inequalities in health: report*. London: The Stationery Office.

Ashwell, M. 1997: *Diet and heart disease: a round table of factors*. 2nd edn. London: Chapman and Hall for British Nutrition Foundation.

Barker, D.J.P. (ed.) 1992: *Fetal and infant origins of adult disease*. London: British Medical Journal.

Beck, U. 1992: *Risk society*. London: Sage.

Cannon, G. 1987: *The politics of food*. London: Century Hutchinson.

Clark, G.M., MacLellan, M., McKie, L. and Skerratt, S. 1995: *Food availability and food choice in remote and rural areas*. Edinburgh: Health Education Board for Scotland.

Committee on Medical Aspects of Food Policy 1984: Diet and cardiovascular disease. *Department of Health and Social Security, Report on Health and Social Subjects* No. 28. London: HMSO.

Cummings, J.H. and Bingham, A. 1998: Diet and the prevention of cancer. *British Medical Journal* **317**, 1636–40.

Cummins, S. and Macintyre, S. 1999: The location of food stores in urban areas: a case study in Glasgow. *British Food Journal* **101**, 545–53.

Fischler, C. 1980: Food habits, social change and the nature/culture dilemma. *Social Science Information* **19**, 937–53.
Gracia, A. and Albisu, L.M. 1999: Moving away from a typical Mediterranean diet. *British Food Journal* **101**, 701–14.
Hutt, M.S.R. and Burkitt, D.P. 1986: *The geography of non-infectious disease*. Oxford: Oxford University Press.
James, P. 1998: Setting food standards. In Griffiths, S. and Wallace, J. (eds) *Consuming passions: food in the age of anxiety*. Manchester: Mandolin, 41–8.
Lacey, R. 1994: *Mad cow disease: the history of B.S.E. in Britain*. St Helier: Cypsela Publications.
McKie, L., Clark, G.M., MacLellan, M. and Skerratt, S. 1998: The promotion of healthy eating: food availability and choice in Scottish island communities. *Health Education Research* **13**, 371–82.
Skerratt, S. 1999: Food availability and choice in rural Scotland: the impact of 'place'. *British Food Journal* **101**, 537–44.
Smans, M., Muir, C.S. and Boyle, P. (eds) (1992): Atlas of cancer mortality in the European Economic Community, *IARC Scientific Publications* No. 107. Lyon: International Agency for Research on Cancer.
Yudkin, J. 1988: *Pure, white and deadly*. Harmondsworth: Penguin.

17

From the Green Revolution to the Gene Revolution

Introduction

The authors have files bulging with press cuttings relating to biotechnology in general and genetically modified food in particular. Much of this material comes from the years 1998 and 1999 when public attention in the UK was much taken with a number of scares about the environmental and health consequences of so-called 'Frankenstein foods'. Even royalty, in the person of HRH the Prince of Wales, has become involved and the story has been headline news in a way rivalled only by 'mad cow disease' in the recent history of food and agriculture in the UK. We feel that a chapter on this area of the food debate is amply justified by this level of interest.

This chapter explores the history of plant breeding over a period of approximately 50 years, starting with a reprise of the Green Revolution. This was, in itself, controversial, mainly because of the argument, from some quarters, that it increased the poverty of small farmers and landless labourers. The section that follows deals with biotechnology, and attempts to tease out the positive as well as the negative arguments.

The Green Revolution

The Green Revolution was the result of an intensive plant breeding programme that relied on brilliant applied science and some luck. It was dubbed 'green' to distinguish it from any 'red' revolutionary politics, to which it offered an alternative, and less disruptive, path to rural development. The Green Revolution has attracted a lot of praise but also some bitter criticism. In this section we look at both sides of the argument.

The Green Revolution was based on the spread of new wheat HYVs from Mexico and rice from the Philippines. It is an example of south–south transfer of technology, although much of the Research and Development funding and expertise originally came from the advanced countries. The story began in 1943 when the Rockefeller Foundation in collaboration with the Mexican government set up a research project in northern Mexico to improve the local variety of wheat. The research station came to be known by its Spanish acronym, CIMMYT, and its director was Norman Borlaug, who was later (1970) awarded a Nobel Peace Prize for his work. CIMMYT began releasing seeds to the rest of the world in the early 1960s, to much acclaim and a fanfare of publicity.

In 1960 the International Rice Research Institute (IRRI) was founded at Los Baños in the Philippines with financial backing from the Rockefeller and Ford Foundations. Part of their job was to collect varieties of rice from all over the world and keep them in a seed bank for possible future use in breeding. In 1966 their 'miracle rice', IR8, was released. It had been a lucky cross between two strains of rice (*Petan*, a tall, vigorous Indonesian variety, and *Dee-geo-woo-gen*, a short, stiff-strawed Chinese rice) and the resulting plants displayed several desirable characteristics (Table 17.1).

By 1988 there were 500 semi-dwarf rice varieties, of which 150 were bred by IRRI. IR64, for instance, which was released in 1985, has proved to be one of the most successful, with high yields and good resistance to pests and diseases. It was produced after 18,000 breeding experiments over 5 years.

High yielding varieties (HYVs) have spread widely (Table 17.2) and by now have passed the global production of traditional varieties. The Green Revolution has not been a simple, single event, however. It now has a more than 40-year history, and we can identify a number of phases.

Table 17.1 The improvements in modern rice varieties

- A shorter stem and a narrower leaf, which increases the strength of the plant and prevents 'lodging' (falling over, weakening the plant). Also there is an increase in the ratio of the weight of the useful grain to that of the rest of the plant, which has less value.
- Rice plants of standard heights, which minimizes the problem of mutual shading.
- Insensitive to 'photoperiod'. Traditional tropical (*indica*) rice flowers when day length shortens, allowing the plant to synchronize with seasonal weather conditions. Temperate (*japonica*) rice on the other hand pays no attention to day length (photoperiod) and can therefore be grown across seasons at any time of year. Putting this characteristic into tropical rice has allowed up to three crops to be grown each year.
- Rapidly maturing. Traditional varieties of tropical rice had ripened in about 160 days, but IR8 was ready in 130 days, and IR28 in 105 days. This aided multiple cropping.
- Higher yields, showing hybrid vigour. Yields, however, were dependent upon a package of inputs that included the application of fertilizer (especially nitrogen), irrigation water, and the control of weeds and pests. However, F1 hybrids do not come true to type after the first generation and, for the full effect, seeds therefore have to be bought each year.

Table 17.2 Percentage of wheat area planted to modern varieties in the developing world

Region	1970	1977	1983	1990	1994
Sub-Saharan Africa	5	22	32	52	59
Middle East	5	18	31	42	57
China	42	69	79	88	91
Rest of Asia	n/a	n/a	n/a	70	70
Latin America	11	24	68	82	92
All LDCs	20	41	59	70	78

Source: Pingali and Rajaram 1998.

Phase One was a period of early euphoria, during the 1960s. The 'miracle' varieties spread quickly in areas of suitable climate and the benefits for the poor seemed clear. Output was increasing at double the rate of population growth. There were also less severe annual harvest fluctuations than before. This is a very important point because moisture stress often affected traditional varieties (TVs) and made employment opportunities unstable for poor labourers, who suffered most because they could not afford to store grain against a possible future shortage. In contrast, HYVs demand more work (10–40 per cent more per hectare for wheat and 30–60 per cent more for rice) for weeding, fertilizer application and moisture control. Consumers also benefited, from lower retail prices, and those small farmers who had previously specialized in growing TVs to be self-sufficient could now devote a smaller area to the HYVs and have some land left to grow a cash crop.

Phase Two: By the early 1970s the evidence was less positive, and some writers even argued that poor people were likely to become relatively poorer. This was said to be because the Green Revolution is a 'package' (Table 17.1), and smallholders were unable at first to adopt all of the elements as quickly as their larger neighbours. Small and marginal farmers are less well informed, less able to take risks, have less cash or credit to invest, and poorer access to water. In theory the technology was 'scale-neutral', in other words it helped everyone equally, but in practice the impact was to make the rich richer and to leave out the poor. Socio-economic polarization was, therefore, a distinct possibility.

The Green Revolution was planned as a technological short-cut to development which would not have to involve any painful social and political restructuring. It therefore reduced pressure to change the existing order, which in many countries was, and still is, unjust and oppressive. The original Mexican wheat breeding project, for instance, was encouraged by a government that was hoping to avoid land reform.

There was also the problem that rice is host to 150 diseases and pests.

TVs, having been developed by trial and error over centuries, were immune to many of these but HYVs were not. In the 1970s and 1980s, for instance, the brown leaf hopper, which carries the grassy stunt virus, caused widespread damage to IR8 crops in Indonesia and several other countries. As a result poor farmers, who are necessarily risk minimizers, were very reasonably reluctant to adopt varieties that were so vulnerable. Combating such vulnerability meant the greatly enhanced use of pesticides, with negative environmental consequences resulting from any chemicals which persisted in the soil.

Green Revolution varieties had characteristics that were inappropriate in some areas. In Bangladesh, for instance, there are regions that are flooded to a considerable depth every year. Here a long stemmed, or even a floating variety, is better adapted than a short-strawed HYV. Also, some Green Revolution varieties were unacceptable to the consumer because they did not mill or cook satisfactorily.

Associated with the Green Revolution have been other trends in modernization that have an in-built threat for the poor. Tractors are the classic example. They may not increase yields or reduce overall costs, but they do improve the timeliness of farming operations such as ploughing and harvesting. As a side-effect they are labour-displacing and are mainly adopted by those larger farmers who can afford such a 'lumpy' investment. The same is true of threshers and herbicides.

Modern varieties have become geographically concentrated in those countries and regions (Table 17.3) which have environmental conditions that resemble those of north Mexico for wheat and the Philippines for rice, and which have ready access to irrigation and fertilizers. Africa, the hungriest

Table 17.3 A spatial typology of the early Green Revolution

- *Type I*. The leading innovative regions, well irrigated or reliably rainfed, some evidence of poverty alleviation. Much of the early literature was written on such areas, giving a biased impression.
- *Type II*. Backward areas with few possibilities to introduce HYVs, due to poor or exhausted soil, few water resources, or a dry or cold climate. The main hopes here are forestry or non-farm rural development.
- *Type III*. Second generation breakthrough areas, in crops other than rice and wheat, for example kharif sorghum in Maharashtra, ragi in Karnataka, maize in Malawi and Zimbabwe, and rice in west Africa. In the last case the experimentation with new varieties was by small farmers.
- *Type IV*. Reasonably favoured areas that are not suited to HYV rice or wheat and where farmers have therefore switched to cash crops that have been abandoned in type I areas, such as pulses and fodder. Examples are Gujarat (India) where there was a change from wheat to mustard, rapeseed and groundnuts, and North Arcot in Tamil Nadu state (India) where rice gave way to groundnuts and sugar.

Source: based on Lipton (1978).

continent, has benefited very little. In India, the irrigated areas of the Punjab were well suited to HYV wheat cultivation, as were parts of Tamil Nadu to HYV rice. Both areas were already relatively prosperous and their success in the Green Revolution increased regional disparities of rural wealth in India. The less favoured areas suffered because the price of their wheat and rice was undercut by the plentiful HYVs, although labour migration to the areas of Green Revolution did introduce a self-correcting mechanism.

Phase Three: in the later 1970s and 1980s, evidence emerged that smaller farmers *were* adopting HYVs. The general view that the poor were losing in absolute terms was modified to a recognition of their relative decline by comparison with their richer competitors. Even the 'package' view of the Green Revolution was abandoned, as it was realized that many small farmers were using HYVs but few could afford the expensive inputs of chemicals and fertilizer.

In the 1970s and 1980s more investment and a great deal more research effort was put into diversifying the original mission of the plant breeders. The initial push had been in yield enhancement of wheat and rice in order to increase food availability. Less thought had been given to the 'entitlements' of poor farmers and to variations in environment and socio-economic context. For example, HYVs were labour-intensive, which was appropriate in densely populated south Asia, but inappropriate in Africa where, in some countries, there is a rural labour shortage.

The Consultative Group on International Agricultural Research (CGIAR) was created in 1971 by a consortium of the World Bank, various regional banks, several UN agencies, charitable foundations and some national governments. It has gradually built a network of centres (Table 17.4), which together have provided a wealth of knowledge about farming systems and their improvement. Crops other than the major cereals have been targeted, along with work on livestock, water management, soil, pests, and farming on slopes. More attention has been given to robustness, such as pest and disease resistance, and the ability to cope with moisture stress. Above all, a greater recognition has been given to the variety of ecological contexts found in the tropics, and researchers have made the effort to understand the complexities and constraints of the enterprises of poor farmers.

An important point about the Green Revolution is that the efforts of the CGIAR centres, and the various national agricultural research stations, were largely in the public and charitable sectors. The sums of money deployed were relatively small (Table 17.5). This was to change dramatically later, in the 1990s, when commercial biotechnology companies became the key players in crop and livestock improvements.

Phase Four: in the 1980s and 1990s there was a realization that the traditional breeding methods, which had been the basic underpinning of the Green Revolution, were nearing their ceiling for increasing production.

Table 17.4 The International Research Centres funded by the CGIAR

- CIAT (Centro Internacional de Agricultura Tropical) (1967), Colombia. Cassava, beans, forage.
- CIFOR (Centre for International Forestry Research) (1992), Philippines. Sustainable forest research.
- CIMMYT (Centro Internacional de Mejoramiento de Maiz y Trigo) (1966), Mexico. Wheat, maize, triticale, barley.
- CIP (Centro Internacional de la Papa) (1972), Peru. Potatoes and sweet potatoes.
- ICARDA (International Centre for Agricultural Research in Dry Areas) (1975), Syria. Mixed animal-crop production systems, especially sheep, durum wheat, barley, lentils, chickpeas and beans.
- ILCLARM (International Centre for Living Aquatic Resources Management) (1995), Philippines. Fisheries.
- ICRAF (International Centre for Research in Agroforestry) (1977), Kenya. Tropical deforestation, land degradation, agroforestry.
- ICRISAT (International Crop Research Institute for the Semi-Arid Tropics)(1972), India. Sorghum, pearl millet, pigeon peas, chick peas, groundnuts, cropping systems in the semi-arid tropics.
- IFPRI (International Food Policy Research Institute) (1975), USA. Policy research.
- IITA (International Institute for Tropical Agriculture) (1967), Nigeria. Cropping systems, grain legumes (cowpeas, soya beans, lima beans, pigeon peas), cassava, sweet potatoes, yams, rice and maize.
- ILRI (International Livestock Research Institute) (1995), Kenya. Tropical crop-livestock systems.
- IPGRI (International Plant Genetic Resources Institute) (1974), Italy. Promotion of agricultural biodiversity.
- IRRI (International Rice Research Institute) (1960). Rice and multiple cropping in Asia.
- ISNAR (International Service for National Agricultural Research) (1979), Netherlands. Policy, organization, management of national agricultural research systems.
- IWMI (International Water Management Institute) (1984), Sri Lanka. Improvement of irrigation management systems.
- WARDA (West African Rice Development Association) (1970), Côte d'Ivoire. Swamp rice.

Source: http://www.iclarm.org/

Yield growth had slowed in areas such as the Punjab in India, due to groundwater exhaustion, micronutrient depletion, and the build up of pests, and conventional plant breeding was unable to find the necessary answers.

A number of higher-level technologies have provided a boost but even these have their limits. First, and most important, is tissue culture, which involves the micro-propagation of a plant in order to make thousands of identical reproductions in a sterile jelly, under carefully controlled laboratory conditions. This is a scientific version of taking cuttings and helps to double the speed of traditional plant breeding. Tissue culture also helps to eliminate disease from key genetic stocks.

Table 17.5 Expenditure on agricultural research as a percentage of agricultural gross domestic product

Region	Countries	1961–65	1971–75	1981–85	Latest year
Developing countries					
South Africa	1	1.4	1.5	2.0	2.6
Rest of Sub-Saharan Africa	17	0.4	0.7	0.8	0.6
China	1	0.6	0.4	0.4	0.4
Rest of Asia and Pacific	15	0.1	0.2	0.3	n/a
Latin America and Caribbean	26	0.3	0.5	0.6	n/a
Middle East	13	0.3	0.5	0.5	n/a
Developed countries	18	1.0	1.4	2.0	n/a
USA	1	1.3	1.4	1.9	2.2
Australia	1	1.5	3.6	4.5	4.4

Source: Pardey and Alston 1995.

Second, embryo transfer (ET) is a convenient means of making genetic material transferrable in livestock breeding and, since the 1980s, it has been an important supplement to the long-established method of artificial insemination. In ET a high quality cow is chemically stimulated to 'superovulate', in other words to produce more eggs than she would naturally. The ova (eggs) are collected, and fertilized in test tubes. They are then implanted into a foster mother. The process is costly but the results make it economically worthwhile. It is, for instance, possible to put Aberdeen Angus eggs into a Friesian and so simultaneously stimulate the cow's lactation and produce high quality beef offspring.

Recently it has become possible to remove immature eggs from the ovaries of slaughtered cows. These are divided by microsurgery, fertilized in the laboratory and allowed to grow for one week. They are then frozen and can be implanted without surgery. It is possible to choose the sex of the offspring. A 60 per cent success rate in ET has been achieved by some breeders and one of their aims is to cut in half the average of 6.5 years presently needed for progeny testing. The use of ET is likely to have major geographical impacts in the long term because the specialist breeding districts are unlikely to maintain their full comparative advantage.

The gene revolution

In the 1990s we began to see the impact of biotechnology, where microbiological techniques are used to manipulate crop characteristics. There have been the beginnings of a shift from the 'Green Revolution' to the 'Gene Revolution'.

Biotechnology is unfortunately a vague term. Our understanding of it here is the application of modern, laboratory developed, high technology to food production. There has been an extraordinary explosion of interest in biotechnology since the 1980s. According to Persley and Doyle (1999), biotechnology is 'any technique that uses living organisms or substances from those organisms to make or modify a product, improve plants or animals, or develop micro-organisms for specific uses'. The main aspects of biotechnology are set out in Table 17.6.

Two methods are employed. In the first, an injection of DNA is made with a very fine needle into the nucleus of an individual plant cell by a so-called 'gene gun'. Second, and more common, is the use of a microbe, *Agrobacterium tumefaciens*, to carry the new gene(s) into a plant's cells. Unfortunately neither approach is able to place the new genetic material at a particular point in the chromosome, and many experiments are therefore usually needed before the desired characteristics are observable.

Much of the early work was in the 1970s, with medical applications such as the genetic engineering of human insulin and various vaccines. One development of relevance to agriculture was the production of recombinant BST for the increase of milk yields in cattle (Chapter 16). Others include the long-standing use of GM yeast in baking and the brewing of beer, and GM rennet in cheese-making. A summary of possible applications is given in Table 17.7.

The first transgenic plant, a herbicide-resistant tobacco, was created in 1983 and licensed in the USA in 1986. From 1986 to 1997 approximately 25,000 transgenic field trials were conducted on more than 60 crops with 10 different traits in 45 countries. By the end of 1997, 48 transgenic crop products had been approved in various countries. Calgene's Flavr Savr tomato, with its delayed ripening, was ready in 1994 and in the same year the British government allowed Monsanto to import GM soya. In 1996 Sainsbury's and Safeway began selling purée produced from GM tomatoes created by a team at Nottingham University for the biotech firm Zeneca.

Animals and fish have also been 'improved'. The insertion of growth hormones has increased their size but the outcome of some experiments

Table 17.6 The key components of biotechnology

- Genomics: mapping the genetic make-up of living things.
- Bioinformatics: assembly of genome data into accessible forms.
- Transformation: insertion of new genes with useful traits.
- Molecular breeding: identification and evaluation of desirable genetic traits in breeding programmes.
- Diagnostics: identification of pathogens using molecular characterization.
- Vaccines: development of recombinant DNA vaccines to control disease.

Source: Persley and Doyle (1999).

Table 17.7 Potential biotechnological applications in agriculture

Crop improvement

- Protoplast fusion and somatic hybridization to produce new crosses.
- Disease-free plant propagation.
- Production of genetic maps.
- Biological nitrogen fixation.
- Genetically-engineered male sterility, to produce hybrid varieties.
- Transgenic plants for pest resistance.
- *In vitro* germ plasm conservation, storage and distribution.

Livestock improvement

- Production of growth hormones using engineered bacteria.
- Embryo manipulation to introduce new traits.
- Transgenic animals for better feed efficiency.
- New vaccines.
- Disease diagnosis.

Source: Conway (1997, 145).

has been of concern: large but unhealthy or even disfigured individuals have been produced. Most famous have been the cloning experiments on sheep, producing in 1997 the world's most photographed ewe, 'Dolly'. She and other animals have been modified with human genes in order that they may yield useful products in their milk, such as blood-clotting factors for haemophiliacs, or alpha-1-antitrypsin, to help with cystic fibrosis and emphysema.

In 1999 the market for biotechnology globally was probably over $15 billion. Although most of that was for medical products, 40 million hectares of land had been planted to 40 varieties of genetically modified (GM) crops, worth about $1.8 billion. About 99 per cent of commercial GM crop production was in the USA, Argentina, and Canada, while more than 50 per cent of the maize, soya bean, and oil seed rape grown in these countries was transgenic (Table 17.8).

Table 17.8 Commercial transgenic crops grown in 1998

	Million hectares	Percent
Herbicide-tolerant soybean	14.5	52
Bt maize	6.7	24
Insect-resistant/herbicide-tolerant cotton	2.5	9
Herbicide-tolerant oilseed rape	2.4	9
Herbicide-tolerant maize	1.7	6
Total	27.8	100

Source: James and Krattiger (1999).

Table 17.9 The modified characteristics of transgenic crops released 1992–95

Modification	Number of releases
Herbicide resistance	212
Increased shelf-life	45
Virus resistance	37
Insect resistance	33
Fungal resistance	24
Bacterial resistance	6
Nematode resistance	1

Source: Nottingham (1998).

Modern biotechnology has been the subject of vast investments of capital and scientific effort by private companies in the last 20 years, but it is still in its infancy. The first field trials have been of crops with a single new feature, such as herbicide tolerance or pest resistance (Table 17.9). The economics of the industry has dictated this because companies have wanted to make a profit quickly in order to demonstrate to their shareholders that the research has been worthwhile. The oft-quoted example is Monsanto, which has developed varieties of maize, oilseed rape, cotton and soya bean which are resistant to Roundup (glyphosate), its own proprietary herbicide, and farmers are thereby encouraged to buy two of the company's products together. Monsanto have also been responsible for 'Bt crops' (maize, cotton, potato), with a genetically built-in insecticide (*Bacillus thuringiensis*), which reduces the need for chemical sprays by 80 per cent.

Table 17.10 Issues in the GM debate

- Opposition to genetic manipulation and to cloning on moral grounds, especially if the technology is applied to humans.
- Wish of scientists and governments to hold field trials of crops before they are released commercially. Opposed by anti-GM campaigners.
- Need for labelling of foods with GM ingredients.
- Unintended environmental consequences, for instance herbicide-resistant crops cross-breeding with wild plants to make 'super-weeds'; the killing by the Bt gene of beneficial insects; the reduction of habitat for some species of birds due to the elimination of weeds.
- Concern about unknown effects upon human health, especially the danger of unpredictable allergic reactions and the possible transfer of antibiotic resistance.
- Patenting of genes and GM crops leading to the commoditization of the building blocks of life. This is a subset of a debate about international intellectual property rights.
- Fears that peasant farmers in poor countries will suffer because of the high prices of the seeds and agri-chemicals needed to grow GM crops, especially if 'terminator' genes are inserted to make second generation seeds sterile.

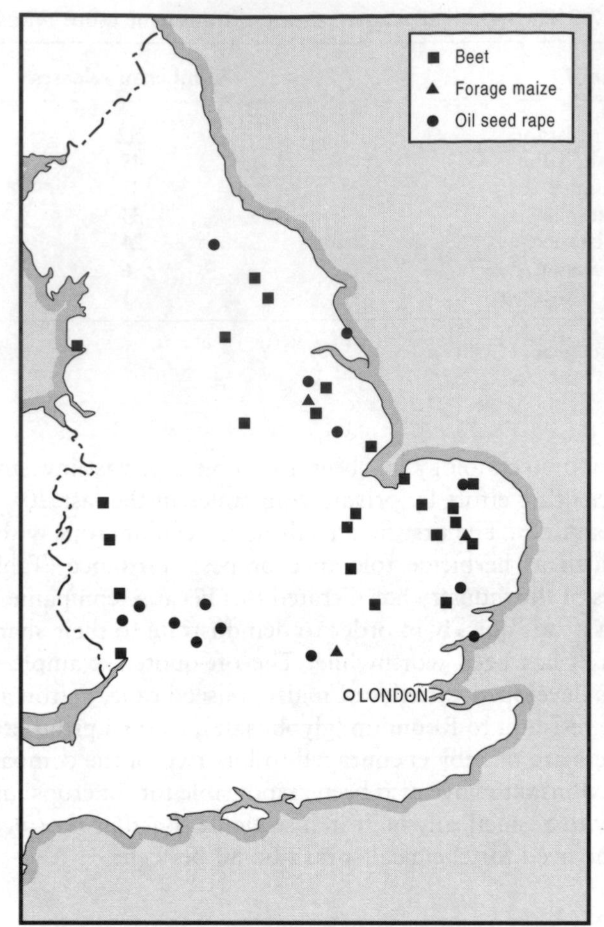

Fig. 17.1 The location of the official GM crop test sites in Britain announced in March 2000. *Source*: http://www.gm-info.gov.uk/

In 1999 a major debate (Table 17.10) began in the UK, and several other countries, about the environmental and human health impacts of GM foods. This has spread and is now undermining political support for the science and marketing of GM products on an international scale. Various environmental groups have made high profile raids on field trials in the UK (Figure 17.1), destroying crops and receiving tacit support from the public.

The question of copyright is a particularly vexed one. Biotechnology companies spend a great deal of money on research and development and they stand to make even more from their successes. It is understandable, therefore, that they should want to protect their products with patents in the same way as any scientific discovery. The Trade Related Intellectual Property Rights (TRIPs) agreement, signed in 1994, met most of the company's needs for the international protection of their discoveries and, since it was negotiated as part of the Uruguay Round of the GATT, it has been signed by 140 countries and has substantial clout (Barton 1999).

In the USA patents have been granted on asexually propagated plants (grafts, cuttings) since 1930, and on seeds since 1970. The first patent on a genetically engineered plant was in 1986, for Molecular Genetics of Minnesota for their maize variety that can make its own tryptophan, an amino acid that is normally deficient in maize. Since then hundreds of patent applications have been filed concerning the DNA sequences of over 50 species (Barton 1999).

One potentially sinister implication of patenting is that pseudo monopolies might arise. The research and development costs are so high that competition is weak. At present five TNCs are dominant: Agrevo/Plant Genetic Systems; Du Pont/ Pioneer; ELM/DNAP/Asgrow/Seminis; Syngenta (a recent merger of Astra-Zeneca and Novartis); and Pharmacia (Monsanto/Calgene/Delkalb/Agracetus/PBI/Hybritech/Delta). Their market power may become another mechanism by which the West might dominate Third World agriculture.

It is clear that GM crops *could* have a major positive impact upon farming in poor countries (Table 17.11). In-bred tolerance to a range of environmental challenges, such as drought or salty soils, would facilitate the expansion of the cultivated area. Biotechnology could also be deployed on poor people's crops (cassava, millet, sorghum, upland rice). This would help hundreds of millions of farm families. However, such developments seem unlikely because all of the effort in breeding GMOs has so far been concentrated on modern-sector crops and animals, where the profitability will be greatest in the short run. Some biotechnology companies have even deliberately engineered their products with 'terminator genes' in order to prevent farmers sowing the seed that they gather from the crop and, therefore, forcing them to return to the market annually.

Table 17.12 implies that the impact of GMOs will vary from country to country in the poor world and therefore generalizations are potentially misleading. To a certain extent this will depend upon the regulation of world trade. There was concern that the World Trade Organization might insist that GMOs must be treated as ordinary commodities and traded without restriction, but the international Biosafety Protocol signed in January 2000 allows individual countries to forbid the importation of those GM foods which are considered to be a threat to human health.

In essence the most advanced biotechnology is about modifying genes in useful plants and animals by the insertion of new genetic material. For the full range of possible improvements to be available it is essential to have a store of germ plasm, and this has been one of the tasks of the CGIAR centres and the numerous national seed banks. Any threat to the diversity of cultivated plants, for instance by the planting of vast areas to single varieties, is also a challenge to the breeding potential of future

Table 17.11 Useful biotechnology for poor countries

At stage of greenhouse or field tests	Laboratory test stage
Input traits	
• Resistance to insects, nematodes, bacteria and fungi in crops such as rice, maize, potato, sweet potato • Delayed senescence, dwarfing, shade tolerance, early flowering in rice • Tolerance of aluminium, submergence, chilling and freezing in cereals • Male sterility/restorer for hybrid seed in rice, maize, oilseed rape, wheat • New plant types for weed control and increased yield potential in rice	• Drought and salt tolerance in cereals • Seedling vigour in rice • Enhanced phosphorus and nitrogen uptake in rice and maize • Resistance to parasitic weed *Striga* in maize, rice and sorghum, to viruses in cassava and banana, and to bacterial blight in cassava • Resistance to nematodes and to black sigatoga disease in bananas • Rice with C_4 photosynthesis
Output traits	
• Increased vitamin A in rice and oilseed rape • Lower phytates in maize and rice to increase bioavailable iron • Modified starch in rice, potato and maize and modified fatty-acid content in oilseed rape • Increased bioavailable protein, essential amino acids, seed weight and sugar content in maize • Lowered lignin in forage crops	• Increased β–carotene, delayed post-harvest deterioration and reduced content of toxic cyanides in cassava • Increased vitamin E in rice • Asexual seed production in maize, rice, millet and cassava • Delayed banana ripening • GM potatoes and bananas which deliver recombinant vaccines • Improved amino-acid content of forage crops

Source: Conway & Toenniessen (1999).

Table 17.12 The possible impact of biotechnology

- High food importers with strong technological potential will benefit the most, since the trends will push their economies towards self-sufficiency.
- High food exporters with strong technological potential could benefit by diversifying their exports.
- Net importers of food with weak technological potential will benefit in the short term from lower world prices, but, in the longer term, domestic food production will suffer.
- Most vulnerable are the net exporters (especially in Africa and the Caribbean) of potentially substitutable products (e.g. sugar) who have low technological potential.

Source: Leisinger (1996).

crops. One estimate suggests that 95 per cent of farm level biodiversity was lost in the twentieth century.

It seems essential to protect the vast variety of genes at present in traditional agriculture in the less developed countries and in wild plants. Ethiopia, for instance, has an extraordinary variety of plant genetic material because of its wide range of ecological niches.

IRRI has collected samples of 86,000 of the 100–120,000 cultivars of rice; ICRISAT has 86,000 holdings of sorghum, millet, chickpea, peanut, and pigeonpea; and so far the International Potato Centre has 12,000 potato varieties in store (Knudson 1999). Other international research centres are responsible for the other major crops. These collections are vital as museums of genetic diversity; but they also have the very practical function of providing scientists with material from which they can breed new strains with useful characteristics. The DNA fragments preserved in these seeds can be used in the ultimate goal of recreating the full genetic scope of each species, known as the genome.

The future

The next steps in the application of technology to agriculture are not easy to predict. The rush by companies, such as Monsanto, to have their GM crops approved by governments around the world has backfired. The public do not want such a *fait accompli* to be imposed upon them at short notice and they are suspicious both about the profit obsession of international capital and about the health consequences of consuming GMOs. A gradual, incremental policy of crop/food testing under a strong régime of regulation looks to be the only one likely to succeed.

Even more revolutionary biotechnologies will be within our grasp in the next 20 years. The next generation of GM foods will have more than one gene changed and the practical possibility of 'designer crops' for specific agri-ecologies will then follow naturally. The necessary investment in research will be immense but the rewards will be also be large and will support a significant portion of the world's economic growth over the coming decades.

Much biotechnology in future will not be of the controversial farm-based variety but will be applied in factories. Goodman *et al.* (1987) predict that agriculture will gradually be reduced in its function to little more than the provision of feedstocks to the bio-industrial complex. This is the very rapidly growing manufacturing sector devoted to the conversion of biomass into food products. Agriculture will become only one amongst several sources of such organic matter, which can be transformed by fermentation and other methods into an unlimited range of synthetic outputs such as mycoprotein, the basis of the vegetarian product Quorn. Bio-industry will add its value in the technology of texturing and flavouring these de-natured

Table 17.13 Some of the uses of maize

- Meat and milk are produced from animals largely fed on maize.
- Frozen meat and fish are coated with a maize starch coating to prevent excessive drying.
- The brown and golden colouring of some soft drinks and puddings comes from maize.
- All canned foods are preserved in a liquid containing maize.
- All modern paper and cardboard products, except for newsprint and tissue, are coated in maize.
- Maize oil is used widely in cooking and in margarine and salad dressings. It is also found in soap and insecticides.
- Monosodium glutamate (MSG), a common flavour enhancer in processed foods, is commonly made of maize protein.
- Maize syrup is used to sweeten ketchup, ice cream, and sweets. It is also found in condensed milk, soft drinks, beer, gin and vodka.
- Maize starch is an essential ingredient in baby foods, jam, pickles, vinegar and yeast.
- All textiles and leathers are coated in maize.

Source: Visser (1986, 22–4).

foods, and the sourcing of the raw materials will become trivial in cost and highly flexible in strategic terms. In effect the farmer's skills will have been appropriated by the manufacturer.

This process is already well underway. Soya beans and maize are two crops which are used generically in the food processing. Corn syrup is used as a sweetener, but according to Visser (1986), all parts of the cob are useful and virtually every item, including some non-food goods, in American supermarkets have been touched by maize in one way or another (Table 17.13).

Third World farmers are the most likely to lose from such substitutionism. Corn syrup and artificial sweeteners are undermining the market for cane sugar, for instance, and biotechnology may make it possible to substitute a genetically modified rape oil for coconut oil, without any loss of the desirable characteristics of the final product. Tissue culture techniques have been used to develop synthetic cocoa flavourings, and research is proceeding on the treatment of low-grade vegetable oils to reproduce the qualities of palm oil.

Conclusion

Biotechnologies, from traditional breeding methods to genetic manipulation in the laboratory, have been shown to have both positive and negative potentialities. Scientists have a responsibility for the knowledge and technical innovations that they produce but the outcomes mainly depend upon economic, political and social factors. It is essential that individual governments and the international community regulate both the science and the commercial applications.

In our opinion, two points are important here. First, it is doubtful that western countries *need* GM foods. There are some benefits to farmers in terms of reduced costs and to the environment if lesser quantities of pesticides are used, but yields do not appear to be increased and there are unquantified dangers for the environment and human health. In a part of the world where farming has achieved embarrassing food surpluses, it might be better rather to put research effort into sustainable food production.

Second, in the poor countries biotechnology has a great deal more relevance to the practical needs of peasant farmers and hungry consumers, yet the commercial companies accord these needs a very low priority. Arguably, the international community needs to take a greater degree of control and encourage public–private partnerships, perhaps with the involvement of the CGIAR, which would be most likely to deliver relevant advances.

Further reading and references

Gordon Conway's (1997) book is by far the best introduction to this topic, having both a broad overview and an up-to-date account of specific issues. There is a burgeoning literature on biotechnology, covering many issues that we have not had time to develop here. O'Mahony's collection, for instance, investigates the discursive context of the debate about genetic modification.

Altieri, M. 1999: Ten reasons why biotechnology will not ensure food security, protect the environment and reduce poverty in the developing world. http://www.foodfirst.org/resources/biotech/altieri-11-99.html.

Busch, L., Lacy, W.B., Burkhardt, J. and Lacy, L.R. 1991: *Plants, power, and profit: social, economic, and ethical consequences of the new biotechnologies*. Oxford: Blackwell.

Conway, G. 1997: *The doubly green revolution: food for all in the twenty-first century*. London: Penguin.

Conway, G. and Toenniessen, G. 1999: Feeding the world in the twenty-first century. *Nature* **402**, Supplement, C55–8.

Goodman, D., Sorj, B. and Wilkinson, J. 1987: *From farming to biotechnology: a theory of agro-industrial development*. Oxford: Blackwell.

James, C. and Krattiger, A. 1999: The role of the private sector. *Focus 2, Brief 4, 2020 Vision*. Washington, DC: International Food Policy Research Institute.

Leisinger, K.M. 1996: Sociopolitical effects of new biotechnologies in developing countries. *2020 Brief 35*. Washington, DC: International Food Policy Research Institute.

Lipton, M. 1978: Interfarm, interregional and farm–non-farm income distribution: the impact of the new cereal varieties. *World Development* 6, 319–37.

Mannion, A. 1995: *Agriculture and environmental change: temporal and spatial dimensions*. Chichester: Wiley.

Nottingham, S. 1998: *Eat your genes: how genetically modified food is entering our diet*. London: Zed Books.
O'Mahony, P. (ed.) 1999: *Nature, risk and responsibility: discourses of biotechnology*. Basingstoke: Macmillan.
Pardey, P.G. and Alston, J.M. 1995: Revamping agricultural R&D. *2020 Brief* No. 24. Washington, DC: International Food Policy Institute.
Persley, G.J. and Doyle, J.J. 1999: Biotechnology for developing country agriculture: problems and opportunities. Overview. *Focus 2, Brief 1, 2020 Vision*. Washington, DC: International Food Policy Research Institute.
Pingali, P. and Rajaram, S. 1998: Technological opportunities for sustaining wheat productivity growth toward 2020. *2020 Brief No. 51*. Washington, DC: International Food Policy Research Institute.
Visser, M. 1986: *Much depends on dinner: the extraordinary history and mythology, allure and obsessions, perils and taboos, of an ordinary meal*. New York: Grove Press.

18

Food ethics, food policies and civil society

'Take a red cock that is not too old and beat him to death'. Opening line of an eighteeneth century recipe of William Kitchiner, quoted in Stead (1985).

Introduction

At first sight it might be difficult to imagine that something as mundane as food might have a connection with the abstract realm of ethics. However, the juxtaposition of extensive malnutrition alongside great wealth in developing countries, such as India, is only one among many points that one might deploy in arguing for an ethical dimension in food studies. In this chapter we cover a range of issues that are not usually grouped together. First, there is a discussion of the 'human right to food' that has been proposed in an international legal framework as a solution to hunger. Second, we introduce the ethics of the relationship between humans and the animals we use for food. Third, we then argue that food regulation by the state arose, in part at least, from notions about what is 'good' for the consumer, and that food policy retains the important function of protecting the public from poor quality and diseased foodstuffs. Finally, there is a discussion of the role of non-governmental organizations in campaigning on the subject of food and the efforts of private citizens in deepening the food awareness of civil society. This arises from a definition of responsibility for food that goes well beyond the state as a formal and centralized institution.

The right to food

Article 25 (1) of the Universal Declaration of Human Rights (1948) insists that 'everyone has the right to a standard of living adequate for the health and well-being of himself [sic] and his family, including food'. The language may not have been politically correct by today's standards, but the sentiments have been shared by many subsequent statements on human rights, notably the International Covenant on Economic, Social and Cultural Rights and the International Covenant on Civil and Political Rights, both opened for ratification in 1966. About 70 per cent of countries have signed up to these Covenants, although the USA has resisted the first and China has set its face against both. The ICESCR is particularly important because it asks that states 'recognizing the fundamental right of everyone to be free from hunger, shall take, individually and through international co-operation, the measures, including specific programmes, which are needed'.

In the 1970s and 1980s the FAO began to lay a greater emphasis upon 'food security' (Chapter 12). This was less a stress upon improved food production than a desire to make sufficient basic foods available at reasonable prices throughout the year and in all regions. In 1984 the World Food Security Compact made food security the responsibility of all humankind, through the articulation of an explicit moral obligation and through international co-operation. The 1992 Barcelona Declaration and the 1995 Quebec Declaration further affirmed the international institutional framework of the Right to Food and the World Food Summit in Rome in 1996 was a very high profile event which publicized it further.

Telfer (1996) discusses the moral obligations that seem to follow from the identification of a Human Right to Food. She makes a distinction between actions for the relief of suffering and the promotion of good. The former obliges us to help starving people, and few would have a problem with that in principle. If taken up in practice, the second would engage us in an open-ended commitment to establish programmes of aid to mitigate the full range of problems associated with malnutrition and dysfunctional food behaviour. By no means everyone agrees that the second set of actions is either necessary or desirable.

Regrettably the Right to Food remains more of a theoretical ethical concept than a practical programme. It establishes the morality of adequate food supply but the international community still has a very long way to go before the objective of a universally well-nourished population is achieved.

Poultry and factory farming: not a leg to stand on?

Farmers in the developed world no longer use oxen to pull the plough or take their produce to market in horse-drawn wagons. Nevertheless, farm animals are made to work very hard in producing meat, milk, eggs and wool, not least because the genetic selection of many breeds to generate profit has outrun the strength of their bodies. Arguably there is a moral issue involved here because there is evidence that the animals suffer discomfort or severe pain throughout their short lives so that we, the consumers, can buy cheap food in the shops.

Take the case of the UK's broiler chicken industry, for instance, which produces 700 million birds a year. Most come from flocks of up to 50,000 birds, which are kept in large, dimly lit sheds and are fed and watered automatically. The broiler has been specially bred to put on weight quickly. What used to take over 80 days in the 1960s for the bird to be saleable now takes half that time. This places great strain on the birds' legs. For the last week or two of the fattening period, anything up to 90 per cent of birds in some flocks have bone deformities which hamper their ability to walk and 26 per cent are likely to suffer chronic pain. This does not affect the quality of the meat for the customer in the supermarket but it does cause suffering for the birds. Turkeys suffer hip joint deformities and pigs are prone to heart disease for the same reasons. They are bred for their meat but the health of their bodies' functions, such as the calcification of bones and the strength of the heart and lung system, has been neglected.

The average milk yield of cows has doubled in the last 50 years. Again breeding has been important, as has the feeding of concentrates. The animal's body is probably reaching the limit of its biological capacity, beyond which its welfare could be threatened. Increased disease is a problem, with up to a third of cows contracting the unpleasant udder disease mastitis, and a further quarter having chronic lameness where the hind legs cannot bear the pressure of the enlarged udder. Cows are killed after about four nine-month lactations (at age six or seven), although their natural lifespan might be 20 years. Some are culled because their yield is inadequate, others are simply worn out by their hard life.

Increasing output is the first way to increase income from farming. A second method with livestock is to enhance their reproduction. For instance, the wild jungle fowl produces about a dozen eggs each year but its domesticated sister lays 300. Since each egg requires calcium for a strong shell, the hen becomes a small factory for processing that element. In her lifetime she produces 25 times more calcium than is contained in her own skeleton, a burden which many birds find intolerable. Their bones are affected by osteoporosis, the brittle bone disease.

Pigs also deserve consideration. The average intensively reared sow produces 23 weaned piglets a year, which may be a strain for some animals. The welfare issue for pigs, however, has become focused on the tethering of sows in stalls that prevent them from turning round. They are kept there for several weeks to allow careful attention but many observers dislike this method because it is so restrictive on the movement of the animals. As a result of much lobbying, sow tethering is now illegal in Britain and will become so throughout the EU in 2005. Once the piglets are born, the sow is again restricted in a farrowing crate to prevent her from rolling on to and crushing her offspring.

In 1993 the UK Farm Animal Welfare Council revised what have become

Table 18.1 The Five Freedoms

- Freedom from thirst, hunger and malnutrition: animals should have easy access to fresh, clean water and adequate, nutritious food.
- Freedom from discomfort: animals should live in an environment suitable to their species, including adequate shelter and a comfortable rest area.
- Freedom from pain, injury and disease: by prevention, rapid diagnosis and treatment.
- Freedom to express normal behaviour: by the provison of sufficient space, proper facilities and company of their own kind.
- Freedom from fear and distress: by ensuring that living conditions avoid mental suffering.

Source: Penman (1996).

known as the Five Freedoms (Table 18.1). These are not legal requirements but they form a basis upon which the notion of animal rights can be discussed. The pressure for action has been building in recent years, with much media attention in 1995 given to a series of demonstrations against the export of live calves to the Continent for the veal industry there.

There has also been press comment on the dirty conditions of production, which mean that many eggs are infected with salmonella, and upon the slaughter of animals, where cruelty has been alleged. The latter is probably rare but profit in abattoirs is to a certain extent a function of the speed of killing and corners on animal welfare are therefore cut. The example of chicken slaughter is a case in point. Live birds are hung from a moving line, which first takes them to be electrically stunned, then to have their throats cut. After that there is the evisceration room where the internal organs are blasted out of the carcase with water jets and, finally, they go into a tank of hot water where their feathers are loosened for plucking.

The usual argument for the continuation of factory farming is that such intensive production is the only means of ensuring cheap food. Animal suffering and dirty conditions are, implicitly, the price we have to pay. But is there really 'no alternative'? Webster (1997) questions the cost point by looking at three production issues with animal welfare implications:

- The use of BST in Britain (see Chapter 16) would increase milk output and reduce dairy farmers' unit costs by 8 per cent. If this cost saving were to be passed on to the consumer there would be an average saving of 2.2 pence per person per week.
- If the present system of broiler chicken production was banned, costs would rise by 30 per cent but, again, the average consumer would pay only 3 to 4 pence more each week.
- The abolition of battery egg production would push costs up by an estimated 28 per cent and each consumer would be 2.8 pence a week worse off.

A number of authors have extended these points about factory farming to the more general argument about our moral obligations to domestic animals. Larrère and Larrère (2000), for instance, want us to see domestication as a form of contract between people and animals, the implicit ethical terms of which are presently breached by intensive rearing and some other livestock husbandry practices. Jensen and Sørensen (1999) suggest the idea of 'ethical accounting', which would provide a basis for ethical decision-making about particular elements of livestock production process.

Vegetarianism: an ethical issue?

It is not a long step from a discussion of animal welfare to a debate about animal rights (Linzey 1976) and even animal liberation (Singer 1990). Do humans have any moral responsibility to the animals they rear for food (Clark 1977; Midgley 1983)? Should animals have rights equal to those of humans? The answer to both questions seems increasingly to be 'yes' in the mind of the general public, and the evidence for this lies in the increasing trend to vegetarianism, especially amongst young women in developed countries. Singer (1998, 71) even predicts that eating meat may disappear over the next 20 years and 'follow smoking into disrepute'.

Vegetarians reverse the traditional hierarchy of foods which privileges meat as a source of potency and animal vitality (Fiddes 1991), but their motivations are complex (Tables 18.2 and 18.3). Some see vegetarianism as closer to nature and less destructive of the environment. For others their repugnance for animal suffering and exploitation is the driving force and they may see a moral issue about the relationship between people and the natural world.

Vegetarianism is really an umbrella term for a wide range of food practices. About half of the people who classify themselves as vegetarians do in fact eat meat occasionally, usually chicken or fish, and it seems to be common for many of us to experiment with vegetarianism or at least sympathize with its principles without fully committing to a meat-free diet. At the stricter end of the spectrum are vegans, who avoid the use of any animal products, including milk and eggs, and will not wear leather, and frutarians, who object also to killing plants and therefore live on what can be harvested without permanent damage.

It has been suggested that the decline in meat-eating (Chapter 20) is not only a function of concern for the ill-health problems associated with animal fat, but is also linked to a growing dislike for the notion of death, blood and viscera. Butchers no longer display the severed heads of animals and strive hard to make their shops different from their image as an annexe to the abattoir. Most consumers of meat now prefer to purchase pre-packaged and disembodied cuts of meat from the supermarket, with as little resemblance

Table 18.2 A hierarchy of foods

	Strength	Food	Cooking
Taboo	Too strong	Human beings Carnivores Uncastrated animals	Raw meat
Dominant culture's boundary			
Meat	Blood, powerful	Red meat	Roasted, stewed
	Non-blood, less powerful	Poultry	Roasted, boiled
		Fish	Fried, steamed
Vegetarian boundary			
Animal products	Less strong	Eggs	Fried, boiled, cooked as dish
		Cheese	Raw, grated
Vegan boundary			
Fruit and vegetables	Too weak	Fruit Leaf vegetables Root vegetables Cereals	

Source: Adapted from Twigg (1983).

Table 18.3 Vegetarians' principal motive for their dietary choice

Moral/spiritual	43
Health	13
Taste	9
Ecological	1
Health + taste	3
Health + moral	4
Moral + taste	3
Total	76

Source: Beardsworth and Keil (1991).

as possible to a dismembered carcase. Fat, bone and sinew are removed in order to eliminate any connexion with a living animal, and there seems to be emerging in the public mind a hierarchy of meats from beef, as the least acceptable when served as a rare, bloody steak, through chicken and fish which are acceptable as 'white meat', to burgers and sausages where the connexion with living animals seems to be most tenuous.

The most common reasons for meat avoidance given in large-scale surveys in developed countries are health, animal rights, ethics, and environment, although we should not forget that the vast majority of the world's vegetarians live in poor countries, where the major factors are

Table 18.4 The percentage of vegetarians in selected Indian states

State	Rural	Urban
Andhra Pradesh	10	20
Assam	5	5
Bihar	23	27
Gujarat	60	60
Jammu and Kashmir	18	8
Kerala	29	10
Madhya Pradesh	47	55
Tamil Nadu	17	20
Maharashtra	30	32
Karnataka	8	24
Orissa	5	0
Punjab	50	62
Rajasthan	60	62
Uttar Pradesh	49	42
West Bengal	5	12
Union Territories	10	12
All India	30	32

Source: Gopalan *et al.* (1971).

religion and insufficient purchasing power. The geographical distribution of vegetarianism in India (Table 18.4) shows regional concentrations that are a function of these factors.

The number of vegetarians doubled in the UK between 1985 and 1995, but about 97 per cent of the population still eat meat at some time or another, while 92 per cent of households serve meat dishes at least twice a week. Girls and young women are the most enthusiastic vegetarians, along with the most affluent socio-economic group. In Australia, eight per cent of teenage girls were found currently to be vegetarian, with a further 14 per cent who have tried it. The corresponding figures for teenage boys were 3 per cent and 3 per cent.

There are several explanations for the gender bias in vegetarianism. The first is perhaps rather too easy: the stereotypical nurture instinct of young females which causes anxiety about the suffering of animals, especially young animals such as lambs and piglets, and also projects anthropomorphic values on to the larger, sentient animals. If the latter point is true, then it does explain why cold blooded fish may remain in the diet of some vegetarians. A second argument makes a link between the concern of adolescent girls with menstruation and their dislike of the thought of the blood that flows in abattoirs. Thirdly, there is evidence that the supportive nature of female networks, among mothers, daughters, sisters and friends, is a strong factor in encouraging and maintaining high levels of vegetarianism. Finally,

Adams (1990) presents a feminist case that violence to animals and meat-eating are expressions of patriarchy and that vegetarianism among girls is therefore a form of rebellion against male-centred values. This last point is even more significant when one couples it with George's (1994) claim that classical vegetarianism was also patriarchal.

State regulation and food policy

Although issues of hunger, animal exploitation and vegetarianism were all discussed in the nineteenth century, none attracted as much attention by the state as the idea of the consumers' right to purchase food and drink of good quality. Two major problems on that theme arose which attracted a great deal of press coverage and political controversy. The first arose from the spread of infectious disease through the food supply (Atkins 1992), an issue which was discussed and regulated through the sanitary discourse which dominated the thinking of the Victorians about the environmental pathologies they saw around them. The second was the problem of food adulteration, such as the fraudulent watering of milk (Atkins 1991). This outraged the moral sensibility of the public and led to the adoption of a number of Acts of Parliament, each in turn more tightly framed, which by the early twentieth century enabled Local Authorities to exercise some measure of control over the food supply.

If one couples these two nineteenth century issues with a third, then the foundation of modern food policy is revealed. This third concern was the call for state intervention to assist farmers and to control the international trade in food. By abolishing the Corn Laws in the 1840s, Victorian legislators had sought to encourage the movement of food across borders with minimal tariff barriers; but the Free Trade gospel of that era became increasingly unpopular as cheap imports of grain and other foodstuffs undermined domestic production. Several long depressions in British agriculture forced the state gradually to modify its policies from the 1890s onwards. This issue is mentioned here because, in modern times, food policy in the advanced world has incorporated the recognition that farmers require subsidies and support in various forms. The CAP of the EU is one example of such a policy framework and this has had major implications for agricultural profitability, international relations, and the cost of a basic diet for the consumer. We are now entering yet another phase. The WTO is looking to reduce barriers to food trade and subsidies to producers, and there has also been a noticeable shift in the balance of political leverage away from producers and towards consumers. Further food policy changes therefore seem inevitable.

A fourth thread of policy might also be mentioned, although we would argue that legislators and regulators have shown less interest in it. This is the advice given on human nutrition by successive generations of scientists throughout the twentieth century. Their evolving knowledge of vitamins and of properly balanced diets did influence decisions about school milk, school dinners and a number of other state initiatives but these never influenced the public's attitudes to food in the way that government food price policy did. Nutrition policy has become much more significant in recent decades but, as Table 18.5 shows, it still has a long way to go in many countries.

Following the long-established example of the Food and Drug Administration in America, many European governments have introduced equivalent agencies or clusters of scientific committees to advise on the increasingly complex science of food. This has become an urgent priority as a result of the political friction caused in the 1990s by food scares (Chapter 16). Table 18.6 lists the committees established in Britain, although it should be noted that many of these have predecessors that operated long before the given date of foundation, and the table does not include either the many *ad hoc* committees that report from time to time on food issues or the several agencies which have an interest in this area. The table makes it clear where the concerns of contemporary state food policy lie: contamination, adulteration, diet, and disease. None of these issues are new but they have a much more powerful currency now than ever before.

Table 18.6 makes food policy look like a dry, technical process of politicians listening to objective scientific advice. In the 1980s and 1990s politics in this previously invisible area of governance came more into the limelight, first as the independence of some of the members of these committees was questioned and, second, as the Conservative government decided to share some of its regulatory powers with the food industry, to the disgust of some independent observers (Marsden *et al.* 2000). Tim Lang is one of a very few Professors of food policy. He is especially worried about the democratic deficit in his area, where it is often not at all clear who is ultimately responsible for policy (Lang 1999). In order to improve this, and other aspects of food policy, he issues seven challenges for the future (Table 18.7).

Table 18.5 National plans of action for nutrition, May 1998

	Finalized/draft prepared		Under preparation		Not yet started	
Region	**Number**	**Per cent**	**Number**	**Per cent**	**Number**	**Per cent**
Africa	39	85	5	11	2	4
The Americas	19	54	11	32	5	14
E. Mediterranean	11	50	2	9	9	41
Europe	30	59	9	18	12	23
S.E. Asia	9	90	0	0	1	10
W. Pacific	24	89	3	11	0	0
World	132	69	30	16	29	15

Table 18.6 UK government food committees and working parties

Title of committee/working party	Founded	Area of expertise
Food Advisory Committee	1983	Labelling, composition and chemical safety of food, as regulated in Food Safety Act (1990).
Advisory Committee on Novel Foods and Processes	1988	Food irradiation, novel foods, novel production processes.
Advisory Committee on Animal Feedingstuffs	1999	Animal feed, e.g. GMOs, sewage sludge, dioxins as contaminants.
Advisory Committee on the Microbiological Safety of Food	1990	Risk to humans of micro-organisms in food.
Committee on Toxicity of Chemicals in Food, Consumer Products and the Environment	1978	Toxic risk of foods and food additives.
Committee on Mutagenicity of Chemicals in Food, Consumer Products and the Environment	1978	All mutagenic hazards in food.
Committee on Carcinogenicity of Chemicals in Food, Consumer Products and the Environment	1978	Cancer risk from foods.
Committee on Medical Aspects of Food and Nutrition	2000	Replaces Committee on Medical Aspects of Food and Nutrition Policy. Will advise on scientific aspects of nutrition and health.
Spongiform Encephalopathy Advisory Committee	1990	Major task to advise on BSE and nvCJD.
Expert Group on Vitamins and Minerals	1998	Vitamin and mineral supplements.
Working Party on Chemical Contaminants in Food	1997	Surveys on chemical contaminants.
Working Party on Chemical Migration from Materials and Articles in Contact with Food	1984	Contamination from packaging, cutlery, preparation surfaces.
Working Party on Dietary Surveys	1991	Dietary surveys.
Working Party on Food Additives	1986	To monitor additive consumption.
Working Party on Food Authenticity	1992	Adulteration and authenticity.
Working Party on Nutrients in Food	1991	Provision of data on nutrient composition of foods.
Working Party on Radionuclides in Food	1988	Radioactivity in food.

Source: http://www.foodstandards.gov.uk/scientific.htm

Food and civil society

Why should consumers be prevented from eating genetically modified foods, or intensively reared poultry meat, if they choose to do so on

Table 18.7 Food policy challenges

1	Accountability in the restructuring of the GATT, requiring a rethinking of the WTO.
2	Democratization of institutions such as the Codex Alimentarius Commission.
3	To steer food culture away from consumerism towards a deeper practice of citizenship.
4	The fusion of environmental and public health.
5	A re-localization of the food supply.
6	Tackling the problems of food poverty.
7	Better co-ordination in food policy circles.

Source: Lang (1999).

grounds of price or some other criterion? This is not an easy question to answer because discussion of the ethical issues raised in this chapter rarely achieves a consensus. It is possible to argue that well-informed citizens should not be subjected to a paternalistic food policy enforced by the regulatory state. Rather they should be empowered to make their own decisions by a system of detailed labelling that would reveal the necessary background about quality and origin of the food, and a description of the farming system (use of chemicals, livestock rearing practices, etc.).

Consumer groups and food campaigners have had a significant impact in the last decade or so. In the British context, Marsden *et al.* (2000) list the Consumers' Association, the National Consumer Council, the National Federation of Consumer Groups, Consumers in Europe Group and the National Food Alliance. Apart from the Consumers' Association, their office staff are few and their cash resources modest. They might not look a match for the might of the food industry and the state but such groups are often very effective in getting their message across. In this they have been assisted by the media, for instance through BBC Radio 4's 'The Food Programme' and some members of the Guild of Food Writers, who publish in newspapers and popular magazines.

Monitoring groups, such as the Food Commission (formerly the London Food Commission), have also been important sources of information about food scandals, and a number of court cases have publicized anxiety about the food system. The most notorious of these was the so-called 'McLibel trial', in which two activists were sued by McDonald's over their claims about the company's food and its attitude to the environment (Vidal *et al.* 1997).

Spontaneous and organized consumer boycotts of certain food products and food companies have also had some effect. The so-called 'baby milk scandal', for instance, about the role of a few international dairy companies in encouraging mothers in poor countries to switch from breast-feeding to the use of their dried milk, produced bad publicity for Nestlé. Another example was the long-standing boycott in Britain of fruit and wine from South Africa, Chile and Israel from the 1970s to the 1990s as a result of the political flavour of their régimes and their oppressive policies.

More positively, consumer demand has made it possible for 'fair trade' products, such as coffee, to be marketed in supermarkets. This alternative channel provides links with small growers and gives an increase in the price that they would otherwise receive from fully profit-motivated food capitalism (Chapter 5).

In addition, the various animal rights societies have brought the attention of the public to the export of live calves, inhumane practices of slaughter, and the many problems associated with factory farming. British civil society has a very long tradition of concern about animals and vegetarianism but the public nerve has only recently been touched on the subject of high quality, organic foods. German consumers and voters are more exercised by green issues in general, and food is one subset of a debate that runs deeply in their consciousness. This helps to explain their reluctance to allow imports of British beef that may in any way have been in contact with BSE (Chapter 16). In America food activism is also significant, but Belasco (1993) shows that there are limits to the success with which ordinary citizens have taken on the giant food corporations.

A final aspect of popular campaigning about food is the street-level demonstration (*see also* Chapter 13), for instance a number of anti-WTO and anti-capitalist protests which have recently hit the headlines. The first was in the City of London in June 1999, when the police were taken off-guard by a march that became violent, soon followed by a large-scale street event in Seattle in November which was timed to coincide with a high-profile meeting about world trade. The latter was so successful in disrupting the WTO talks that it seems likely to encourage further action by the rainbow coalition of demonstrators, drawn from various anarchist, socialist and ecological groupings. Food trade and food capitalism were important foci of these protests, as evidenced by the symbolic wrecking of a branch of McDonald's during the Mayday2k demonstration in central London.

Conclusion

In this discussion of ethics, policy and social action, we have not addressed the issue of biotechnology and the ethical doubts that have been raised extensively about genetically modified organisms. The topic was considered in Chapter 17 but interested readers are also referred to the ethical literature that is beginning to form in this area.

In conclusion, we assert that ethics does have a part to play in food studies. The 'good' of the consumer was a basis for food policies in the late nineteenth century and the notion of morality has since become a focus of discussions about human hunger and animal welfare. Environmental ethics are another strand of ethical debate, which overall seems likely to grow further, in parallel with public awareness.

Further reading and references

For a fuller introduction to ethics see Simmons (1993), Singer (1993) and Mepham *et al.* (1995). This important and expanding area of interest is also well represented in journals such as *Environmental Ethics*; the *Journal of Agricultural and Environmental Ethics*; *Agriculture and Human Values*; and *Ethics, Place and Environment*.

Adams, C.J. 1990: *The sexual politics of meat: a feminist-vegetarian critical theory*. Cambridge: Polity Press.

Atkins, P.J. 1991: Sophistication detected: or, the adulteration of the milk supply, 1850–1914. *Social History* **16**, 317–39.

Atkins, P.J. 1992: White poison: the health consequences of milk consumption. *Social History of Medicine* 5, 207–27.

Beardsworth, A. and Keil, T. 1991: Health-related beliefs and dietary practices among vegetarians and vegans: a qualitative study. *Health Education Journal* **50** (1), 38–42.

Beardsworth, A.D. and Keil, E.T. 1992: The vegetarian option: varieties, conversions, motives and careers. *Sociological Review* **40**, 253–93.

Belasco, W.J. 1993: *Appetite for change: how the counterculture took on the food industry*. Ithaca: Cornell University Press.

Flynn, A., Harrison, M. and Marsden, T. 1998: Regulation, rights and the structuring of food choices. In Murcott, A. (ed.) *The Nation's Diet: the social science of food choice*. London: Longman, 152–67.

George, K.P. 1994: Should feminists be vegetarians? *Signs* **19**, 405–34.

Gopalan, C., Balasubramanian, S.C., Rama Sastri, B.V. and Visweswara Rao, K. 1971: *Diet atlas of India*. Hyderabad: National Nutrition Institute, Indian Council of Medical Research.

Jensen, K.K. and Sørensen, J.T. 1999: The idea of 'ethical accounting' for a livestock farm. *Journal of Agricultural and Environmental Ethics* **11**, 85–100.

Lang, T. 1999: The food policy challenge: can we really achieve public, environmental and consumer protection? Paper for Eating into the Future, the first Australian Conference on Food, Health and the Environment, Adelaide, April 11–13th.

Larrère, C. and Larrère, R. 2000: Animal rearing as a contract? *Journal of Agricultural and Environmental Ethics* 12, 51–8.

Marsden, T.K., Flyn, A. and Harrison, M. 2000: *Consuming interests: the social provision of foods*. London: UCL Press.

Mepham, T.B., Tucker, G.A. and Wiseman, J. (eds) 1995: *Issues in agricultural bioethics*. Nottingham: Nottingham University Press.

Penman, D. 1996: *The price of meat*. London: Gollancz.

Singer, P. (ed.) 1993: *A companion to ethics*. Oxford: Blackwell.

Singer, P. 1998: A vegetarian philosophy. In Griffiths, S. and Wallace, J. (eds) *Consuming passions: food in the age of anxiety*. Manchester: Mandolin, 71–80.

Spencer, C. 1994: *The heretic's feast: a history of vegetarianism*. London: Fourth Estate.

Stead, J. 1985: *Food and cooking in 18th century Britain: history and recipes.* London: English Heritage.
Telfer, E. 1996: *Food for thought: philosophy and food.* London: Routledge.
Twigg, J. 1983: Vegetarianism and the meanings of meat. In Murcott, A. (ed.) *The sociology of food and eating.* Aldershot: Gower, 18–30.

PART V

FOOD CONSUMPTION SPACES

19
Introduction

Food shapes us and expresses us even more definitively than our furniture or houses or utensils do. Visser 1986, 12.

The purpose of Part V is to investigate various aspects of food consumption. There is some continuity from Part IV because in Chapter 21 we look at the role of the environment in terms of the ecology of consumption but, on the whole, the next three chapters adopt a social science approach. The emphasis will shift from environmental and biological considerations to notions about the economic and social constructions of diet. As prefigured in Chapter 1, this Part has been influenced by the 'cultural turn' in food studies in the 1990s, but our excursions into the further realms of post-structuralism are limited by available space. Bell and Valentine (1997) have already made inroads in this direction and in future we expect a significant expansion of post-structuralist work on food.

The cliché 'you are what you eat' and its geographical sibling 'you are where you eat' may have become tedious in the repetition but they do contain grains of truth. There are clear correlations between people's diet (quantity, variety, quality) and their characteristics in terms of social class, income, age, sex, ethnicity, household composition, religion and culture. Chapter 20 unpacks these relationships and argues that they change across both time and space.

In Chapter 21 we extend the temporal point about the evolution of diets by seeking the origins of taste. Although taste does have biological and psychological aspects to it, our main thrust is cultural. We acknowledge that a 'regional geography' of food habits has much to offer but we reject the spatial determinism of the 'super-organic' cultural geography of food in favour of a more flexible approach that recognizes the influence of factors such as cookery books, exotic cuisines, eating out, health considerations, and the media and advertising.

Chapter 22 looks at the anthropo-sociological interpretation of food habits and beliefs. We dip our toes into the water and find it to be invitingly warm. There are many extraordinary insights to be gained from the ethnographic approach to food studies. In a short chapter it is difficult to do justice to this rapidly expanding field, but we comment on the grammar of meals and food choice; on why certain foods are avoided; problems of obesity and anorexia in societies where much of an individual's identity is seen to be body-centred; and the role of knowledge in shaping food culture.

Finally, Chapter 23 deals with gendered aspects of food. This is very definitely not an afterthought or an intellectual ghetto but rather a recognition that women have historically had a pivotal role in food preparation and presentation. The literature is still developing in this area but there is already a strong foundation laid through studies of women's agency in provisioning western families, and also research on household food security in LICs.

References

Bell, D. and Valentine, G. 1997: *Consuming geographies: we are where we eat*. London: Routledge.

Visser, M. 1986: *Much depends on dinner: the extraordinary history and mythology, allure and obsessions, perils and taboos, of an ordinary meal*. New York: Grove Press.

20

Factors in food consumption

Introduction

It was Paul Vidal de la Blache (1926) who noted that among the connexions between nature and culture, 'one of the most tenacious is food supply'. Nineteenth and early twentieth century geographers emphasized this mainly in discussions on the importance of physical geography as a key set of constraining variables in human activity. Soils and climate were identified as fundamental to food production, and so indeed they were in economies dominated by subsistent peasant agriculture. Since then the steady progress of urbanization and industrialization has put paid to such deterministic assumptions, in the developed world at least, and modern social scientists have different perspectives on the environmental context of the economy in general and consumption in particular.

Table 20.1 transcends the idea of environmental determinism in food habits to encompass a wider range of socio-economic and physiological factors. In this and the following two chapters we start with a recognition that the ecology of food production does still have some direct and indirect relevance for a study of consumption patterns, and we then turn to a more conventional discussion of other factors, such as prices and incomes, social class, taste, tradition and lifestyle.

Ecology

In truly subsistence economies the diets of farming families are limited to own-production, plus some local trade. The ingredients are, therefore, constrained by the range of farming systems practised within a particular adaptation to physical conditions, such as climate, soils, topography and the availability of water for irrigation. Such fully independent agri-ecosystems are increasingly rare in the modern world of regional and international

Table 20.1 Factors in food choice

Geo-environmental	Socio-economic	Physiological
• Agro-ecosystem	• Religion, taboo, social custom	• Heredity
• Time of day, season	• Ethnicity	• Allergy
• Regionality of food culture	• Income, social class	• Therapeutic diets
• Spatio-temporal and hierarchical diffusion of food habits	• Household composition	• Taste, acceptability
	• Knowledge of nutrition	• Sex, body size
	• Attitude to food-related health risk	• Age
	• Advertising, mass communications, travel	
	• Retail system	
	• Moral values	

Source: modified and extended after McKenzie (1979).

trade, but the foods eaten by the bulk of the world's population are still dominated by locally produced staples and by the inertia of dietary tradition. The global pattern of food régimes (Figure 20.1) demonstrates this, as does the spatial distribution of regional dietary staples in India (Figure 20.2), although Chapter 21 will show why such maps do not tell the full story.

Table 20.2 The protein quality of the main foods consumed in West Africa

	Amino acids			
Food	**Isoleucine**	**Lysine**	**Methionine and cystine**	**Tryptophan**
Fish	72	141	73	70
Beef	73	137	72	70
Cow's milk	71	121	60	88
Fonio	60	40	152	88
Millet	62	53	87	122
Rice	57	59	61	78
Yam	56	64	50	80
Manioc	42	64	49	72
Sorghum	59	31	52	76
Maize	55	41	63	44

Note: The protein score is calculated as a percentage of the respective amino acids in egg protein.
Source: Annegers (1974).

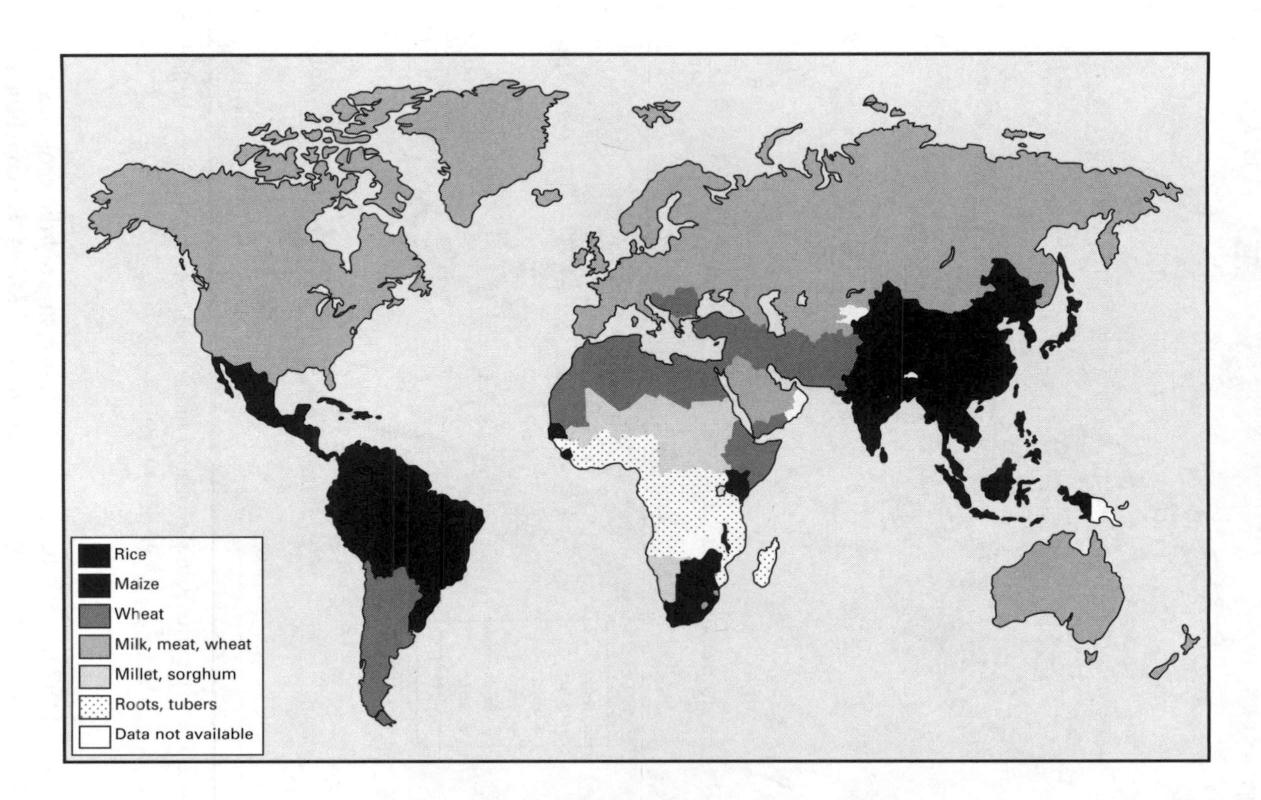

Fig. 20.1 World dietary patterns, by major food groups. *Source*: World Food Summit website (accessed January, 1999).

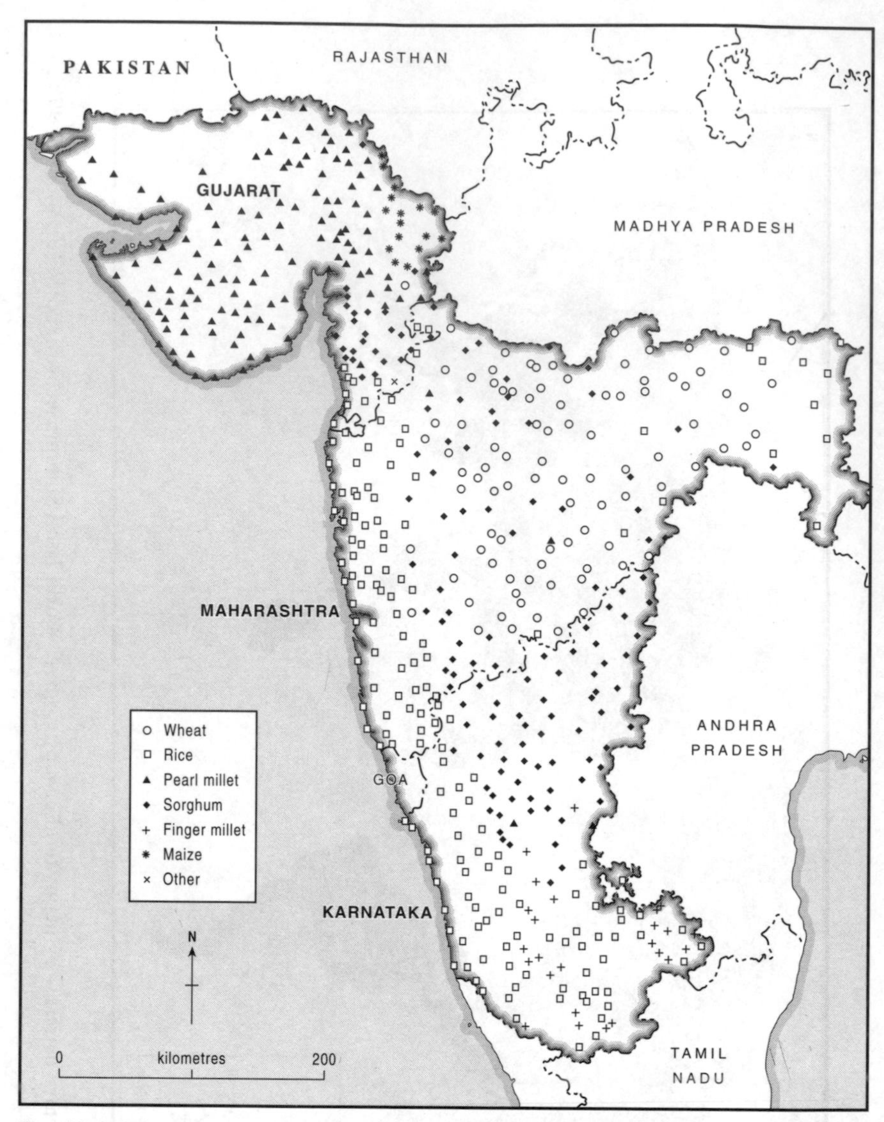

Fig. 20.2 Dietary patterns in the west of India. *Source*: compiled from the Indian Population Census, 1981.

Annegers (1973a, 1973b, 1973c, 1974) investigated the ecological background to certain nutritional imbalances in West Africa. He concluded that, while historical factors have been important in the spread of particular crops and farming systems, it is the agri-climatic environment that has had most significance in the evolution of regional dietary patterns (Figure 20.3). Although he found that the region generally has a sufficiency of calories, the

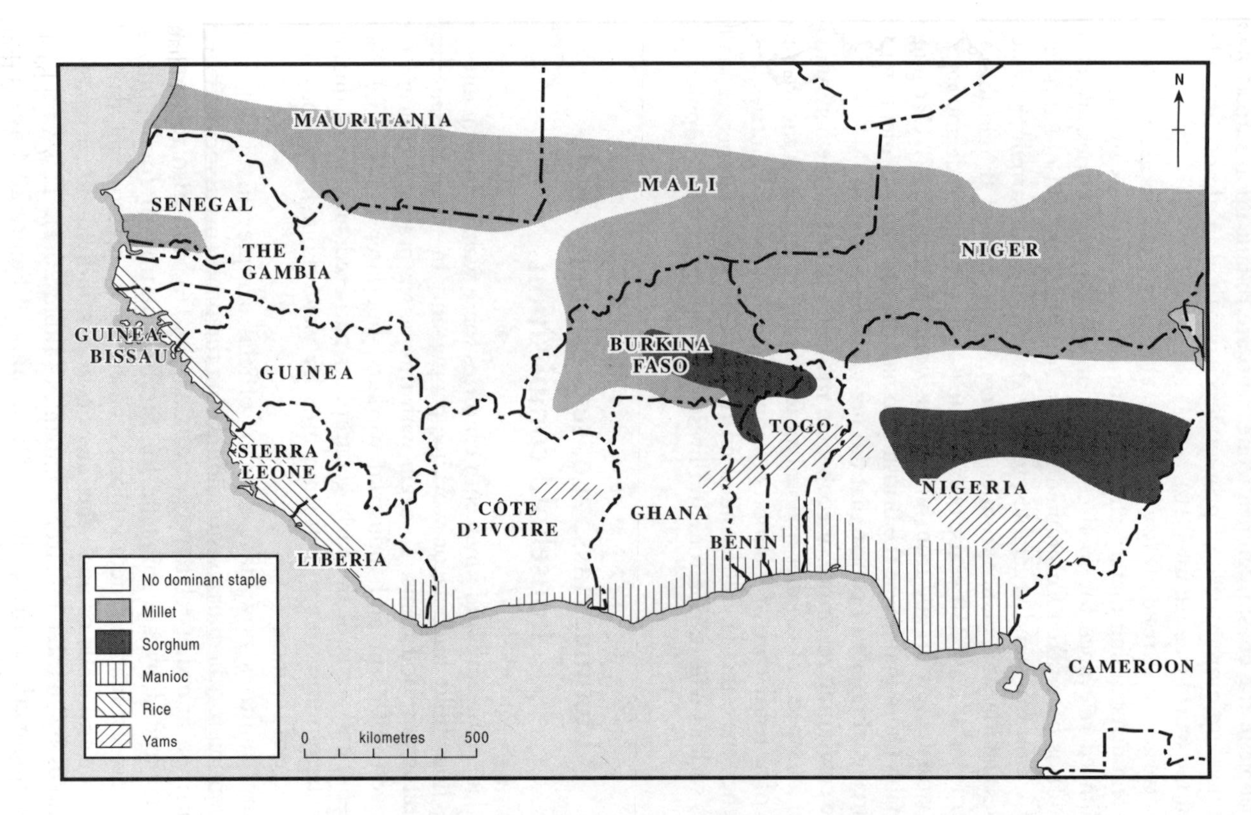

Fig. 20.3 Dominant starchy staples in West Africa. *Source*: based on Annegers (1973a).

protein quality of local foods varies considerably, with certain root crops and coarse grains falling short of requirements (Table 20.2). The result is that people in the coastal 'guinea' zone, who are reliant upon staples such as manioc, are more at risk of protein deficiency than in almost any other part of the world (*see* Chapter 10).

Lest we get too carried away with the neat symmetry of a local agricultural ecosystem determining dietary choice, let us remind ourselves that this has only limited relevance for food consumption in richer countries. Trade in foodstuffs is now taken for granted and supermarkets are full of exotic and out-of-season fruits and vegetables, to the extent that local farming products elicit less and less allegiance from consumers. In the case of Britain this has been true for at least 250 years, because the policies of free trade and imperial expansion drew in food from all over the world (Walvin 1997). One might reasonably claim that it was the British who had the first 'global' diet in terms of sourcing, although of course that did not guarantee quality.

The continuing relevance of ecology in the HICs is not in the determination of all aspects of the diet by local agricultural output but in the maintenance of a steady core of the food economy. In many Asian countries, even the richer ones, this is still founded on rice, and in many temperate lands staples such as wheat, potatoes and livestock products remain popular.

Income, class, age, sex, ethnicity and household composition

In the late eighteenth and nineteenth centuries there was a growing interest in the household budgets of working people and in particular in their expenditure on food items. To begin with this was a practical concern for the poverty of farm and industrial workers, but later Frédéric Le Play (1806–62) and some of the early geographers gathered substantial quantities of data in an attempt to form a broader view about livelihoods in a wide range of settings.

Many countries now have regular, official sample surveys of food consumption for the purposes of monitoring the problems associated with malnutrition and diet-related disease (Chapter 16), and also to provide the data needed by governments for the formulation of food policies. Others use estimation methodologies based upon macro-level data of production and trade or upon the interviewing of individuals or panels of households by market research companies (Table 20.1). This variety of compilation methods makes it sensible to think of Figure 20.4 as impressionistic because of the problems of comparability (see also Grigg 1993a and 1993b).

Some of the early researchers realized that a number of factors influenced food consumption beyond the context of the regional food culture. Income

Table 20.3 Approaches to generating data on food consumption

Data from	Method
Sample survey	A representative sample is drawn from a certain population, bearing in mind the need to have accurate information on consumers of all income groups, and also sex, age and household composition. Questionnaires are used to prompt people's memory of recent meals, or the respondent may be asked to fill in a food diary. The most accurate method is for skilled surveyors to weigh all food portions and analyse typical foods chemically but this is too time-consuming and labour-intensive for large-scale research.
Panel survey	Here the respondent group may be retained over a period of time.
Balance sheets	This method is commonly used at the national scale. If a country's agricultural production and international food trade statistics are available, then it is possible to generate estimates of per capita consumption after allowances have been made for seeds, wastage and storage carry over.
Scanning data	Most modern supermarkets have automatic technology to read and log the bar codes on the food items they sell. As a result there are data on the aggregate sales of each store and also on each consumer if they have a loyalty card. Commercial and individual confidentiality mean that this data is not readily available.

and class, for instance, could not be ignored. Higher income groups were observed to purchase higher quality items rather than increase their intake of basic staples. This is also true today, with the adoption of tastier, higher status, and easy-to-prepare foods as incomes rise, and this helps to explain the long-term rise of animal foods (meat, fish, cheese) and the decline of certain of the cheaper carbohydrates (potatoes, bread) (Table 20.4).

As long ago as 1857, Ernst Engel recognized that family income was important. He found that in low-income families a larger proportion of expenditure goes on food than in higher income groups. This common sense relationship holds good for cross-sectional data (comparisons between families, groups or countries at one time – see Table 20.5) and time series data.

Sociologists have argued that socio-economic class is a persistent factor influencing food habits. Calnan and Cant (1990), for instance, found the following variations between middle class and working class households in a sample drawn in south east England:

- Information about healthy eating was more readily available to the middle class households, whereas working class expenditure, although constrained by the need for careful budgeting, was more directed to the purchase of high status meats.

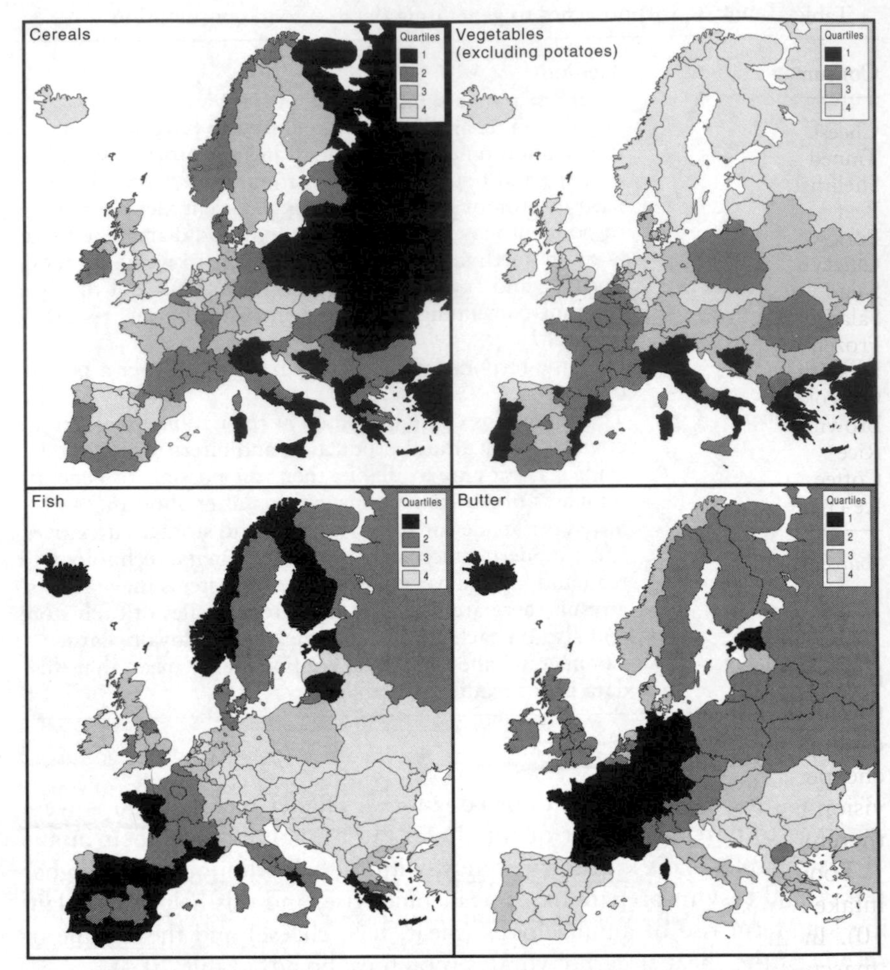

Fig. 20.4 European food consumption by EU region and non-EU countries. (high = 1, low = 4) *Source*: Eurostat.

- Middle class husbands played a greater part in decision-making about shopping lists, although they made little more effort in the shopping or cooking.
- Working class recipes and ingredients were more likely to be conservative and repetitive.

The income/diet relationship may create problems for the needy. In LICs food poverty may lead to hunger and malnutrition (Chapter 10) but the rich countries are by no means exempt. In the USA part of the state welfare provision is paid as 'food stamps' in order to create a safety net for the poor and unemployed, but there is a stigma attached to such handouts. Even families with a worker on regular (but low) wages are at risk because there

Table 20.4 Foods whose consumption is strongly related to income in the UK

Consumption rises with income	Consumption falls when income rises
Cheese	Tinned meat
Tinned salmon	Sausages
Shell fish	Herrings
Beef	Margarine
Pork	Lard
Chicken	Tinned milk puddings
Salad vegetables	Potatoes
Salad oils	Dried pulses
Frozen vegetables	Tinned vegetables
Fresh fruit	Tea
Chocolate biscuits	White bread
Brown and wholemeal bread	Oatmeal products
Rice	
Coffee	
Ice-cream	

Source: Ritson and Hutchins (1991).

is a strong drive in all of us to emulate the diets of the cultural norm and this may lead to over-expenditure on some items and the neglect of the essentials. Expensive 'junk foods' may be consumed in preference to cheaper and more nutritious alternatives, especially in families where television-inspired pressure comes from young people, and food poverty may therefore be perpetuated.

Food consumption differences between the sexes are not only a matter of intake to match different body weights and nutrient requirements (Chapter 10). In most societies there are foods that are seen to be 'feminine' or 'masculine' (Chapter 23). For example, in Britain the consumption of beer

Table 20.5 The percentage of EU national incomes spent on food purchase in 1994

Country	%	Country	%
Belgium	12.04	Netherlands	12.08
Denmark	13.51	Portugal	21.22
Finland	16.00	Spain	22.82
Greece	17.81	Sweden	19.52
Italy	21.12	United Kingdom	15.76
Luxembourg	12.33		

Source: Eurostat.

Fig. 20.5 Trends in household composition in the United States. *Source*: Senauer *et al.* (1991, 85).

is commonly accepted as a largely male preserve, whereas among teenagers a predominance of vegetarians are female because girls and young women have become concerned with methods of animal rearing and slaughter (Chapter 18).

The age and sex composition of the population is also significant for the overall structure of demand. This was discussed as early as the 1860s by one investigator, Dr Edward Smith, and he even incorporated into his analysis what we would now call 'reference adults'. This allowed him to calculate the average consumption per head within a family in terms that were then comparable between families.

According to Senauer *et al.* (1991), modern demographic and societal trends, such as families with fewer children, the steadily ageing population, and an increasing number of single person households (Figure 20.5), are all fundamental to an understanding of a nation's diet. One example of this is that the older age cohorts are most likely to cling to lifelong food habits and, as a result, there are several foodstuffs that are age-dependent for their consumption (Figure 20.6). Since senior citizens will be forming a larger proportion of the population in future in most countries (for instance the proportion of people over 65 in the USA will increase from four per cent in 1900 to 22 per cent in 2050), this conservative element of taste is important. It is also worth remembering that the elderly need 15–20 per cent less energy intake than they did when they were aged 35, and in consequence their spend on food is reduced (Senauer *et al.* 1991, 91).

Finally, one might assume that the connexions between ethnicity and food are 'obvious'. The popular stereotype is that Asian, Caribbean or other immigrant groups in Britain, for instance, will continue to eat their

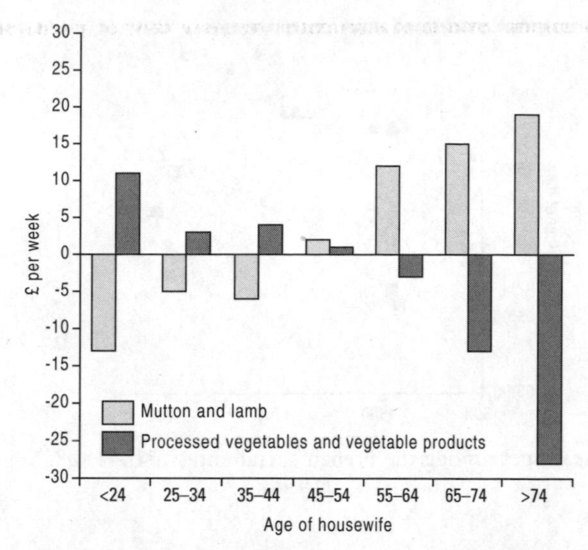

Fig. 20.6 Expenditure on lamb and processed vegetables by age of housewife (£ per week compared with national average). *Source*: Ritson and Hutchins (1991).

'own' types of food, even in second and subsequent generations (Chapter 21). There seems to be some truth in this for Italians and south Asians in Glasgow (Williams *et al.* 1998), but Caplan *et al.* (1998) found in Lewisham (south London) that among younger people of West Indian parentage, although they did still regard Caribbean recipes as important to their identity, many had slipped away from using them every day because they are time-consuming to prepare and too 'heavy and starchy' to fit with weight watching. Rather, dishes such as rice and peas were reserved for Sundays when eating with relations. In 'modern' societies there is a convergence of food experience in which most groups are willing to experiment and perhaps to adopt the dishes of others, although usually in a modified form which is acceptable to the local palate.

Time

Any serious study of diet should consider the dimension of time. Stephen Mennell (1996), citing Norbert Elias, sees time as crucial to the process and structure of civilization, and argues that the social construction and regulation of time owes much to the question 'when shall we eat?' The answer to this question has varied according to the era and location. Figure 20.7 illustrates this by showing the changing number and timing of meals among the social élite in France.

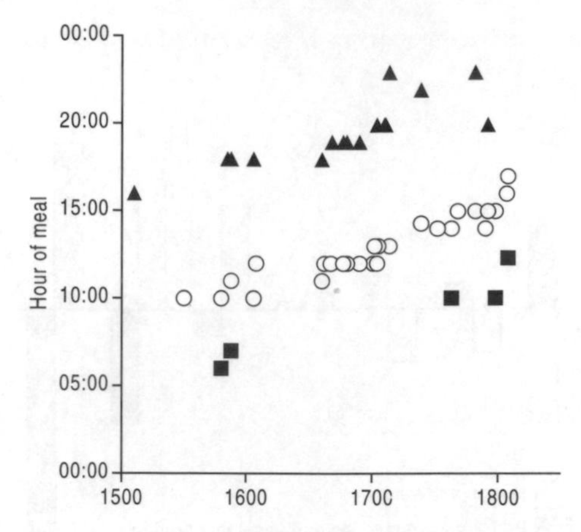

Fig. 20.7 Mealtimes among the French social élite, 1510–1808. *Source*: Flandrin (1996).

Among the most immediate observations we can make is our own experience of the diurnal rhythms of hunger, eating and digestion (Tables 20.6 and Figure 20.8). It seems from Table 20.7 that there are geographical variations in this, some of them quite surprising. In Poland, for instance, among the elderly at least, there is a greater intake of energy at breakfast time than from the evening meal.

There are also seasonal shifts in the availability of certain foods, and in some societies even seasonal fluctuations in energy intake (Chapter 10), but it is the fundamental changes over longer time periods that concern us in this section. They tend to mirror the process of economic development, with greater prosperity and social dynamics encouraging the evolution of dietary patterns.

In Britain, The National Food Survey provides an historical data series that allows us to describe these changes since World War II. As Figure 20.9

Table 20.6 Percentage nutrient intake at different meals of 899 men in Augsburg, Austria

Nutrient	Breakfast	Morning snack	Lunch	Afternoon snack	Dinner	Supper
Energy	17.2	7.4	29.2	6.8	33.1	6.4
Protein	13.6	7.2	36.3	4.1	35.5	3.3
Fat	16.9	7.0	32.5	5.5	35.1	3.1
Carbohydrates	22.9	7.7	25.1	8.8	28.5	7.0
Fibre	19.9	7.4	29.4	5.2	32.7	5.3
Calcium	21.3	7.5	24.3	7.8	32.1	7.1
Alcohol	2.1	7.6	21.0	8.9	39.8	20.5

Source: Winkler *et al.* (1999).

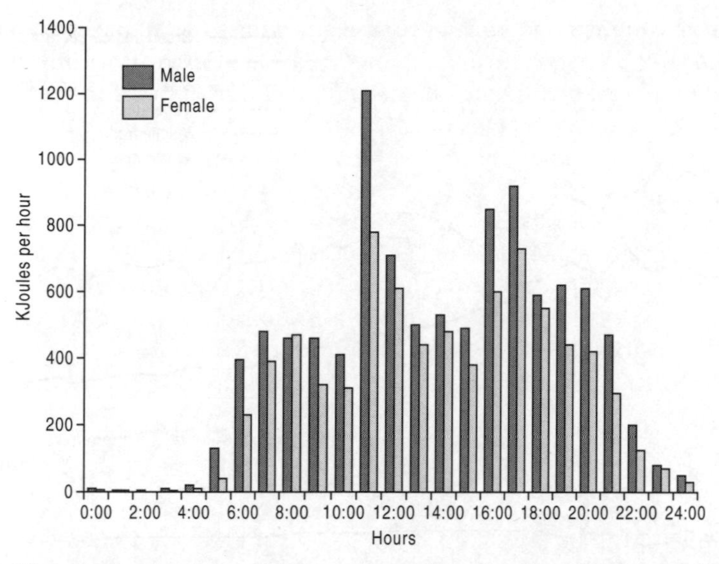

Fig. 20.8 Energy intake at different times of the day. *Source*: Roos and Prättälä (1997).

shows, food purchases have been fluid. Rising incomes are a factor (e.g. in the falling consumption of bread and potatoes), but health considerations have also been important (e.g. in the decline of full cream milk), as have secular changes in taste. Nelson (1993) argues that such national averages hide social class variations, especially before the war, when there were remarkable differences in nutritional health between the top and bottom earners.

Convenience foods have also become increasingly important in the diets of people in western, advanced countries. This seems to be a result partly of shifts in the working patterns of women, the traditional preparers of

Table 20.7 Energy intake at different meals as a percentage of total consumption among the elderly in eight European towns

	Breakfast	Midday	Evening	Snacks
Padua, Italy	11	45	37	7
Romans, France	18	45	30	7
Haguenau, France	19	39	30	12
Yverdon, Switzerland	19	39	33	9
Marki, Poland	28	33	27	12
Ballymoney, N. Ireland	22	32	30	16
Roskilde, Denmark	19	25	35	21
Culemborg, Netherlands	15	21	33	31

Source: Schlettwein-Gsell *et al.* (1999).

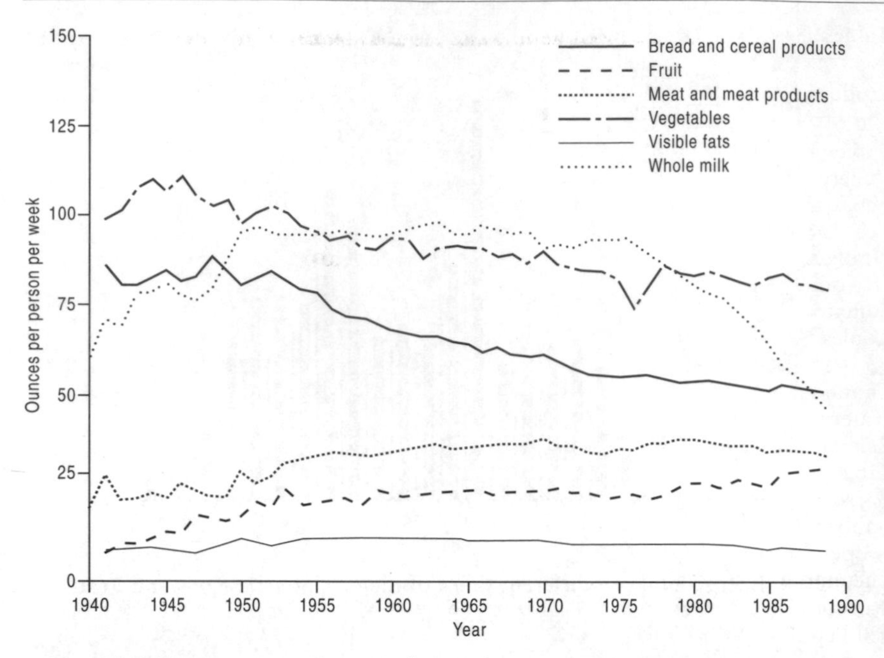

Fig. 20.9 Food consumption trends in Britain, 1940–90. *Source*: Buss (1991).

food, who are now participating more in the workforce and therefore have less time available for complex recipes. Ritson and Hutchins (1991) refer to one study of Newcastle families, which found that 94 per cent of meals now take less than ten minutes of preparation and 51 per cent no time at all. Only 7 per cent take more than 20 minutes of cooking time. Such trends have been facilitated by the explosive growth of sales in ready-to-eat meals, coupled with the kitchen technology of freezer and microwave.

Warde's (1997) content analysis of women's magazines from 1968 to 1992 has confirmed a significant increase in the mention of convenience in food preparation, but there are other trends. Recommendations about health also became more prominent, as did tradition in a nostalgic sense, and indulgence. The role of economy in meal preparation remained constant, but less emphasis was given to novelty, extravagance and care.

Conclusion

The USA is a country in which one might expect food consumption habits to show relatively few regional trends. First, a majority of the population is sufficiently wealthy to purchase an adequate and varied food basket, with a proportion of non-local items high enough to rule out ecologically-

Table 20.8 Consumption* of fresh fruit and vegetables in 18 American cities, 1985

Produce	Average	Lowest	Highest
Table potatoes	62.6	Baltimore-Washington (27.9)	Dallas (141.8)
Iceberg lettuce	41.1	Baltimore-Washington (21.2)	Boston (91.9)
Bananas	35.9	Philadelphia (10.2)	San Francisco-Oakland (54.6)
Oranges	23.7	Baltimore-Washington (9.7)	Boston (51.2)
Dry onions	22.0	Detroit (8.2)	Columbia (50.7)
Tomatoes	21.7	St Louis (8.7)	Columbia (103.2)
Apples	21.5	Baltimore-Washington (8.7)	San Francisco-Oakland (62.1)
Cantaloupes	14.9	Detroit (8.2)	Columbia (35.6)
Watermelons	14.7	San Francisco-Oakland (8.5)	Columbia (137.3)
Celery	13.6	Detroit (6.1)	Boston (34.7)
Table grapes	13.2	Baltimore-Washington (6.8)	Boston (26.4)
Carrots	11.2	Baltimore-Washington (4.8)	Buffalo (27.6)
Cabbage	10.0	Detroit (4.8)	Columbia (45.3)
Grapefruit	9.0	New Orleans (4.6)	Boston (23.0)
Cucumbers	8.2	Chicago (4.4)	Boston (24.6)
Broccoli	6.4	Detroit (2.7)	Boston (14.4)
Bell Peppers	5.6	Baltimore-Washington (3.1)	Boston (16.4)
Sweet corn	5.4	Chicago, Detroit (3.2)	Columbia (16.1)
Honeydew melons	5.2	New Orleans (1.5)	Los Angeles (10.6)
Pears	5.2	Atlanta, Balt-Wash (2.4)	Boston (11.5)

* Thousand cwt of deliveries per 100,000 population.
Source: Shortridge and Shortridge (1989).

determined diets. Second, although there are spatial concentrations of immigrant groups, one would not expect these to have developed into homogeneous regional cuisines as has happened in countries with a longer cultural history. Third, the commercialization of both the grocery food chain and fast food restaurants has reached such levels in America that the standardizing tendencies of modernity have penetrated deeply into the national food psyche.

However, the Shortridges (1989) have found, for fresh foods at least, that there are significant variations (Table 20.8) between the big cities in America in terms of dietary preferences. The friction of distance from the sources of the various fruits and vegetables is one factor, along with some local traditions and the diffusion of awareness about the health implications of consumption.

The conclusion of this chapter is that such variations are not surprising because, even in America, there are significant regional differences in the ecological, economic and social factors that underpin the consumption of food, and also in the organization of the food economy by manufacturers and retailers. In Chapters 21 and 22 we extend the argument further by investigating cultural factors; we need to give due recognition to the very large and complex literature on the anthropology and sociology of food which adds subtleties of analysis and helps us to analyse the evolution of food habits.

Further reading and references

For a sample of the francophone literature in this area see Sorre (1952), George (1963), Livet (1969), Bernard, Salles and Thouvenot (1980), Favier and Thouvenot (1980).

Annegers, J.F. 1973a: Ecology of dietary patterns and nutritional status in West Africa. 1: distribution of starchy staples. *Ecology of Food and Nutrition* **2**, 107–19.

Annegers, J.F. 1973b: The protein-calorie ratio of West African diets and their relationship to protein calorie malnutrition. *Ecology of Food and Nutrition* **2**, 225–35.

Annegers, J.F. 1973c: Seasonal food shortages in West Africa. *Ecology of Food and Nutrition* **2**, 251–57.

Annegers, J.F. 1974: Protein quality of West African foods. *Ecology of Food and Nutrition* **3**, 125–30.

Atkins, P.J. and Raw, M. 1995: *Agriculture and food*. London: Collins Educational.

Bernard, A.J-M., Salles, P. and Thouvenot, C. 1980: Consommation alimentaire: une orientation interdisciplinaire, *Annales de Géographie* **493**, 258–72.

Buss, D. 1991: The changing household diet. In Slater, J.M. (ed.) Fifty years of the National Food Survey. London: HMSO, 47–54.

Calnan, M. and Cant, S.L. 1990: The social oganization of food consumption: a comparison of middle class and working class households. *International Journal of Sociology and Social Policy* **10**, 53–79.

Caplan, P., Keane, A., Willetts, A. and Williams, J. 1998: Studying food choice in its social and cultural contexts: approaches from a social anthropological perspective. In Murcott, A. (ed.) *'The Nation's Diet': the social science of food choice*. London: Longman, 168–82.

Cummins, S. and Macintyre, S. 1999: The location of food stores in urban areas: a case study in Glasgow. *British Food Journal* **101**, 545–53.

Favier, A. and Thouvenot, C. 1980: Eléments de cartographie alimentaire. *Annales de Géographie* **493**, 273–89.

Flandrin, J.-L. 1996: Mealtimes in France before the nineteenth century. *Food and Foodways* **6**, 261–82.

George, P. 1963: *Géographie de la consommation*. Paris: Presses Universitaires de France.

Grigg, D.B. 1993a: International variations in food consumption in the 1980s. *Geography* **78**, 251–66.

Grigg, D.B. 1993b: The European diet: regional variations in food consumption in the 1980s. *Geoforum* **24**, 277–90.

Livet, R. 1969: *Géographie de l'alimentation*. Paris: Editions Ouvrières.

McKenzie, J. 1979: Economic influences on food choice. In Turner, M. (ed.) *Nutrition and lifestyles*. London: Applied Science Publishers, 91–103.

Mennell, S. 1996: Sociogenetic connections between food and timing. *Food and Foodways* **6** (3–4), 195–204.

Nelson, M. 1993: Social class trends in British diet, 1860–1980. In Geissler, C. and

Oddy, D.J. (eds) *Food, diet and economic change past and present*. Leicester: Leicester University Press, 101–20.
Ritson, C. and Hutchins, R. 1991: The consumption revolution. In Slater, J.M. (ed.) *Fifty years of the National Food Survey*. London: HMSO, 35–46.
Roos, E. & Prättälä, R. 1997: Meal pattern and nutrient intake among adult Finns. *Appetite* **29**, 11–24.
Schlettwein-Gsell, D., Declari, B. and De Groot, L. 1999: Meal patterns in the SENECA study of nutrition and the elderly in Europe: assessment method and preliminary results on the role of the midday meal. *Appetite* **32**, 15–22.
Senauer, B., Asp, E. and Kinsey, J. 1991: *Food trends and the changing consumer*. St Paul, Minnesota: Eagan Press.
Shortridge, B.G. and Shortridge, J.R. 1989: Consumption of fresh produce in the metropolitan United States. *Geographical Review* **79**, 79–98.
Sorre, M. 1952: La géographie de l'alimentation. *Annales de Géographie* **61**, 184–99. Translated as 'The geography of diet'. In Wagner, P.L. and Mikesell, M.W. (eds) 1962: *Readings in cultural geography*. Chicago: University of Chicago Press, 445–56.
Vidal de la Blache, P. 1926: *Principles of human geography*. New York: Holt.
Walvin, J. 1997: *Fruits of empire: exotic produce and British taste, 1660–1800*. Basingstoke: Macmillan.
Warde, A. 1997: *Consumption, food and taste: culinary antimonies and commodity culture*. London: Sage.
Williams, R., Buch, H., Lean, M., Anderson, A.S. and Bradby, H. 1998: Food choice and culture in a cosmopolitan city: South Asians, Italians and other Glaswegians. In Murcott, A. (ed.) *'The Nation's Diet': the social science of food choice*. London: Longman, 267–84.
Winkler, G., Döring, A. & Keil, U. 1999: Meal patterns in middle-aged men in southern Germany: results from the MONICA Augsburg dietary survey 1984/85. *Appetite* **32**, 33–7.

21

The origins of taste

It could plausibly be argued that changes of diet are more important than changes of dynasty or even of religion. Orwell (1937, 82).

Introduction

In matters of food consumption we are at the same time driven and constrained by biological necessity and socio-cultural factors. We may be an omnivorous species but this does not mean that we are either attracted to eat or able to eat every item of food that may be edible. There are perfectly good reasons to be suspicious of food because of possible poisoning or disease, but there are also self-imposed preferences or avoidances that have nothing to do with health. 'Taste', therefore, has both physical and social meanings. Despite some variations (Figure 21.1), all palates are attuned to distinguish the sensations of sweet, sour or salty foods, and to identify common flavours. But tastes are also derived from our culturally constructed inclinations for particular dishes and ingredients, and our socially-derived desire for our consumption habits to show us in the best possible light. To Pierre Bourdieu (1984), the latter sense is especially important. He argues that the distinction between social groups, especially classes, in their tastes for food and other commodities may become a badge of their identity.

Rozin (1976) and Fischler (1980) have discussed this tension between the biological and the cultural aspects of taste. They identify the so-called 'omnivore's paradox' as becoming more of a problem in the modern age because, on the one hand, people are naturally suspicious about any new foods which do not fit with the traditional foodways that have delivered nutritional stability over generations. In the last 20 years such doubts have extended even to staple foods, with scandals such as *Salmonella* in eggs, mad cow disease, and many others (Chapter 16). On the other hand, there is a natural curiosity about novelties, to the extent that adventurous

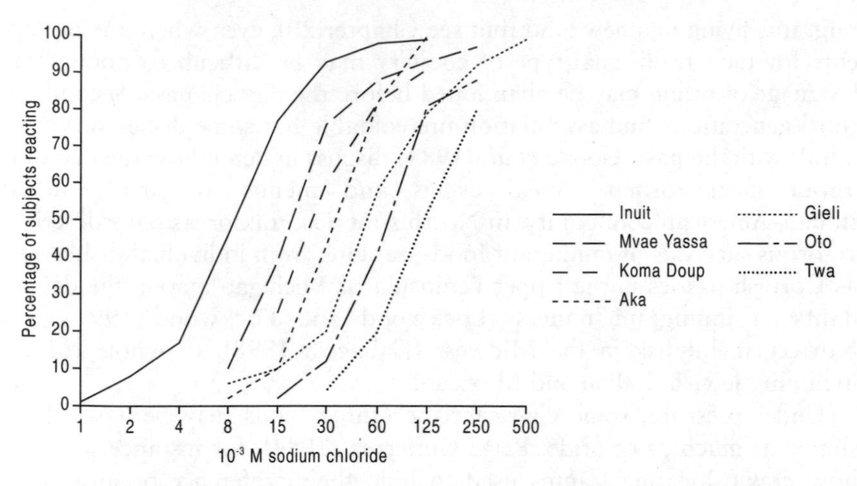

Fig. 21.1 The recognition of saltiness by Inuit (solid line) and forest-dwelling pygmy tribes (other lines). *Source*: after Hladik (1997).

Table 21.1 Three food paradoxes

Positive aspects of food	**Negative aspects**
Provides gustatory pleasure and satiety	Can produce dyspepsia, nausea, vomiting
Required for vigour, energy and health	Can lead to illness and disease
Required for continuation of life	Food production involves death of plants and animals

Source: Beardsworth (1995).

consumers have become familiar with the cuisines of distant lands or are creating their own 'fusion foods' from mixtures of different cuisines. Alan Beardsworth identifies three food paradoxes that summarize contemporary anxieties (Table 21.1)

In this chapter we will look at some of the social and cultural influences on taste and food habits, a discussion that will be continued in Chapter 22.

Food and identity

One factor in the origin and persistence of foodways is that they often represent an important expression of our identity, both as individuals and in reference to a broader ethnic, class or religious grouping. For instance, it has been noted that inertia in food habits is strong amongst first generation migrants living in a new land (but see Chapter 20), even where the ingredients for their traditional type of cookery may be difficult to obtain. The language of origin may be abandoned before the diet changes. Second and third generations find assimilation unavoidable but some dishes survive as a link with the past. Goode *et al.* (1984) discuss in detail how the decisions about meal formats, meal cycles and menus are made in an Italian–American community. In North America, folklorists have described vigorous survivals of immigrant foods, ranging from individual dishes such as Cornish pasties in the Upper Peninsula of Michigan among the descendants of immigrant miners (Lockwood and Lockwood 1991), and Norwegian lutefisk in the Midwest (Gabaccia 1998), to whole cuisines including Jewish, Italian and Mexican.

Under pressure, some characteristic ethnic foods may be a source of shame as much as of pride. Paige Gutierrez (1984), for instance, recounts how crawfish-eating Cajuns used to hide their preference because of its associations with poverty. However, the 'new ethnicity' which has been evident in the USA since the 1960s, has led to a revival of culinary roots and to the celebration of that country's extraordinary diversity. Ethnic speciality restaurants have multiplied and they have sought to negotiate with their customers a version of cultural authenticity which, although reductionist and exaggerated, does satisfy a demand for an exotic consumption experience.

One of the most striking expressions of identity through food is the love affair of the Japanese with rice. Although rice is no longer dominant in her people's diet, Japan has held strongly to the view that it remains central to national life. Rice farmers until recently were heavily subsidized and foreign grains excluded by high import tariffs. This was not just for economic protection but because the highly urbanized and industrialized Japanese value their links with the countryside, and the rural way of life is still seen as a continuing foundation of their culture. Rice has spiritual meaning and in religious terms is believed to be a medium between the earthly and godly realms. It is not surprising then that Ohnuki-Tierney's (1993) book on this topic is entitled 'rice as self', because here we have a commodity that over the centuries has absorbed powerful economic, cultural and political energies, and these are now diminishing only slowly under the impress of westernization and modernization.

Food and place: regional specialities and 'typical' foods

The geographies of food preparation, cooking, recipes, meals and diet have received scant attention from scholars. There are numerous popular books claiming to give insights into regional specialities, and these have their own significance in that they influence the food habits of certain groups in society, but they are usually highly selective in content and discursive in intent. They seek to construct a perception of both place and cuisine which will convey benefits upon the parties to the transaction: the reader, writer, producers of particular foods, the manufacturers of certain items of kitchen equipment and, indirectly, perhaps even the tourist economy of attractive locations.

Folklorists and ethnographers have made major contributions to the understanding of food in particular settings but this type of work has often been innocent of broader social theory and therefore has limited appeal. There are several recent examples of how such place-specific research may be useful, for instance in the description of how new dishes or food cultural complexes may arise. Thus Lloyd (1981) has uncovered the history of Cincinnati chilli, starting in 1922 with the invention of a recipe by a Macedonian immigrant, and later spreading to 200 restaurants in the city and becoming an important Midwest regional dish. Kelly (1983) studied Loco Moco in Hawaii, which in his opinion is a 'folk dish in the making'. This was a local version of fast food, made up of a hybrid concoction of rice with a hamburger patty, a fried egg and gravy. From its origins in 1949, this meal has gradually spread to all of the Hawaiian islands. A British example, again arising from the innovative ideas that come from cross-cultural situations, is the Balti meal. This is a variation on the theme of the curry, which itself has so successfully penetrated popular culture. Birmingham Asian restaurateurs have developed the Balti as a special blend of ingredients that are served in a metal container. Now the dish is found on the menus of many Asian take-aways in Britain, but it is said to be unknown on the sub-continent itself.

While food habits and food systems may not be constitutive of spatially bounded cultures (see below), nevertheless there are regional geographies of food production, food marketing and food preparation in which place and space play an important role. Comparative advantages of climate and soil, coupled with historical traditions based upon particular skills or trade patterns, have given market dominance to the products of particular places. We can recognize three variations of such food-place associations.

First, there are highly specialized production regions, some of which are of relatively recent origin as a result of intensive capital investment. In Britain there are several which are unknown to the public, such as the concentration of rhubarb cultivation in West Yorkshire or of celery growing in Cambridgeshire. Because the place association is weak or non-existent in the mind of the consumer (*see also* Chapter 15), we will leave this category aside.

Second, there are foods which may have originated with a traditional recipe in a particular place, but which over time have become generic food products. Yorkshire pudding, Black Forest gateau, Cheddar cheese, and Eccles cakes are such, now universally known and manufactured without reference to their places of origin other than in the name. This type of

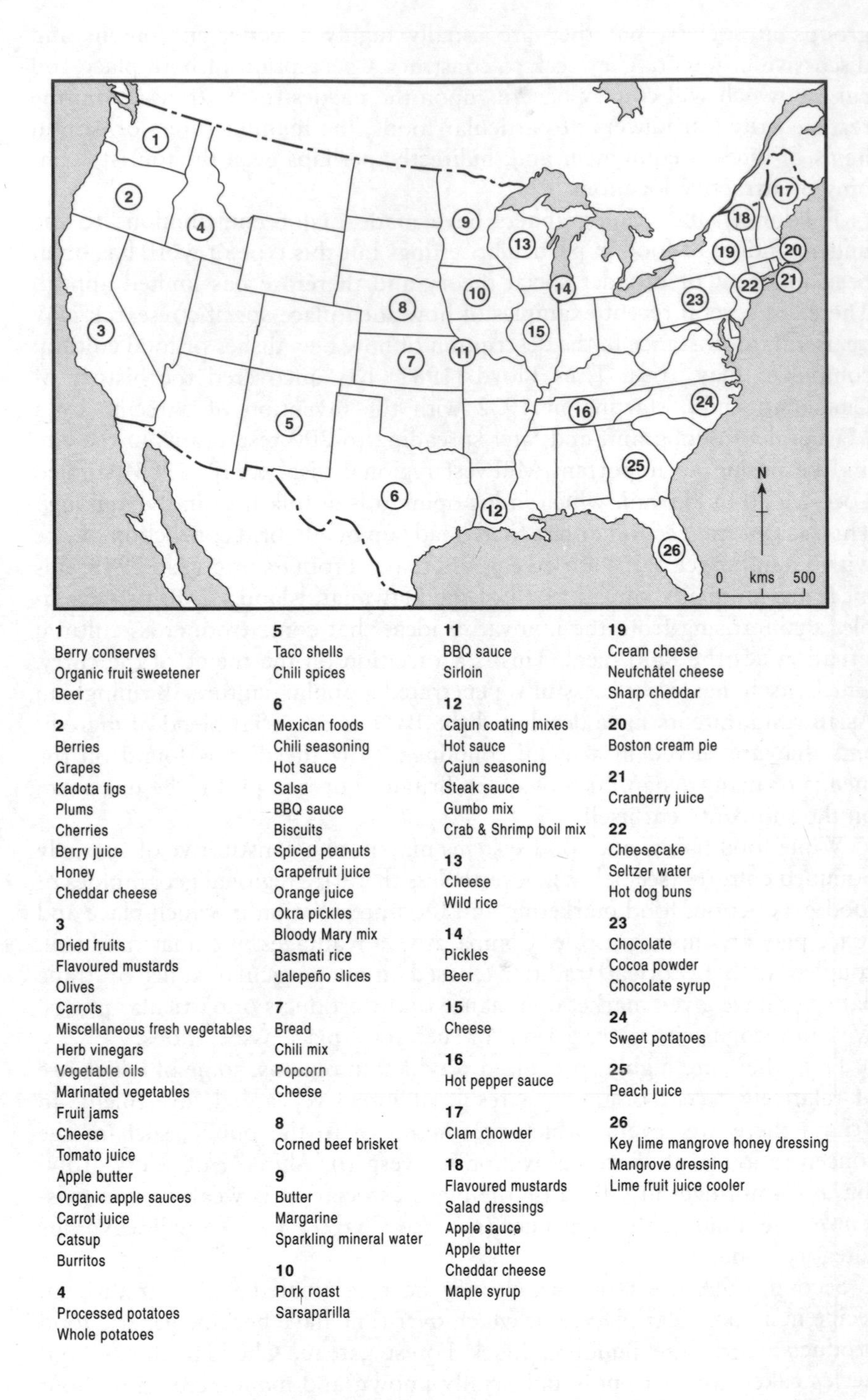

Fig. 21.2 Food-place associations. *Source*: De Wit (1992).

nominal link is becoming more common as consumers wish to sample the foods of other lands, and often the ingredients and the recipe are adapted to suit local tastes.

Cary de Wit (1992) has traced this type of food-place association by looking at the labels on food products displayed in American supermarkets. The resulting map (Figure 21.2) shows that Texas, California, Vermont, Oregon, Idaho and Louisiana have the strongest representation, but some of these links are false. De Wit found New York Seltzer Water that is made in California, Vermont Cheddar from Wisconsin, and Oregon Berry Juice out of Vermont. This contradicts the consumers' wish for labelling that allows the traceability of foods to the site of production.

In the third category we might include foods that have maintained strong links with particular regions in terms of production, quality control and identity (Chapter 15). One thinks here of the high quality artisanal production of Parma ham and many types of Continental cheeses, but the best

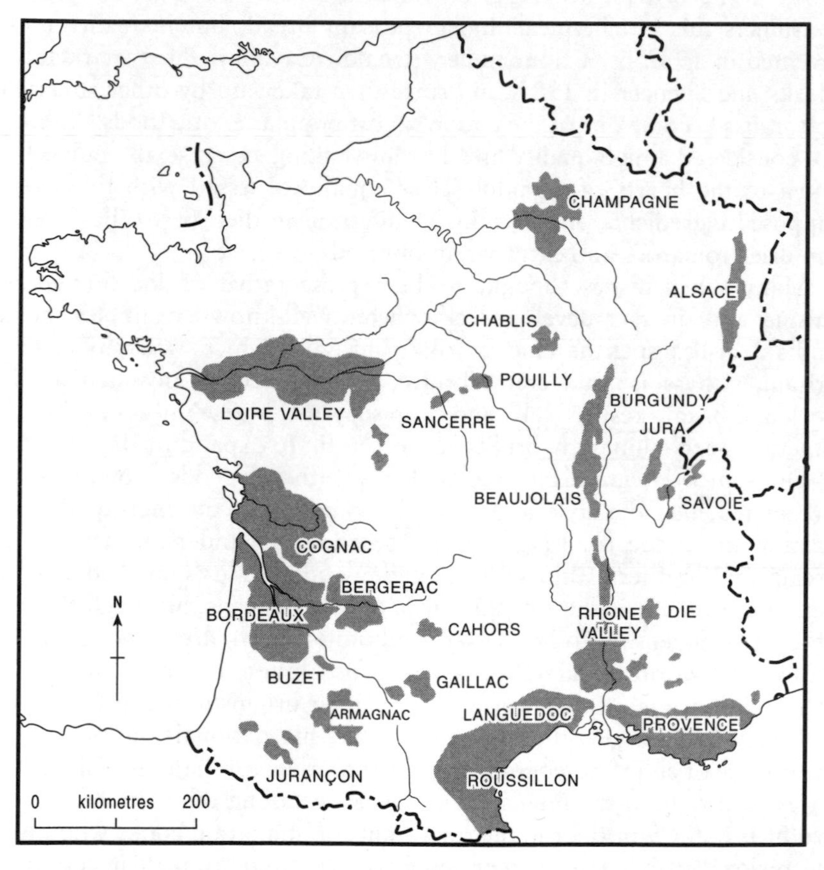

Fig. 21.3 France's major wine producing regions. *Source*: Moran (1993).

example is probably the legal framework of Apellations Controllées in France (Chapter 15) which over decades has protected individual varieties of wine and which has effectively frozen the map of wine regions (Figure 21.3).

Ironically, there is no guarantee that foods which have place associations in the mind of the consumer actually are true regional specialities. There are several examples of invented regional traditions. Lymeswold cheese, for instance, was launched in 1982 by the then Minister of Agriculture, Peter Walker, in an effort to persuade consumers that British soft cheese was preferable to imported products. After the novelty wore off, Lymeswold failed to maintain more than a small part of the market and was abandoned in 1992. More successful has been Wensleydale cheese, which arose in the 1890s from a need of Dales farmers to sell their surplus milk. Through a local initiative, a creamery was built and a new mass-market cheese was created.

Among the most interesting of recent introductions of Mediterranean foods to the British diet has been ciabatta bread. In the minds of many consumers this is a 'typical' Italian peasant bread, but in reality it was invented in 1982 by a flour miller, Arnaldo Cavallari. First marketed by Marks and Spencer in 1985, and since then taken up by other supermarkets, it has become one of the nation's most popular exotic foods. Although not considered a high quality bread by any baking expert, ciabatta has been taken to the hearts of a middle class equally obsessed with those other supposed ingredients of a healthy Mediterranean diet: pesto (basil paste), sun dried tomatoes and extra virgin olive oil.

Many other dishes thought to be representative of location-specific cuisine were in fact developed elsewhere. Well-known examples include 'Indian' dishes such as chicken tikka massala, which was pioneered in Britain; vichyssoise and crème brulée, 'French' recipes invented in New York and Britain respectively; and chop suey, a 'Chinese' meal developed in America. Such culinary hybridity seems certain to expand further in future.

Our fourth category is the regional cuisine that depends upon its distinctive ingredients, the style and skill of cooking, and the high quality and palatability of the resulting dishes. According to Anderson (1988), only French and southern Chinese (Cantonese), and perhaps south Indian cooking, meet the very highest of standards of haute cuisine, although there are of course dozens of other lesser traditions which are now recognized around the world. Figure 21.4 is a map of Chinese regional cuisines that identifies four major traditions that have their origins in the distant past.

Mintz (1996) argues that it is regional (i.e. sub-national) cuisines that are the essence of culinary geography, and not national or 'hautes cuisines'. The latter exist only in the minds of marketeers retailing the so-called 'classic' products of the world, or in the restaurants of immigrant cooks who assemble regional dishes into a 'representative' selection from their country of origin. The regional may even stand for the national, as seems to be the case

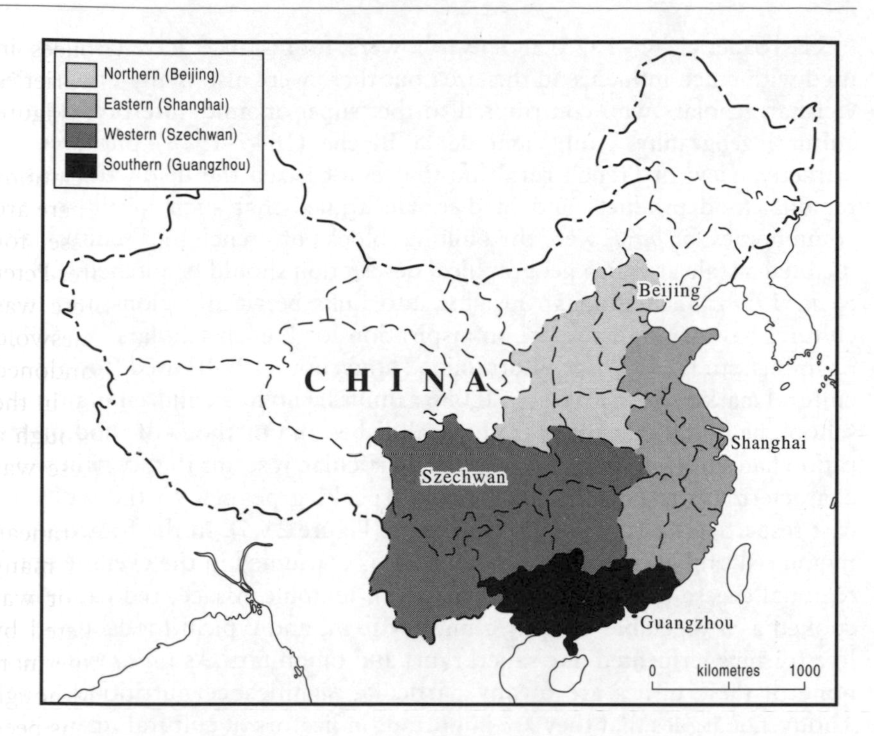

Fig. 21.4 The four major regional culinary traditions of China.
Source: Simoons (1991).

with much of the world-renowned 'Italian' cuisine of pasta and pizza, which is heavily reliant upon traditions from the south, the Mezzogiorno, with Neapolitan dishes especially favoured.

The super-organic cultural geography of food

There has been a tradition in American geography of seeing landscapes, artefacts and human behaviours (e.g. food habits) as outcomes of regionalized 'cultures'. To the modern and especially the post-modern eye some of this scholarship seems naive and uncritical, falling back at times upon description rather than engaging with social process. But something of a defence is possible of the often high quality of research that was involved. Detailed fieldwork and a thorough knowledge of a region was a *sine qua non*, and many cultural geographers of the old school had an excellent feel for historical origins, the diffusion of cultural traits and ideas, and the 'human ecology' of food production and consumption in pre-industrial societies.

Carl Sauer (1889–1975) and his followers, for instance, have been attributed with much influence in this area but there were also many French and German scholars who contributed to the 'super-organic' interpretation of cultural geography. Paul Vidal de la Blache (1845–1918) purveyed an attractive view of French rural life that emphasized the distinctiveness of regional food products and of diet. He argued that such locally rooted communities, or *pays*, were the building blocks of French local culture and dictated a scale at which geographical description should be pitched (Atkins *et al.* 1998). Such Vidalian insights into links between regions and food cultures have continued to be an inspiration for French scholars.

In his study of Alsace-Lorraine, Thouvenot (1978) used food as a cultural marker. In a survey of 30,086 families who had children in primary school, he found that many traditional dishes and methods of food preparation had survived until the 1970s. In particular, it seems that certain foods characterized the French- and German-speaking peoples to the west and east respectively of the linguistic frontier (Figure 21.5). In the francophone region red cabbage is used in salad, soup is consumed in the evening and a regional cheese (cancoillotte) is typical; in teutonic Alsace, red cabbage is cooked as a vegetable, soup is a midday item, and typical foods are naveline (turnips fermented like sauerkraut) and onion tart. As far as we know, none of these dishes are of any particular significance nutritionally, but Thouvenot argues that they are important indicators of cultural realms that have not yet been integrated, despite their juxtaposition within the boundaries of France. The implication is that culinary history carries with it a burden of symbolism that reaches beyond the skills and technology available to the individual cook. In the early twenty-first century we might be tempted to think that such dietary regions will have no significance because the emergence of a global culture will annihilate local differences; but experience suggests that inertia is strong in many aspects of food practice.

English writers have not been immune to super-organic interpretations. Allen (1968) in particular wrote a book on regional tastes that was full of assumptions about the spatial structure of material culture. His raw material was the kind of report produced by consultants on behalf of industry, to inform them about marketing opportunities in different parts of the country. The interpretation he put on data about the demand for goods in the 'North' as opposed to the 'South', however, pandered to prejudices about regional consumers:

> Detectable in the Northerner there is . . . a streak of something almost childlike, to be seen not merely in his artlessness, but in the more miniature scale of some of his tastes and habits: 'fairy' cakes, for example, which almost look as if they came out of dolls-houses . . . The Northerner eats as he lives, plainly and even starkly, without attention to décor or side issues . . . The Southerner's . . . palate . . . is more finely tuned. He goes for rarer flavours.

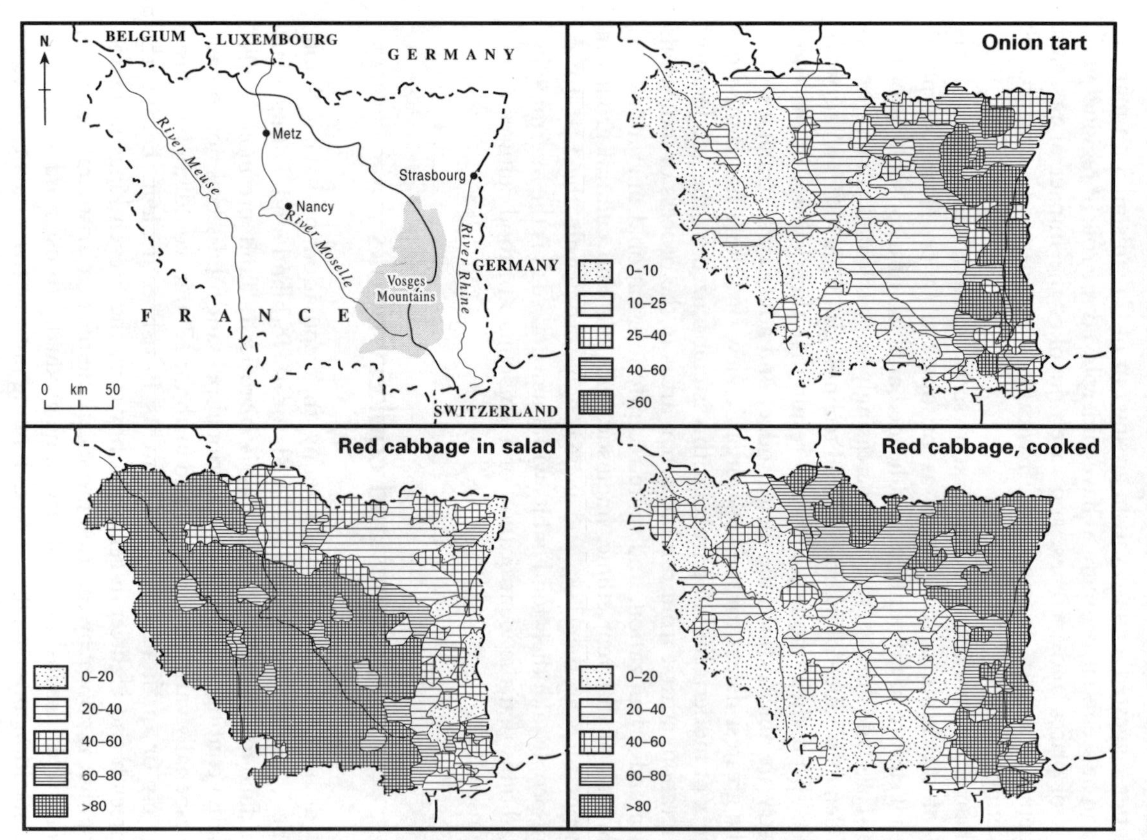

Fig. 21.5 Food as a cultural marker in Alsace-Lorraine. Percentage of surveyed families consuming a particular food. *Source*: Thouvenot (1978).

Allen was unable to account for regional preferences, such as that found in the Midlands of England for sour foods (pickles, acid sauces, bitter beer) rather than sweet, or for the apparent Scottish dislike of pork. He was tempted by a notion of genetically inherited tastes for the former and for the latter suggested a conservatism arising out of Calvinism. For a marketing man he seems surprisingly to have underplayed the role of regional variations of prices and incomes, and rarely invokes the impact of targeted marketing strategies or even, more basically, the variety and structure of marketing channels.

The modern market research literature seems to have totally rejected this kind of caricature of consumers as prisoners of a crude stereotype of regional culture in favour of a hard-headed analysis of disposable incomes, social class and lifestyles. Their increasingly sophisticated surveys nowadays deploy a range of techniques and technologies to investigate the ways people spend their money, with the result that their clients in the food industry are able to fine-tune their product and advertising strategies.

The greatest defeat for the superorganic view of culture has come from the flux of modern life. In particular this is epitomized by the acceleration of migration, at first from rural to urban areas in the process of industrialization and urbanization, and more recently international migrations and the mixing of different ethnic, linguistic, religious and cultural groups in countries such as the USA, and increasingly also within the EU. To an extent people take their food habits with them and such is the intermingling of traditions that the identification of a single map of food cultures would be nonsensical.

Cuisines and cookery books

Why do some countries/regions enjoy the reputation of having 'haute cuisines', while others are castigated for their bland and boring diet? According to Mennell (1985), this is historically (and presumably also spatially) contingent. In France traditional peasant food seems not to have been especially varied or original, and in about 1700 it would not have been creatively or qualitatively superior to its British equivalent. During the seventeenth and eighteenth centuries, however, the French kings increased their political power by centring the social life of their aristocratic subjects at court. This urban-based nobility came gradually to use food as a means of competition amongst themselves, drawing prestige from the richness and elaboration of their meals. After the Revolution, this complex, élitist diet was further developed by restaurateurs, who brought it to the wider attention of a bourgeois clientèle, who at first were unable to use their wealth as conspicuously as was possible for the gentry in England, and who therefore welcomed a gustatory channel of display. During the nineteenth century attitudes to food diffused down the French social hierarchy and were solidified at the local level by the development of location-specific recipes and food products. Peltre and Thouvenot (1989) have edited a collection that sheds light on these regional food habits.

Throughout this period French agriculture was protected by import tariffs and the ingredients of meals therefore remained securely local in character. By comparison, the nineteenth century diet in Britain saw the annihilation of regional difference under the press of industrialization and the importation of widely available, cheap ingredients from the Empire and the wider world. The local roots of this diet became increasingly tenuous and consumers relied upon mass-produced, processed and manufactured foods such as roller-milled and refined flour, margarine, condensed milk and manufactured pickles, jams and preserves. In Britain food has therefore not been a means of establishing difference. This has been achieved in other ways, such as through accent, education, type and location of residence, or leisure.

It seems that the uniformity and blandness of the British diet is an historical legacy of early industrialization and globalization. Any pride derived from favourite dishes such as fish and chips or roast beef and Yorkshire pudding is rather marginal to the national psyche and, if the truth were told, these foods are something of an international joke, alongside the British weather. Other nations with a more fractured identity have deemed food to be more central to their cultural goals, and perhaps even a way of forging some commonality.

India is one such, with an extraordinary range of ethnic, linguistic, religious, caste and other cultural traditions. The culinary idiom is also varied of course, but it has been possible to discern a trend towards universalism in recent decades as cookery books have penetrated the Indian mass market for the first time. The readership of such books and cookery columns in newspapers and magazines is mainly female and middle class and, interestingly, much of this literature is published in English. Certain dishes have emerged as being especially popular, particularly those influenced by the Mughlai and colonial traditions, and the presumption is that these now have a much wider currency in urban literate households than used to be the case. Regional traditions have not disappeared but a hierarchy of cuisines is beginning to emerge at the national level. Appadurai (1988) suggests a 'processual model' of change that might be tested in other countries, where the standardization of consumers' habits is contributing to a more centralized 'national' culture.

One could extend Appadurai's argument to a more general one about the impact of the media upon food consumption habits. Women's magazines are one source of information about contemporary messages on both practicable and desirable food habits (Warde 1997, 44–9), along with a recent surge of food and drink and cooking programmes on television. But perhaps even more significant is the vast scale of food advertising that is pumped through television and the printed media.

The diffusion of food habits

Diets evolve through time in response to the changing circumstances of supply and demand (Chapter 20). One example is the spread of exotic foods, which may become acceptable to new consumers in a number of ways. They may, for instance, be cheaper or more convenient to prepare than traditional staples, as has been the case with wheat food aid in Africa which has out-competed local coarse grains and root crops (Chapter 13); there may be some perceived status associated with the consumption of a particular item, as was originally the case with tea-drinking in Britain; or the commodity may be heavily promoted in the media, for instance soft drinks such as Coca-Cola and Pepsi. The growth in international trade has encouraged the adoption of new foods, both in its mercantilist and capitalist phases, and other forces such as imperialism have also been at work.

An example of the latter process is the introduction of dairy products to Indonesia by the Dutch from the late nineteenth century onwards (Hartog 1986). This was not necessarily a very promising region for milk marketing because many Asian people have a lack of the stomach enzyme, lactase, which breaks down lactose (milk sugar) (Chapter 22). As a result, the consumption of milk may cause them some discomfort. Not deterred, the colonizers set up dairy farms to supply their own needs and also imported tinned, condensed milk from Europe. Gradually the use of milk spread to those among the indigenous people who were in closest contact with the Dutch, and especially among the educated urban élite. From 1910 Nestlé were active in targeting Indonesian consumers, initially by distributing free tins of sweetened condensed milk and then by its promotion for infant feeding. Thus Java became a centre of milk production and consumption, quite contrary to its traditional non-milk food culture before colonization.

Eating the 'other'

'Consuming' the exotic has a variety of meanings. These may range from the creation of images of other cultures, as a means of stereotyping and taming their otherness, to the importation and incorporation of the products of other economies. In the latter case, exotic, usually tropical, fruits and vegetables have become a common feature of the retail experience in rich countries in the last 10–15 years. What used to be called 'queer gear' is now acceptable for many consumers and sales are increasing steadily. According to Cook (1994), the large supermarket chains have carefully nurtured this trade in two ways. First they have ruthlessly imposed conditions of quality and timeliness upon their suppliers in the tropics through systems of contracts that have been responsible for the restructuring of many farm enterprises. The negative side of this is that the poor farm and factory workers in these countries experience low rates of pay and unregulated conditions of work. Second, consumers have been persuaded of the attractiveness of exotic lines by in-shop leaflets and advertising campaigns. These help each company to establish a distinctiveness for their product lines, which they hope will encourage shoppers to return.

May (1996) argues that the consumption of exotic food has another function. It is a (relatively cheap) way of establishing distinctions between social groups. Demonstrating a knowledge of other cuisines, other cultures, and of culinary authenticity, thus has a status currency in those social groups who are perhaps less likely to be able to compete in purely materialistic terms. Cook and Crang (1996) add to this an analysis of restaurants in London which serve exotic food. They see the city as a site of 'global miniaturization' where links are created with many countries and cultures through a range of consumption activities, of which food is one of the most important. Such geographies of displacement are important because they form one source of cultural mixing that may lead to hybridized forms.

Eating the 'other' is therefore not an 'innocent' activity. It has economic, social and cultural implications to add to the political echoes from the past. British colonies for centuries supplied sugar, bananas and other exotic produce to the metropolitan country. Even the economies of the 'white' colonies were partly founded on exports of agricultural produce such as butter, cheese, mutton and beef.

Awareness of tropical cuisines has a long history in Britain, the best example undoubtedly being the 'curry'. 'Indian' cuisine has long been a feature of the restaurant and take away markets in Britain. The number of dishes is limited according to the capacity of each outlet but the similarities between menus are striking, using words that virtually every diner would recognize: curry, dal, bhaji, nan, tandoori, tikka, vindaloo, korma, biryani, mosala. Ironically perhaps, this collapsing of the vast range of South Asian gustatory delights into a comfortingly short list, with only occasional surprises, has been achieved by an industry dominated by restaurateurs from one small district of Bangladesh, Sylhet. Their entrepreneurship has moulded the British view of 'Asian' food.

One interpretation of ethnic restaurants is that they are a reproduction of colonial-style cultural hierarchies. Diamond (1995) finds this in her discussion of restaurant reviews. The non-dominant culture, in this case the dishes, decor and service provided by the restaurant, are exoticized, fetishized, and consumed. Reviewers in turn patronize and racially stereotype their subjects, while always seeking the 'authentic' experience through which they can encapsulate, simplify and therefore tame the culture. One Greek restaurant, for instance, was found to be

> . . . a milieu, a world. They are simply being themselves, in effect getting paid for what they like to do best: talk; sing; argue; munch on olives, cheese, and bread . . . What makes the Minerva special is that

it is unaffectedly, unabashedly, wholly Greek, and that attracts a constant Greek clientèle.

Eating out

Eating out in Europe has its origins in the inns which accommodated travellers from the middle ages and in the eighteenth century coffee houses, which were meeting places for financiers and entrepreneurs. Restaurants as providers of meals purely for pleasure came later, and in France fulfilled the important function of democratizing luxury (Mennell 1985).

In Britain eating out has gone through two phases. The first was the era of fish and chips, a working class food that was (and still is) fried and sold hot in a neighbourhood shop. The availability of this convenience food at unsocial hours fitted well with the shift pattern of industrial workers' lives. Factory canteens were also a feature of this working class experience. In the second phase restaurants, both sit-down and takeaway, have become an increasing source of both fast food and more leisurely eating. European cuisines such as French and Italian were well represented from the outset, but from the 1960s there has been a sustained challenge from Asian foods, mostly prepared by immigrants from South Asia and Hong Kong. From the 1980s, throughout the western world and now increasingly also in the large cities of developing countries, there has been a strong trend for American owned/franchized and American-inspired stores selling burgers, fried chicken and pizzas under highly standardized conditions (Chapter 7).

Warde and Martens' survey (1998b) (Table 21.2) concentrated on the proportion of their respondents who had eaten in a particular type of outlet

Table 21.2 The proportion of a sample of 1,001 people in London, Bristol and Preston who had eaten out at particular types of restaurant in a 12-month period

	%		%
Specialist pizza house	41	Hotel restaurant	25
Fastfood restaurant/burger bar	49	Other British-style restaurant	6
Fish and chips (eat in)	18	Indian restaurant	33
Wine bar	17	Chinese/Thai restaurant	29
Roadside diner or service station	31	Italian restaurant	31
In-store restaurant or food court	31	American-style restaurant	12
Café or teashop	52	Other ethnic restaurant	21
Traditional steakhouse	19	Vegetarian restaurant	9
Pub (bar food)	49	Any other establishment	1
Pub (restaurant)	41	None	7

Source: Warde and Martens (1998b).

Table 21.3 Eating out in Britain in 1997 (grams per person per week)

Chinese dishes	17	Fish	23
Curry and Indian dishes	19	Chips	68
Other ethnic foods	19	Other vegetables	124
Rice, pasta and noodles	27	Sandwiches and rolls	89
Hamburger or cheeseburger	15	Ice cream, desserts, cakes	56
Meat pies	15	Crisps, nuts, snacks	11
Chicken or turkey	22	Chocolate and sugar confectionery	19
Sausages and sausage rolls	19	Other foods	106
Other meat and meat products	36	Tea, coffee and other beverages	406 ml
Pizza	11	Soft drinks and milk	348 ml
		Alcohol	490 ml

Source: MAFF (1998).

in the previous year. The National Food Survey (Table 21.3) quantifies the wide variety of foods and drink away from home. In 1997 this accounted for 28.3 per cent of their expenditure on food and drink and in nutritional terms it contributed approximately 11–12 per cent of total energy and protein. Interestingly, in India the proportion of calories derived from outside the home, mainly from street food, is even higher at 18–21 per cent for low income men in Hyderabad (Sujatha *et al.* 1997), so eating out is important even in poor countries.

Eating out has also increased in the USA. The classic roadside diner which dominated the catering trade from the 1940s, and which made a significant landscape impact, is now fading and the baton has been taken up by ethnic restaurants and by fast food outlets with a variety of corporate affiliations. Among the ethnic restaurants, the most popular cuisines are Chinese, Italian and Mexican (Figure 21.6). In the fast food sector the greatest competition is between burgers, pizzas, chicken, barbecue, and hot dogs, with some striking regional preferences revealed in Figure 21.7.

By 1987 in the USA there was one commercial eating establishment for every 2700 people. This away-from-home food market was supported by over half of the population making a purchase every day. The American people spent $60 billion on fast food alone in that year. This experience is not necessarily the future for other developed countries because consumers have become increasingly aware of the health issues associated with the high fat and salt content of these foods, and because microwave ovens in the household have brought even greater convenience within easy reach.

Warde and Martens (1998a) found that 60 per cent of respondents to their survey wanted to eat out more often and in a variety of ethnic styles (Warde *et al.* 1999), but that the choice is much more limited than we sometimes imagine. Some diners limit themselves according to culturally learned

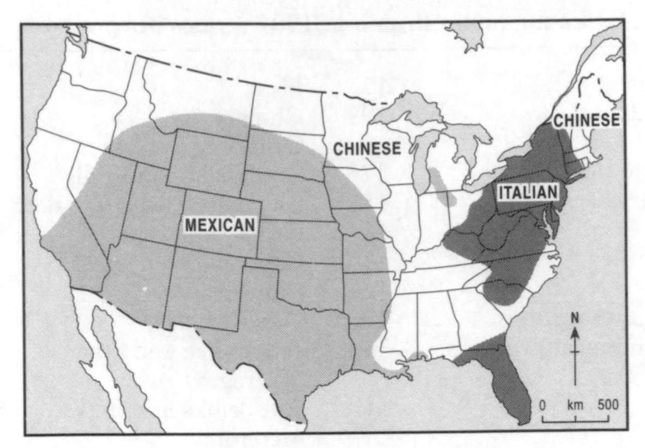

Fig. 21.6 Principal ethnic restaurants by region in the USA, 1980. *Source*: after Zelinsky (1987).

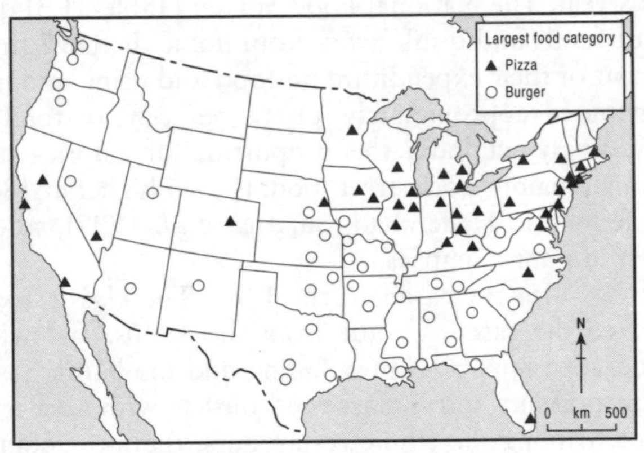

Fig. 21.7 The dominant fast food in American metropolitan areas. *Source*: Roark (1985).

preferences which may rule out anything more adventurous than a generic 'curry', but also the restaurant environment is constraining in terms of the specialized (often ethnic) food served or in the structure of a 'set menu'. Most Asian and Chinese restaurateurs source their raw materials at a limited number of wholesalers and many also work to set formulae in terms of menus and cooking methods. At the lower end of the market there is often little to choose between individual outlets even though the ownership may be different.

Finkelstein (1989) emphasizes the negative aspects of dining out. She sees the restaurant experience as limiting and disengaging from an aware sociality. Diners are manipulated by their surroundings into conventional,

Table 21.4 Satisfaction with eating in restaurants

	A lot	A little	Disliked	Other
How much did you enjoy the food?	80	13	3	3
How much did you enjoy the service?	65	22	6	7
How much did you enjoy the atmosphere?	72	18	3	7
How did the place rate as value for money?	67	17	7	10
How much did you enjoy the occasion overall?	81	14	2	4

Source: Martens and Warde (1999).

mannered behaviours and they become lazy and uncritical. Martens and Warde (1997, 1999) agree with this to a certain extent but their interviewees made it clear that the experience is generally enjoyable (Table 21.4), despite the restrictions. According to Martens and Warde, Finkelstein makes an error in treating the consumer's view as a kind of false consciousness and seeks to impose her own interpretation.

Health considerations

We can show that risk-minimization has played a part in food-selection from an early date, but British consumers have only recently had a sufficient cushion of economic prosperity to allow them a real choice in their diets and a wide range of foods available as a result of the retail revolution ushered in by the large supermarket chains. Thus, from the 1970s there has been mass-awareness of and participation in a health-food market. The messages have come from government bodies such as COMA (Chapter 16) and from manufacturers making claims for their foods, but the public has responded in several ways. White meat (poultry) consumption has increased at the expense of red meat (beef, veal, mutton, lamb); skimmed milk and margarine are replacing full-cream milk and butter. Breakfast cereals and cakes have switched places in their importance in cereal consumption. Other strong growth sectors that have health connexions are fresh fruit and vegetables and wholemeal bread. On the other hand, convenience foods such as frozen chips are also popular and some items recommended on health grounds, such as fresh and processed fish, have declined. Most worrying has been the steady increase in the last 50 years of the proportion of energy derived from fats, from about 35 to 42 per cent: this has implications for coronary heart disease (Chapter 16).

Successive governments have attempted to influence eating habits in an attempt to bring diet into line with contemporary scientific advice about nutrition and health. This was originally attempted through health education campaigns, of a somewhat didactic nature, and more recently by providing information about lifestyles and health. In reality the efforts of the British state, which invested £0.7 million for nutrition education in 1996–97, seem totally inadequate when compared with the substantial amounts spent annually on advertising by the food industry (Keane 1997).

The research that has been published on the reactions of consumers to the food/health messages in government pronouncements, supermarket leaflets, television and the printed media, and various other sources, suggests a substantial and significant amount of resistance to change. Keane (1997), for instance, found that healthy eating advice was treated with scepticism by many of her interviewees. Some, especially young men and the elderly of both sexes, felt that it was of no relevance for their particular circumstances, while others were either confused or irritated by the apparently contradictory messages and even by the motives of the information sources:

> 'I don't really trust doctors and nutritionists 'cos they tend to contradict themselves.'
>
> 'One day zinc is good for you, the next it's bad. It's a minefield. No one could possibly be expected to understand it all.'
>
> 'I tend not to believe any of it really. I think that food research is often financed by people who have an interest in a certain outcome' (Keane 1997, 181–87).

The media, advertising and the world of signs

Those of us who live in the wealthier countries have in recent years become used to a barrage of information about food disseminated through a variety of media. Some of this is educational, for instance the efforts of governments to influence the eating habits of their citizens; but the overwhelming majority is commercial advertising by food manufacturers and retailers.

An average American, it is estimated, sees 150,000 advertisements on television in her or his lifetime, and advertising is increasing worldwide, faster than population or incomes. The global advertising spend, by the most conservative reckoning, is now $435 billion annually (all goods), and its growth has been particularly rapid in developing countries. In the Republic of Korea it increased nearly threefold in 1986–96, for instance, and in the Philippines by 39 per cent a year in 1987–92. In 1986 there were only three developing countries among the 20 biggest spenders in advertising. A decade later there were nine. And in spending on advertising relative to income, Colombia ranks first with $1.4 billion, 2.6 per cent of its GDP.

Food advertising is carefully targeted. In the Norwegian print media, for instance, only 7 per cent of food-related adverts are to be found in men's publications, while 63 per cent are placed in women's magazines (Lien

1995). The messages of these adverts relate mostly to health (words such as natural, safe, light, healthy), women's role as food providers and carers (convenience, variety, children), and the imaginary dimension of nostalgia, quality and exoticism.

The available evidence suggests that food advertising does indeed affect people's diet. Helmer (1992) concluded that McDonald's 'won the burger wars' in 1980s America as a result of an advertising campaign which concentrated on familial images and so persuaded customers that their restaurants were 'a potential source of love and human happiness . . . a place for *being* a family'. By then the company had been successful in making the golden arches an icon of American culture, which it has since gone on to export to the rest of the world.

In a different vein, Chang and Kinnucan (1991) found that bad publicity about cholesterol has been a significant factor in the long term decline of butter consumption in Canada, but they also concluded that an advertising campaign by the Dairy Bureau of Canada reduced the rate of decline and gave butter an edge over margarine in the minds of some consumers in terms of flavour. One should not overestimate the effect of advertising, however: price is still the main consideration in people's food buying habits.

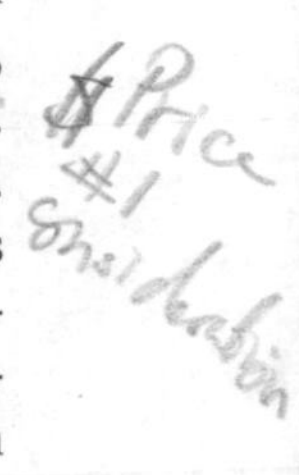

Food messages are received even when there is no advertising intent. British viewers of American TV sitcoms and police and hospital dramas have apparently started asking at local supermarkets for soda pop and blueberry muffins. These people are buying into a lifestyle and the manufacturers of brand leaders such Coca-Cola, and McDonald's, have benefited in general from the world-wide popularity of American culture.

The messenger is frequently blamed for the (food) message. The media are frequently vilified, particularly by politicians and those involved in the

Table 21.5 The top food commodities in UK advertising, 1998

Food	**£ million spent**
Ready-to-eat cereals	95.6
Chocolate	88.7
Fast food and chain restaurants	77.2
Tea and coffee	62.8
Potato crisps and snacks	36.9
Sugar confectionery	29.6
Cooking sauces	28.8
Ice creams and lollies	28.4
Cheese	27.6
Margarine and low-fat spreads	27.5
Ready-to-eat meals	27.1
Total	770.2

Source: Advertising Association (1999).

food industry, for irresponsible 'scaremongering' and exaggeration with regard to food issues, such as the health consequences of consuming eggs contaminated with salmonella or beef infected with BSE (Chapter 16). Although media campaigning on food is certainly not new, there do seem to have been more column inches devoted to food in general, both by investigative journalists and those concerned more with entertainment, style and leisure. The public seems to be ready for this, despite the scepticism noted above.

Miller and Reilly (1995) and Reilly and Miller (1997) have analysed the management of the egg and beef scare stories. Their conclusion is that the media played a vital role in the way both issues unfolded and were dealt with by the political machine and the civil service. The MAFF's original cloak of secrecy was forced aside but full information has probably not yet been revealed on either, despite an official Public Enquiry on BSE. Miller and Reilly see the media as an arena that is unevenly structured, with inequities of access to the various parties, and this helps to explain how stories evolve and why some rather than others become so important that policy may change.

One worrying aspect of food advertising is its persuasive impact on children to desire items that are unhealthy. British and American children aged 5–12 watch about three hours of television a day, which means that they see an average of 11,000 food-related advertisements a year, mostly for breakfast cereals, fast food restaurants and sweets. Consumers International (1996) has monitored these adverts in the UK and found that 62 per cent of the foods promoted were high in fats, 50 per cent were high in sugar, and 61 per cent high in salt. Research has demonstrated that televised food messages do significantly influence children's immediate dietary choice and there is concern that they may also alter the patterns of a lifetime. Some countries have therefore introduced controls, such as Austria where no advertising of any kind is allowed during children's programmes, and the UK where there are several restrictions (Table 21.6).

Table 21.6 UK regulations concerning food advertising to children

- Slimming products may not be aimed at people under 18.
- Children must not be encouraged to eat frequently throughout the day.
- Consumption of food or drink (especially sweet, sticky food) near bedtime must not be promoted.
- Advertiser cannot suggest that confectionery or snack foods can be substituted for balanced meals.
- Advertiser must not discourage eating of fruit and vegetables.
- Regard should be paid to oral hygiene.
- Health claims must be based on sound scientific evidence.

Source: Consumers International (1996).

Conclusion

Changes to taste in the 1990s have, in the opinion of James (1996), four discursive dimensions. First, she finds the globalization process to have been significant both economically, through rapidly expanding world trade in branded goods such as Coca Cola, and symbolically through the spread of McDonald's to Moscow and Beijing. Second, there is 'expatriate food' mediated through literature and other media as the consumption of 'the other'. Third, she identifies a discourse of 'food nostalgia' in which distinctive, traditional and supposedly authentic products and recipes are being rediscovered as part of the national heritage. This is a process of localization, a counterpoint to the homogenization of global diets that seems to be an aim of corporate capital. Finally, there is 'food creolization', the mixing of foods in a mélange of tastes that suits the post-modern palate.

The efforts of James, and other writers in this area, have demonstrated the complexity of the origins and evolution of taste. But our conclusion is that much more research remains to be done before we can claim to have a satisfactory theoretical depth to the literature.

Further reading and references

There is no single textbook devoted to all of the material in this chapter, but the following are helpful for further reference: Lloyd; Lockwood and Lockwood; Manzo; and Zelinsky. Pierre Bourdieu's work has been influential, as have the publications of Fischler.

Advertising Association 1999: *Advertising statistics yearbook*. 17th edn. Henley-on-Thames: NTC Publications.

Allen, D.E. 1968: *British tastes: an enquiry into the likes and dislikes of the regional consumer*. London: Hutchinson.

Anderson, E.N. 1988: *The food of China*. New Haven: Yale University Press.

Appadurai, A. 1988: How to make a national cuisine: cookbooks in contemporary India. *Comparative Studies in Society and History* 30, 3–24.

Atkins, P.J., Simmons, I.G. and Roberts, B.K. 1998: *People, land and time: an historical introduction to the relations between landscape, culture and environment*. London: Edward Arnold.

Beardsworth, A. 1995: The management of food ambivalence: erosion or reconstruction. In Maurer, D. & Sobal, J. (eds) *Eating agendas: food and nutrition as social problems*. New York: Aldine de Gruyter, 117–42.

Bourdieu, P. 1984: *Distinction: a social critique of the judgement of taste*. London: Routledge and Kegan Paul.

Chang, H.-S. and Kinnucan, H.W. 1991: Advertising, information, and product quality: the case of butter. *American Journal of Agricultural Economics* 39, 1195–1203.

Consumers International 1996: *A spoonful of sugar. Television advertising aimed at children: an international comparative survey*. London: Consumers International.

Cook, I. 1994: New fruits and vanity: symbolic production in the global food economy. In Bonnano, A., Busch, L., Friedland, W.H., Gouveia, L. and Mingione, E. (eds) *From Columbus to ConAgra: the globalization of agriculture and food*. Lawrence, Kansas: University Press of Kansas, 232–48.

Cook, I. and Crang, P. 1996: The world on a plate: culinary culture, displacement and geographical knowledges. *Journal of Material Culture* **1**, 131–54.

De Wit, C.W. 1992: Food-place associations on American product labels. *Geographical Review* **83**, 323–30.

Diamond, R. 1995: Become spoiled Moroccan royalty for an evening: the allure of ethnic eateries, *Bad Subjects* **19**, http://eserver.org/bs/19/Diamond.html.

Finkelstein, J. 1989: *Dining out: a sociology of modern manners*. Cambridge: Polity Press.

Fischler, C. 1980: Food habits, social change and the nature/culture dilemma. *Social Science Information* **19**, 937–53.

Fischler, C. 1988: Food, self and identity. *Social Science Information* **27**, 275–92.

Gabaccia, D.R. 1998: *We are what we eat: ethnic food and the making of Americans*. Cambridge: Harvard University Press.

Goode, J.G., Curtis, K. and Theophano, J. 1984: Meal formats, meal cycles, and menu negotiation in the maintenance of an Italian–American community. In Douglas, M. (ed.) *Food in the social order: studies of food and festivities in three American communities*. New York: Russell Sage Foundation, 143–218.

Hartog, A. den 1986: *Diffusion of milk as a new food to tropical regions: the example of Indonesia, 1880–1942*. Wageningen: Stichting Voeding Nederland.

Helmer, J. 1992: Love on a bun: how McDonald's won the burger wars. *Journal of Popular Culture* **26**, 85–96.

Hladik, C.M. 1997: Primate models for taste and food preferences. In Macbeth, H. (ed.) *Food preferences and taste: continuity and change*. Providence: Berghahn Books, 15–26.

James, A. 1996: Cooking the books: global or local identities in contemporary British food cultures. In Howes, D. (ed.) *Cross-cultural consumption: global markets, local realities*. London: Routledge, 77–92.

Keane, A. 1997: Too hard to swallow? The palability of healthy eating advice. In Caplan, P. (ed.) *Food, health and identity*. London: Routledge, 172–92.

Kelly, J.L. 1983: Loco Moco: a folk dish in the making. *Social Process in Hawaii* **30**, 59–64.

Lien, M. 1995: Fuel for the body, nourishment for the dreams: contradictory roles of food in contemporary Norwegian food advertising. *Journal of Consumer Policy* **18**, 157–86.

Lloyd, T.C. 1981: The Cincinnati chili culinary complex. *Western Folklore* **40**, 28–40.

Lockwood, Y.R. and Lockwood, W.G. 1991: Pasties in Michigan's Upper Peninsula: foodways, interethnic relations, and regionalism. In Stern, S. and Cicala, J.A. (eds) *Creative ethnicity: symbols and strategies of contemporary ethnic life*. Logan: Utah State University Press, 3–30.

Martens, L. and Warde, A. 1997: Urban pleasure? On the meaning of eating out in a northern city. In Caplan, P. (ed.) *Food, health and identity*. London: Routledge, 131–50.

Martens, L. and Warde, A. 1999: Power and resistance around the dinner table. In Hearn, J. and Roseneil, S. (eds) *Consuming cultures: power and resistance*. Basingstoke: Macmillan, 91–108.

May, J. 1996: A little taste of something more exotic: the imaginative geographies of everyday life. *Geography* **81**, 57–64.

Mennell, S. 1985: *All manners of food: eating and taste in England and France from the Middles Ages to the present*. Oxford: Blackwell.

Miller, D. and Reilly, J. 1995: Making an issue of food safety: the media, pressure groups, and the public sphere. In Maurer, D. and Sobal, J. (eds) *Eating agendas: food and nutrition as social problems*. New York: Aldine de Gruyter, 305–36.
Ministry of Agriculture, Fisheries and Food 1998: *National Food Survey, 1997*. London: The Stationery Office.
Mintz, S. 1996: *Tasting Food, tasting freedom*. Boston: Beacon Press.
Moran, W. 1993: The wine appellation as territory in France and California. *Annals of the Association of American Geographers* **83**, 694–717.
Ohnuki-Tierney, E. 1993: *Rice as self: Japanese identities through time*. Princeton: Princeton University Press.
Orwell, G. 1937: *The road to Wigan Pier*. London: Gollancz.
Paige Gutierrez, C. 1984: The social and symbolic uses of ethnic/regional food ways; cajuns and crawfish in south Louisiana. In Brown, L.K. and Mussell, K. (eds) *Ethnic and regional food ways in the United States*. Knoxville: University of Tennessee Press.
Peltre, J. and Thouvenot, C. (eds) 1989: *Alimentation et regions*. Nancy: Presses Universitaires de Nancy.
Reilly, J. and Miller, D. 1997: Scaremonger or scapegoat? The role of the media in the emergence of food as a social issue. In Caplan, P. (ed.) *Food, health and identity*. London: Routledge, 234–51.
Roark, M.O. (1985) Fast foods: American food regions, *North American Culture* **2**, 24–36.
Rozin, P. 1976: The selection of food by rats, humans and other animals. In Rosenblatt, J.S., Hinde, R.A., Shaw, E. and Beer, C. (eds): *Advances in the study of behaviour*. London: Academic Press.
Simoons, F.J. 1991: *Food in China: a cultural and historical enquiry*. Boca Raton: CRC.
Sujatha, T., Shatrugna, V., Narasimha Rao, G.V., Krishna Reddy, C.K., Padmavathi, K.S. and Vidyasagar, P. 1997: Street food: an important source of energy for the urban worker. *Food and Nutrition Bulletin* **18**, 318–22.
Thouvenot, C. 1978: Studies in food geography in France. *Social Science and Medicine* **12**, 43–54.
Warde, A. 1997: *Consumption, food and taste: culinary antimonies and commodity culture*. London: Sage.
Warde, A. and Martens, L. 1998a: A sociological approach to food choice: the case of eating out. In Murcott, A. (ed.) *'The Nation's Diet': the social science of food choice*. London: Longman, 129–44.
Warde, A. and Martens, L. 1998b: Eating out and the commercialisation of mental life. *British Food Journal* **100**, 147–53.
Warde, A., Martens, L. and Olsen, W. 1999: Consumption and the problem of variety: cultural omnivorousness, social distinction and dining out. *Sociology* **33**, 105–27.
Zelinsky, W. 1987: You are where you eat. *American Demographics* **9** (7), 30–33.

22

Food habits, beliefs and taboos

> *A meal is, above all, a phase in a much- but not all-embracing process, a node in a network, not exhausting but still touching or embracing the totality of all existence. Otnes (1991).*

Introduction

Food habits and foodways concern us in this chapter. Some have the curiosity value of the culturally specific, but others are indicative of broader meanings in society and perhaps even of biologically determined processes of thought.

CFL or M+S+2V?: the patterns in meals

Food habits are among the most deeply ingrained forms of human behaviour. Sharing food in the sociable setting of the meal is not only enjoyable but also has fundamental symbolism for the solidarity of families and the reproduction of relationships. Douglas (1972) noted a symbolic threshold between food and drinks:

> Drinks are for strangers, acquaintances, workmen, and family. Meals are for family, close friends, honoured guests . . . Those we only know at drinks we know less intimately.

Table manners, the ritualized forms of behaviour that regulate these occasions, are usually first inculcated in childhood at the family table. The Victorians decreed that children were to be 'seen but not heard' at dinner, and for them 'every meal is a lesson learned' (Visser 1991, 48).

There are also wider social norms that come under the heading of taken-for-granted practices. None of Warde and Martens' (1998, 131) interviewees, for instance, ate roast beef with custard, because of 'internalized constraints which effectively prohibit people from such courses of action on pain of embarrassment or being considered socially deplorable or grossly incompetent'. Nor, one supposes, did they engage in cannibalism, because of an even deeper-rooted taboo that predates organized religion and which now embraces most of the world's cultures.

Food is the focus of many of our festivals, both the regularly occurring events such as those in the religious calendar and occasional events such as weddings and funerals. Special foods may be served, often rich in ingredients, complex in preparation, and usually generous in quantity. The occasion may only be annual, as with the roast turkey, mince pies, cake and pudding of the British celebration of Christmas, but the significance is well-known to all, including those outposts of British culture where the feast may be consumed on a hot beach rather than in the depths of winter.

Even more frequent than 'meals' are the so-called 'food events' or snacks that are becoming an essential part of modern life. Teenage grazing is a popular stereotype but it seems to have some substance in reality if Table 22.1 is to be believed. Here we see the establishment of a pattern of adolescent eating where only 58.4 per cent of an American sample were eating three 'square meals' a day. One interesting finding of recent research is that consumers of snack or 'junk' food are well aware of the health consequences of eating items high in fat, sugar and salt, but they find the pull of the associated lifestyle irresistible, especially the teenagers who can thereby demonstrate their peer group loyalty and their independence of parental influence.

Time-honoured food habits and meal patterns are said to be dissolving under the pressure of modern life. On the one hand, time has become such a constraint for busy people that the leisured cooking of complex recipes is now less of an option than it was only a few decades ago, and meals of a predictable composition served at set times have also declined. On the other hand, there have been powerful shifts in family structure, with the growth

Table 22.1 The pattern of meal intake in a sample of American adolescents (1989–91)

Daily pattern of meals	Percentage
Three meals, ± snack(s)	58.4
Breakfast, lunch, ± snack(s)	5.5
Lunch, dinner, ± snacks(s)	13.4
Breakfast, dinner, ± snack(s)	14.4
One meal and snack(s)	5.4
One meal or snack(s)	2.9

Source: Siega-Riz *et al.* (1998).

Table 22.2 Gastro-anomy

- Self-sufficient producers operating in a constrained but certain world have been replaced by 'consumers'.
- The traditional rhythm of eating is changing with a shift from set meals to 'snacking' or 'grazing'.
- Eating, formerly a social practice, in family groups, is now becoming individualized.
- Choice of food is no longer constrained by ecology/season.
- Cookery has lost its role as the means of transforming potentially dangerous raw foods into 'civilized' meals because much food is now purchased in processed form.
- The technology of food processing is making it increasingly difficult to identify the ingredients. Required labelling is one reaction.
- Familiar, local ingredients and recipes are challenged by imported goods and ideas that come from an increasingly globalized food culture.
- Eating out is an increasingly attractive and affordable alternative to self-provisioning.
- Consumers worry in general terms about food-related risks such as those from mad cow disease or genetically-modified organisms.
- The emergence of food sectarianism: vegetarian, vegan, macrobiotic.
- Advice about food choice is now common, from governments and the media. This may amount to 'dietary tyranny'.

Source: Modified after Fischler (1980, 1988).

in single person households undermining traditional collective meals. For some this is a liberating experience but Fischler (1980) has described this new era as one of 'gastro-anomy', echoing Durkheim's concept of 'anomie' (normlessness) (Table 22.2). Dickinson and Leader (1998) report one survey that found that two-thirds of evening meals in Britain are now consumed in front of the television. It seems that both the family and the family meal are under threat.

However, most food cultures still recognize the continued importance of the group sharing of food. Murcott (1997) reports evidence from the UK and Germany that family meals remain the norm. The idea of a 'proper' meal is particularly strong. This is a reflection of core values and beliefs, some of which may have religious overtones. The components of a meal are usually a core item of animal protein (meat or fish), a staple of plant origin (potato, rice, maize), vegetable trimmings, and a dressing of sauce or gravy, but there are almost infinite variations on this theme. Mintz (1996) identifies a primitive grammar of the meal that he calls CFL, comprised of a core item (such a rice), a fringe item (such as a sauce), and a legume. With industrialization this has changed to M+S+2V, meat plus a staple (such as potatoes), plus two vegetables.

Mary Douglas and her collaborators codified the structure of British working class meals (Table 22.3) and found that certain strong themes have survived. First they classified four forms of eating, and then three types of

Table 22.3 A codification of British working class meals

Different forms of eating

1 Meal: a structured event, a social occasion when food is eaten according to rules concerning time, place and sequence of actions.
2 Snack: unstructured food event without rules of combination and sequence.

Types of meal

1 A major meal or the main meal. The first course is based on a staple (potato), a centre (meat, fish or egg), vegetables and dressing (e.g. gravy). The second course is sweet.
2 A minor or secondary meal, e.g. breakfast.
3 Less significant meal, e.g. tea and biscuits.

Source: Douglas (1972).

Table 22.4 A classification of food and meal types

(a) Foods

Category	Foods	Nutrients
a	Meat, fish, poultry, egg, milk, cheese	Animal protein and fat, iron, zinc, calcium
b	Rice, pasta, bread, dried legumes, seeds, potatoes	Starch, plant protein, dietary fibre
c	Green vegetables, fruit, berries, roots	Starch, carotenoids, ascorbic acid
d	Nuts, olives, avocado	Plant fat, plant protein
e	Cooking fat, spreads, cream, fatty sauces	Fat
f	Sugar, alcohol, ice cream, sweets, chocolate, biscuits, sweet desserts	Sugar, fat, alcohol
g	Water, coffee, tea, unsweetened light beverages	No nutrients

(b) Meals

Categories	Types of meals/snacks	Examples
a+b+c	Complete meal	Meat, potatoes or bread, carrots
a+b	Incomplete meal	Meat, potatoes or bread
a+c	Less balanced meal	Meat, carrots
b+c	Vegetarian meal	Potatoes or bread, carrots
a or b or c	High quality snack	An apple
any of a or b or c and/or d, and/or e and/or f	Mixed quality snack	An apple and some chocolate
e and/or f	Low quality snack	Chocolate
g	No energy snack	Diet coke

Source: Lennernäs and Andersson (1999).

Table 22.5 Features of a 'proper meal' in South Wales

- Meat, such as beef, lamb, pork or chicken
- Cooked by roasting or grilling
- Potatoes
- A green vegetable
- Plated, each component in a separate pile
- Gravy
- Substantial portions
- Eaten with knife and fork

Source: Murcott 1982.

meal were identified. This structural approach has been influential and has been widely applied (Table 22.4). In South Wales, Murcott (1982, 1983) found that the ideal cooked dinner comprised meat, potatoes and vegetables, with some elaborations at the weekend and at Christmas (Table 22.5). The housewife, however, rarely has autonomy in the choice of ingredients, her role as 'gatekeeper' being constrained by patriarchal authority and by the 'choosiness' of children.

Getting married seems to be a key threshold in establishing patterns of consumption, with the adolescent junk food phase being left behind. Thereafter food is used as a means of establishing and reaffirming a family's identity and of drawing external boundaries. The tastes of single males and females are domesticated into a new joint diet, although Table 22.6 indicates that differences remain over individual items. Food can also be a means of articulating linkages with the outside world.

Table 22.6 Male, female and partner foods in a 1985 Australian survey: percentages of respondents consuming items at least once a week

Food item	Partnered men	Single men	Partnered women	Single women
Beer/cider	61	67	18	17
Peas	90	84	87	69
Hot chips	73	61	62	34
Low fat yoghurt	9	15	28	27
Pears	61	48	67	64
Crispbread	23	15	43	36
Mince meat	75	51	77	52
Cheese slices	45	24	44	23
Pasta	49	33	56	31
Grapefruit	18	16	25	34

Source: Worsley (1988).

Food choice and food preferences

Neophobia is a dislike of the new, and in the case of food this may be manifested in its taste, odour and appearance. Young children are especially prone to the rejection of food for this reason but adults may also refuse novel foods or dishes that seem to lie beyond the limits of their

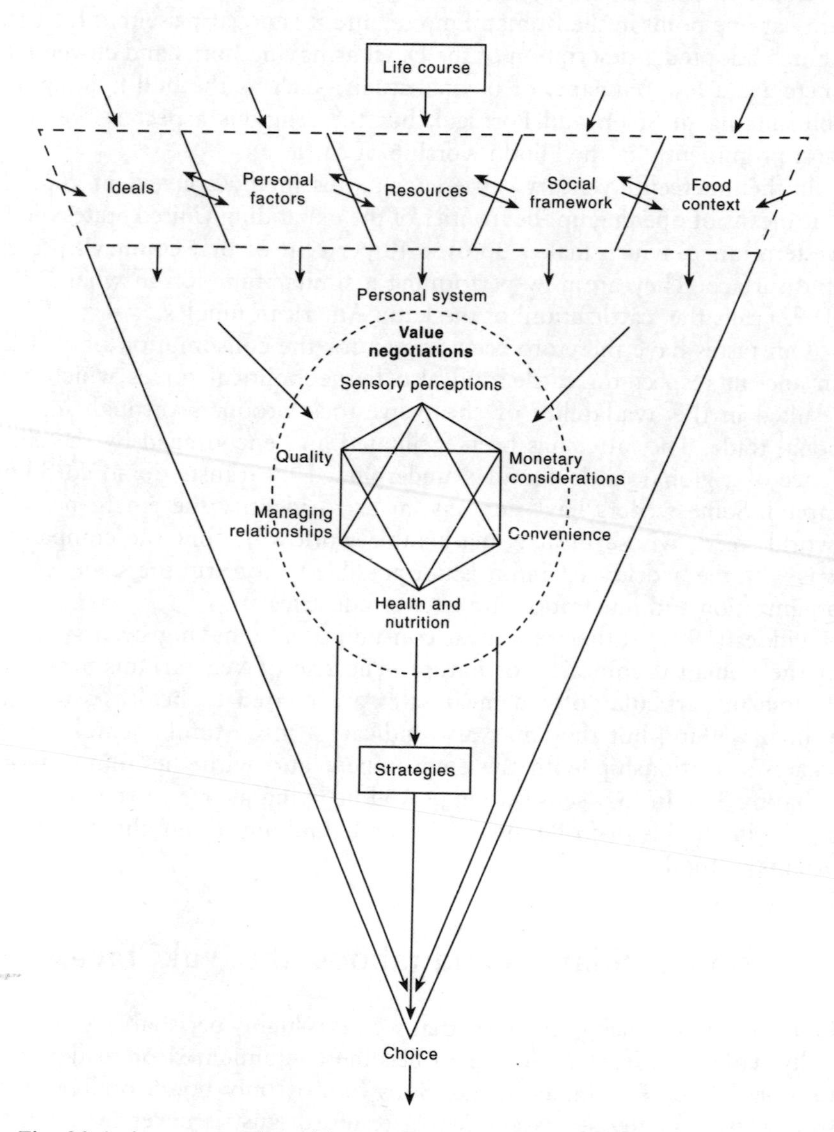

Fig. 22.1 A conceptual model of the process of food choice. *Source*: after Furst *et al.* (1996).

socially-constructed taste. The social context of such tastes is extremely complex and we have evidence that it may sometimes have historical origins which go back for several thousand years. An example is our attitude to cattle, beef and milk.

There was a deeply ingrained religious reverence for cattle in many neolithic cultures, from the Mediterranean to India, no doubt arising from their status as one of the first forms of moveable wealth (Harris 1987). Bull-reverence in the cult of Mithras actually challenged Christianity for popularity at one point in the Roman Empire, and it is no surprise that the early Church adopted a description of the Devil as having horns and cloven feet. There are a few remnants of bovine rituals, such as the bull-fighting and bull-running in Spain and Portugal, but the religious aspect is preserved most prominently in the Hindu worship of cattle.

In the nineteenth century cattle later represented a convenient and reliable means of opening up the frontier of the expanding United States on the western range, which makes up over 40 per cent of that country's present land surface. They are now performing a similar function in what Rifkin (1992) calls the 'cattlization' of the Latin American jungles.

Our tastes have therefore been primed for the consumption of meat by an ancient respect for cattle, and also by geopolitical forces which have resulted in the availability of cheap livestock products through international trade. The latter has been facilitated and encouraged by the emergence of a global cattle complex underpinned by transnational corporate capital. Some writers have gone as far as to identify the emergence of a 'world steer', whose meat is marketable worldwide, but the comparison was with the 'world car' and it is not possible to confirm the same level of organization and integration in the two industries.

Fiddes (1991) rather sees meat consumption as having been symbolic of the human domination of nature. The rise of vegetarianism and the decline in particular of red meat sales are related to health issues and animal welfare, but they may also indicate a more fundamental shift in society's relationship with the environment and with the animal world (Chapter 18). In that sense we may well be living at one of the key hinge points in the history of human diet and thinking about the acceptable origins of food.

Food avoidances and taboos: the 'yuk' factor

In its raw state, much of the food that we eat is highly perishable and potentially dangerous if it is allowed to become contaminated or to decay. As Lupton (1996, 3) comments, 'delicious food is only hours or days away from rotting matter, or excreta. As a result, disgust is never far from the pleasures of food and eating'.

Rozin *et al.* (1997) suggest that such revulsion may be classified into 'core disgust' (very bitter tastes, faeces, certain animals and insects), and 'animal nature disgust' (poor hygiene, gore, contact with death). But they also find that there are remarkable degrees of cultural variations in disgust responses. This is because our behaviour is affected by our perception of the polluting power of 'unclean' foods, which may originate from a religious taboo, as with pork for Muslims and Jews, or from a disgust generated by custom. Many Britons may abhor the notion of eating horseflesh, snails or dogmeat, but some of their traditional foods such as black pudding (dried blood), tripe (cow's stomach) and cheese veined with blue mould are equally nauseating for others.

Cannibalism is of course the ultimate taboo. The popular belief seems to be that it was once a widespread practice that has yielded under the press of civilization. Many of us have colourful mental images of human sacrifice by the ancient societies such as the Aztecs, in which blood or flesh is consumed, either as a celebration of the defeat of enemies and the absorption of their strength (exo-cannibalism) or as a means of honouring relatives and giving them new life (endo-cannibalism). But authors such as Arens (1979) have controversially sought to disappoint us by claiming that cannibalism has in reality always been rare. For Arens it is little more than a myth, one that incidentally has served the disciplines of cultural anthropology and archaeology well in their search for a topic of popular interest (but *see* Rawson 1999).

Simoons (1994) has written the classic text on food avoidances. Through close scholarship of texts, archaeological and ethnographic evidence, he reconstructs the spatial extent of taboos on the consumption of pork, beef, chicken and eggs, horseflesh, camel, dog and fish. Perhaps his most interesting conclusion (Simoons 1970, 1978) is that the correlation between the traditional areas of dairying and lactose tolerance is the best example we

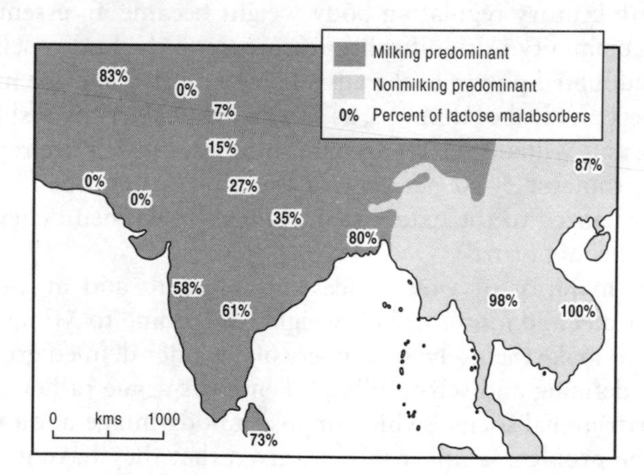

Fig. 22.2 Lactose intolerance and the milking and non-milking regions of Asia. *Source*: Simoons (1981).

have of a long-standing economic/cultural practice influencing human genetics. The non-milking areas of the old world (Figure 22.2) coincide with a reduction after infancy of lactase, the enzyme in the body that helps us to absorb milk sugar (lactose). This suggests that pastoralists elsewhere gradually became adapted genetically to the nutritional bounty that their female livestock supplied.

Harris (1986, 1987) also looks at food taboos. He analyses the Hindu prohibition on eating beef and concludes that it is not as paradoxical as it might first appear. The outsider may wonder how it is that a sub-continent, which for hundreds of years faced occasional famines and chronic hunger amongst its many poor people, could abstain from using such an abundant protein resource, but Harris reasons that Indian cattle are functional in providing the motive power for ploughing. There is no taboo on consuming their milk, so there is no blanket aversion to cattle products, and even scruples about slaughter are overcome when a blind eye is turned to sales of cattle to Muslim and Christian dealers. His overall thesis (Harris and Ross 1987) is that human food habits serve an evolutionary purpose and that they are best understood in that light.

Food and the body

From Foucault to the postmodernists, there has been much writing in recent years on the body as a locus of social discipline, as an arena of risk and health, as raw material for the politics of agency/self, and as a focus of identity creation. We do not have the space here to cover any of these topics in depth, but let us consider one interesting aspect: dieting and eating disorders in the creation of body image.

Although more men are obese, it is mostly women and girls who diet. In the twentieth century regulating body weight became an essential element of modern femininity. A significant pressure upon the bodily self-image has been the undoubted power of the advertising industry and the media generally to project an ideal of thinness, as shown particularly by fashion models. As Germov and Williams (1996) remind us, these models are representative only of the thinnest 5–10 per cent of the population, but their look has become normalized to the extent that millions of women believe that they themselves are 'abnormal'.

The vast majority of women feel this pressure and at various times restrict their diet and monitor their weight. According to Mintz (1996, 80), they do so 'to make themselves members of a gender-defined group of abnegators, self-defining and self-fulfilling'. Feminists argue rather that it is the values of patriarchal society which impose a body image upon women and therefore the problem is *not* insoluble, given that they have greater power over their own lives than in the past.

According to the Body Mass Index (weight in kilograms divided by height in metres squared), a desirable status is 20–25, overweight 25–30, and obesity >30. Recent data suggest that obesity in the UK has risen to 20 per cent of women and 17 per cent of men in 1999, compared with 36 and 31 per cent in the USA. In the latter country 52 per cent of women have a self-image of being overweight, and 64 per cent of those with a BMI of below 25 have dieted at some point even though they are demonstrably not overweight according to medical criteria.

The main beneficiaries of what Germov and Williams (1996) call this 'epidemic of dieting women' are the companies which annually sell $30 billion worth of diet foods in America alone. It seems that about 90 per cent of diets fail, so the benefits are minimal, and there are medically-related dangers of increased anxiety, lowered metabolic rate, and the potential for malnutrition.

It is arguable that binge eating (bulimia nervosa) and self-starvation (anorexia nervosa) (Table 22.7) are merely extreme extensions of this behaviour rather than the mental illness diagnosed by some psychiatrists. If so, then dieting is one of the best examples of how one portion of society (girls and women) can be strongly socialized to a norm of behaviour which seemingly has little impact upon another group (boys and men). The psychiatric argument does hold some water, however, because many sufferers share particular character traits, such as low self-esteem and/or perfectionism, and many are treatable by anti-depressant medication.

Recent medical research has suggested a genetic origin for the eating disorders of some people but there is a large literature which prefers a sociological explanation. Mennell, Murcott and van Otterloo (1992), for instance, argue that the social status of the wealthy in the medieval period was affirmed at extravagant feasts where gluttony was encouraged. Such conspicuous

Table 22.7 Diagnostic criteria of the two main eating disorders

Bulimia nervosa	**Anorexia nervosa**
• Recurrent episodes of binge eating.	• Refusal to maintain body weight at minimally normal level.
• Rapid consumption of large amounts of food in a short period of time.	• Intense fear of gaining weight or becoming fat, even though underweight.
• A sense of lack of control of eating during the episode.	• Undue influence of body weight or shape on self-evaluation.
• Compensatory behaviour to prevent weight gain: vomiting, misuse of laxatives and diuretics.	• Amenorrhoea.
• Episodes occur at least twice a week for three months.	• At times may involve binge eating and purging.
• Self-evaluation unduly influenced by body shape and weight.	• Middle class girls/women.

Source: modified from Stunkard (1997).

consumption was only possible for the few because food shortages were the norm for the majority. In the modern period famine and undernutrition have become so rare in developed countries that the establishment of distinctiveness in a well-fed population has switched to slimming. The supposed growth in the twentieth century of eating disorders such as anorexia and bulimia seems to fit this trend. The logical outcome of this argument would be the eventual decline of eating disorders, because their present rapid rate of growth will ultimately make them common and therefore the distinctiveness gained from being ultra thin will be lost.

Fine (1998) prefers to explain eating disorders within a framework of political economy. He sees anorexia, bulimia and obesity in the same light, as outcomes of the tension between our physical compulsion to eat and our mental control over our weight. Because over-eating is a high risk in societies where there is no shortage of food, it is not at all surprising that the body should have become a battle ground between the different forces in society. Female beauty was commodified during the twentieth century and, with profitability at stake, the fashion and dieting industries, along with the media in all of its guises, have stressed the 'value' of thinness. This 'message' has been received most attentively by those well-educated and relatively prosperous young women, for whom self-control with regard to food is a means of demonstrating their independence and personal identity. Working class women are more prone to obesity and this is also a reaction to the message, using food as a compensatory indulgence for disappointments about missing the impossible targets set by society.

If this political economy model is viable, then two consequences flow from it. First, the fashion and beauty industries are now beginning to extend their influence to young men, until the 1980s a largely untapped market. One might expect, therefore, the number of males with eating disorders to increase in future. Second, the explanation is one that operates at the macro-structural scale and not at the level of the individual. The psychiatric analysis of individual personalities is downgraded in this schema because it is a fundamental assumption of the model that eating disorders will not disappear until there is a shift in social norms and their relationship to the wider economy.

Bell and Valentine (1997) adopt yet another approach. They prefer to analyse body weight issues in terms of the post-structuralist literature, where the body is seen as the site of competing cultural and social processes. For instance, they cite Foucault as one theorist of relevance, particularly for his insights into social discipline and surveillance. Women's bodies are disciplined for their appearance by society's gaze but fat men are also subjected to prejudice, for instance in the workplace where some corporations filter out obese people at the interview stage in order to maintain a positive image in the minds of their clients.

Consumer knowledge

The information available to consumers, and their ability and willingness to seek it out and absorb it, affects their purchasing habits. The top priority in the minds of most of us is information about the prices of particular foods and their variation between shops. This is probably followed by knowledge about their palatability and healthiness, and a minority would also want to know whether a product comes from a politically correct source in terms of fair trade, employment practices or the treatment of animals. Since few people have the time to research the market fully, we tend to rely upon information provided by our previous experience, by other consumers, and also by retailers, the media and by government. Consumer knowledge in practice is always imperfect due to practical constraints, but we also display degrees of resistance to the receipt of additional information due to scepticism, custom, conservatism, and even prejudice.

One aspect of food information is the labelling provided by manufacturers and retailers. Some of this is required by law (Chapter 16) and some is added for public relations purposes or as a marketing ploy. In recent years there has been a growth of health claims made on labels associated with the so-called 'functional foods'. Some countries regulate labels carefully but the international picture is confused and one suspects that misleading claims are now so rife that they amount to 'disinformation'. A recent example has been consumer disquiet in Europe about genetically modified foods (Chapter 17) and the failure of manufacturers to declare GM soya as an ingredient in many products. Eventually global standards will be required but these seem a long way off at present.

Cook, Crang and Thorpe (1998) have found that consumers have wide-ranging knowledges about the geographical origins of foods. They argue that foods have 'biographies' which are traceable and they propose an ethnographic methodology for collecting information on how consumer knowledge is constructed.

Knowledge of consumers

One of the characteristics of capitalism is its flexibility in the face of changing circumstances. It seems to have an ability to learn and adapt to shifts in the market and to manipulate demand where possible. Information Technology has been an important aspect of this increased fine-tuning of the food industry. Barcode scanning at checkouts in supermarkets, for instance, allows the identification of broad trends specific to individual stores and also gives insights into the spending behaviour of customers with loyalty cards. Scanning has also been used in the home by market research companies who compile representative panels of households and monitor their purchasing/consumption behaviour over periods of time. Also important has been the analysis of small area census statistics and the compilation of data profiles for postcode districts, both of which are possible using GIS. These enable marketing strategies to be carefully targeted at households that have particular features in terms of income, age and sex.

Lifestyle is more difficult to investigate and here food companies employ surveyors to sample public attitudes. The results have suggested categories which to an extent cross-cut the classical sociological variables, but which are nevertheless discriminatory in terms of buying behaviour. Senauer *et al.* (1991) illustrate this approach with the example of the US company, Pillsbury. In the late 1980s they identified five groups of customers for their products. The 'Chase and Grabbits' (26 per cent) are young, upwardly mobile people without children, who prefer convenience food and who are willing to try new foods. The 'Functional Feeders' (18 per cent) are working class consumers who prefer plain meals, and the 'Down Home Stokers' (21 per cent) eat traditional and ethnic foods. The 'Careful Cooks' (20 per cent) are older but well educated, and they insist on healthy, nutritious food. Finally, the 'Happy Cookers' (15 per cent) have at least one family member who enjoys cooking and is willing to experiment.

Conclusion

Anthropologists and sociologists have taught us a great deal about food habits and beliefs. These are far from just empirical observations, because there is now a large literature of generalizations and theoretical interpretations. One important aspect that we have not dealt with here is the role that food potentially can take in the power play of social relationships. We will touch on this in the domestic setting in Chapter 23, and Valentine and Longstaff (1998) deal with it in the context of prison. They argue that food is employed by the warders as a tool of reward or punishment and by the prisoners both as a currency amongst themselves and as a means of resistance against the institution. In other words, food habits may not just be cultural curiosities but may be either short-term or long-term means of deploying power in society.

Further reading and references

Much of the literature for this chapter lies in anthropology and sociology. Beardsworth and Keil provide a helpful introduction, along with Germov and Williams (1999), and Mennell *et al.* (1992).

Arens, W. 1979: *The man-eating myth: anthropology and anthropophagy*. Oxford: Oxford University Press.

Beardsworth, A. and Keil, T. 1997: *Sociology on the menu: an invitation to the study of food and society*. London: Routledge.
Bell, D. and Valentine, G. 1997: *Consuming geographies: we are where we eat*. London: Routledge.
Cook, I., Crang, P. and Thorpe, M. 1998: Biographies and geographies: consumer understandings of the origins of foods. *British Food Journal* **100**, 162–67.
Dickinson, R. and Leader, S. 1998: Ask the family. In Griffiths, S. and Wallace, J. (eds): *Consuming passions: food in the age of anxiety*. Manchester: Mandolin, 122–29.
Douglas, M. 1972: Deciphering a meal. *Daedalus* **101**, 61–81.
Fiddes, N. 1991: *Meat: a natural symbol*. London: Routledge.
Fine, B. 1998: *The political economy of diet, health and food policy*. London: Routledge.
Fischler, C. 1980: Food habits, social change and the nature/culture dilemma. *Social Science Information* **19**, 937–53.
Fischler, C. 1988: Food, self and identity. *Social Science Information* **27**, 275–92.
Furst, T., Connors, M., Bisogni, C.A., Sobal, J. and Falk, L.W. 1996: Food choice: a conceptual model of the process. *Appetite* **26**, 247–66.
Germov, J. and Williams, K. 1996: The epidemic of dieting women: the need for a sociological approach to food and nutrition. *Appetite* **27**, 97–108.
Germov, J. and Williams, L. (eds) 1999 *A sociology of food and nutrition: the social appetite*. Melbourne: Oxford University Press.
Harris, M. 1986: *Good to eat: riddles of food and culture*. London: Allen & Unwin.
Harris, M. 1987: *The sacred cow and the abominable pig*. New York: Touchstone/Simon & Schuster.
Harris, M. and Ross, E.B. 1987: *Food and evolution: toward a theory of human food habits*. Philadelphia: Temple University Press.
Lennernäs, M. and Andersson, I. 1999: Food-based classification of eating episodes (FBCE). *Appetite* **32**, 53–65.
Lupton, D. 1996: *Food, the body and the self*. London: Sage.
Mennell, S., Murcott, A. and van Otterloo, A.H. 1992: *The sociology of food: eating, diet and culture*. London: Sage.
Mintz, S.W. 1996: *Tasting food, tasting freedom*. Boston: Beacon Press.
Murcott, A. 1982: On the social significance of the 'cooked dinner' in South Wales. *Social Science Information* **21**, 677–95.
Murcott, A. 1983: 'It's a pleasure to cook for him': food, mealtimes and gender in some South Wales households. In Gamarnikow, E., Morgan, D., Purvis, J. and Taylorson, D. (eds) *The public and the private*. London: Heinemann, 78–90.
Murcott, A. 1997: Family meals: a thing of the past? In Caplan, P. (ed.) *Food, health and identity*. London: Routledge, 32–49.
Otnes, P. 1991: What do meals do? In Fürst, E.L., Prättälä, R., Ekström, M., Holm, L. and Kjærnes, U. (eds) *Palatable worlds: sociocultural food studies*. Oslo: Solum Forlag, 97–108.
Rawson, S. 1999: Unspeakable rites: cultural reticence and the cannibal question. *Social Research* **66**, 167–93.
Rifkin, J. 1992: *Beyond beef: the rise and fall of the cattle culture*. New York: Dutton.
Rozin, P., Haidt, J., McCauley, C. and Imada, S. 1997: Disgust: preadaptation and the cultural evolution of a food-based emotion. In Macbeth, H. (ed.) *Food preferences and taste: continuity and change*. Providence: Berghahn Books, 65–82.
Senauer, B., Asp, E. and Kinsey, J. 1991: *Food trends and the changing consumer*. St Paul, Minnesota: Eagan Press.
Siega-Riz, A.M., Carson, T. and Popkin, B. 1998: Three squares or mostly snacks: what do teens really eat? A sociodemographic study of meal patterns. *Journal of Adolescent Health* **22**, 29–36.

Simoons, F.J. 1970: Primary adult lactose intolerance and the milking habit: a problem in biological and cultural interrelations. II: a culture historical hypothesis. *American Journal of Digestive Diseases* 15, 695–710.
Simoons, F.J. 1978: The geographic hypothesis and lactose malabsorption: a weighing of the evidence. *American Journal of Digestive Diseases* 23, 963–80.
Simoons, F.J. 1981: Geographic patterns of primary adult lactose malabsorption: a further interpretation of evidence for the old world. In Paige, D.M. and Bayless, T.M. (eds): *Lactose digestion: clinical and nutritional implications*. Baltimore: Johns Hopkins University Press, 23–48.
Simoons, F.J. 1994: *Eat not this flesh: food avoidances from prehistory to the present*. 2nd edn. Madison, Wisconsin: University of Wisconsin Press.
Stunkard, A. 1997: Eating disorders: the last 25 years. *Appetite* 29, 181–90.
Valentine, G. and Longstaff, B. 1998: Doing porridge: food and social relations in a male prison. *Journal of Material Culture* 3, 131–52.
Visser, M. 1987: *Much depends upon dinner: the extraordinary history and mythology, allure and obsessions, perils and taboos, of an ordinary meal*. Toronto: McClelland and Stewart.
Visser, M. 1991: *The rituals of dinner: the origins, evolution, eccentricities, and meaning of table manners*. New York: Grove Weidenfeld.
Warde, A. and Martens, L. 1998: A sociological approach to food choice: the case of eating out. In Murcott, A. (ed.) *'The Nation's Diet': the social science of food choice*. London: Longman, 129–44.
Worsley, A. 1988: Cohabitation-gender effects on food consumption. *International Journal of Biosocial Research* 10, 107–22.

23

Food, gender and domestic spaces

Introduction

Much of the book hitherto has been cast in the form of debates about macro-scale structures and processes concerning food. Lest we forget the micro-scale, this chapter seeks to remind us of the importance of the family as a unit of production and consumption. It also focuses upon the role of women as providers and mediators between the raw and the cooked; between nature and culture.

Food and families

In most societies throughout history there has been a seemingly ineluctable connexion between women and the domestic duties of food preparation. So deeply rooted are the associated assumptions about women's roles that they were rarely challenged before the late twentieth century. Even now, the cooks among the supposedly 'new men' of our more enlightened era would no doubt find some ironic amusement in Bridget O'Laughlin's account (cited in Fürst 1991) of a tribe in Chad where cooking pots are seen to be female property to such an extent that any men using them are taken to be transvestites.

There was a kind of power to be drawn from the traditional knowledge and skills that were passed on orally from mother to daughter but such reproductive duties have always been shunned by men and given a lowly status by society at large. Lupton (1996, 2) recalls that Greek philosophers such as Plato were keen to distance themselves from the constraints of their bodies which they regarded as distractions from the search for truth. She therefore concludes that 'philosophy is masculine and disembodied; food and eating are feminine and always embodied'.

The nineteenth century was an important threshold for gendered spaces and roles. The rapid growth of suburbs created areas that were female-dominated, during the daytime at least, and middle class prosperity facilitated an increase in the number of 'housewives' who were specialized home-makers. The invention of 'domestic science' as part of the compulsory education system at the end of that century also formalized the notion of the kitchen as the woman's domain and explicitly excluded boys from any obligation to learn about cooking.

There is an interesting literature on the role of women as gatekeepers in deciding which foods are suitable for their families to eat. Most women admit that the views of men and children heavily influence their choice of meals, ingredients and methods of preparation, but that is hardly surprising in a consensual family setting (Chapter 22). It has been suggested that this deference amounts to reinforcement of the family power hierarchy, and there is certainly evidence that lack of submissiveness over choice of food and timing of meals may lead to domestic violence. When it comes to choosing food for themselves, however, according to Mennell, Murcott and van Otterloo (1992), 'women are expected to deny themselves in order to remain slim and therefore sexually attractive . . . women's body images are products of their structural position in society'. Sharman (1991) also shows how the individual life histories of women and their families can have a profound effect upon their food habits (Figure 23.1)

Technological developments in the last few decades have lightened the chore of preparing and cooking food, and have to some extent reduced the need for traditional knowledge and skills (*see* Chapter 16 for health implications). Family meals made by specialist cooks are on the decline, along with the role of the housewife as the arbiter of taste and nutrition. However, DeVault (1991) argues that women's work in coordinating food provisioning for the family remains crucial and that the effort involved in pleasing everyone is a continuing source of emotional subordination. She even asserts that 'feeding work has become one of the primary ways that women "do" gender'. The implication is that the transference of such tasks to other members of the family or to technological substitutes would undermine the self-image of many women and their gendered relations with the world; but in reality the ways of 'doing' gender are both flexible and dynamic and are certainly not necessarily tied to the kitchen sink.

In their classic text, Charles and Kerr (1988) also found that women remain the key food providers in British society, but their exploration of class added a further dimension. They found that income levels, housing, (un)employment, and proximity to parents are important and influence/constrain women's decision-making about meal patterns and food choice. The patriarchal context of family food ideologies also varies with class. Middle class houses have more space for formal eating occasions, and middle class households are more likely to have a member who is a vegetarian. Working class extended families eat together more often but they have fewer 'dinner parties' with friends. All social classes maintain a view of what is a 'proper meal' (Chapter 22).

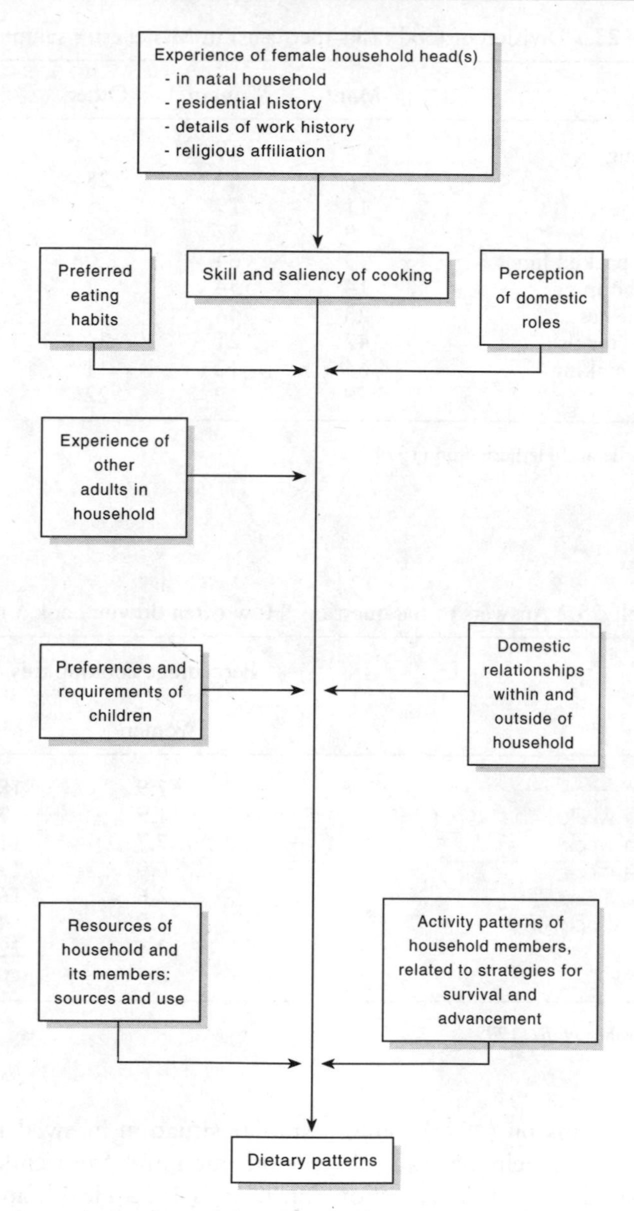

Fig. 23.1 Interrelationships giving rise to dietary patterns. *Source*: Sharman (1991, 194).

Tables 23.1 and 23.2 indicate that gender divisions of food-related labour are far from dead in the British household. The data in Table 23.1 are from a sample of 278 households in Greater Manchester, and they show a continuing feminine dominance in all but home beer-making and the notoriously testosterone-laden task of supervising barbecues. Ekström

Table 23.1 Division of food tasks (per cent) in Manchester sample, 1990

Task	Man	Woman	Other	Shared
Cake baking	5	61	31	4
Jam making	6	63	28	3
Cooking meals	11	79	5	5
Baking bread	9	57	27	7
Preparing packed lunches	13	64	20	3
Grocery shopping	14	54	2	30
Washing dishes	23	46	20	10
Take-away meals	42	21	26	10
Beer/wine-making	64	12	11	14
Barbecues	59	9	22	11

Source: Warde and Hetherington (1994).

Table 23.2 Answers to the question 'How often do you cook a meal?'

	Percentage cooking this often	
	Women	Men
Every day	67.9	18.4
5–6 days a week	11.9	7.0
3–4 days a week	7.7	11.9
1–2 days a week	6.0	19.4
<1 a week	2.6	16.2
Only special occasions	1.0	4.3
Never	2.6	22.4
Don't know	0.3	0.4

Source: Caraher *et al.* (1999).

(1991) and Jansson (1995) found a similar situation in Sweden, although there was more help generally there from men and from children. Table 23.2 is based on Health Education Authority's 1993 Health and Lifestyles Survey of 5,553 people in England.

The discussion in this section has made an implicit assumption that families are nuclear units made up of heterosexual adults and their children. Such assumptions yield only selective insights because they marginalize an increasingly large proportion of the population in western societies and also ignore the complex realities in poor countries. However, the impact of changing household structure on the cultural attributes of food has yet to be researched.

Women and household food security in poor countries

As we saw in Chapter 13, food security depends upon a number of factors. Probably most important are adequate food production, economic access to available food, and nutritional security. The role of women is vital in all three. In sub-Saharan Africa, for instance, they produce and market up to 80 per cent of the food grown locally. In addition they are responsible for over 80 per cent of domestic food processing and storage, 70 per cent of hoeing and weeding, 60 per cent of harvesting and marketing, 50 per cent of livestock care, 50 per cent of planting, and 30 per cent of ploughing. This is in addition to 95 per cent of housework, including the collection of water and fuel wood. In Asia they account for 65 per cent of household food production, and 45 per cent in Latin America and the Caribbean.

Home gardens, usually female territory, are in many countries models of sustainable and intensive multi-cropping. In Indonesia, for instance, they provide on average 20 per cent of household income and 40 per cent of domestic food supplies. A survey in Nigeria found home gardens to be astonishingly diverse, with between 18 and 57 useful plant species. These may be fertilized with household waste, while small livestock are also fed on scraps.

One estimate compared the weight and distance of food-related materials carried by men and women in the rural 'South'. This included water and fuel wood collection, crop harvesting and marketing. The conclusion was that able-bodied girls and women annually carry an average of 80 tonnes for a distance of 1 km, by comparison with one tonne over the same distance by men and boys.

Female labour and skill therefore underpin the whole food system and policy-makers must consider this in their attempts to reverse the decline in the food security of certain low-income countries. One reason for the high levels of feminization of agriculture and some downstream parts of the food system, especially in Sub-Saharan Africa, is male migration to job opportunities in cities and industry. Thirty-one per cent of households in Africa south of the Sahara are female-headed for this and for other reasons, by comparison with 12 per cent for low and middle income countries as a whole.

According to Quisumbing *et al.* (1995), women farmers in poor countries face a number of constraints that reduce their productivity. Their land rights are weak and they are allowed only limited access to common property resources such as communal land, fuelwood and water. Traditionally, these are controlled by men (Tables 23.3 and 23.4) but modern land reforms have exacerbated this concentration of rights. Female farmers also have poor access both to modern technology such as HYV seeds and labour-saving equipment, and also to the credit that might assist an intensification of their

Table 23.3 Agricultural decision-making, from survey of Korogwe District, Tanzania

Type of decision	By women (%)	By men (%)	By both (%)
Type of crop to grow	36	48	16
Where to plant	22	56	22
What agricultural techniques to use	20	60	20
Sale of surplus crop	33	46	21
Sale of surplus livestock	18	73	9
Distribution of income from crop sale	27	41	32
Distribution of income from sale of livestock	30	40	27
Total	30	48	27

Source: FAO (1998).

Table 23.4 Access to and control of resources by women in North Wollo, Ethiopia

Resource	Access	Control
Irrigation	25	25
Spring water	75	75
River water	50	50
Land	30	30
Cows	30	0
Oxen	50	40
Sheep and goats	100	100
Poultry	0	0
Horticulture	45	45
Extension	20	0
Crop produce	40	60
Trees	50	40
Credit	0	0
Labour	65	50
Team work	35	0
Farm inputs	0	0
Money	45	50

Source: FAO (1998).

enterprises. They are allowed only limited contact with extension workers (99 per cent of whom are men in Asia and Africa) and their low levels of literacy would anyway restrict the use they could make of any advice given. Some development projects organized by the international community have reproduced the bias, by granting formal land titles only to men in resettlement and irrigation schemes.

Table 23.5 Effects upon household welfare of men's and women's incomes

Country	Effect on	Effect of women's income	Effect of men's income	Ratio effect of women's income to that of men
Kenya	Household calorie level	Positive	Negative	–
Taiwan	Household share of alcohol	Negative	Negative	1.3
Guatemala	Food expenditure	Positive	Positive	2.0
Brazil	Child weight for height	Positive	Positive	4.2
Brazil	Child survival	Positive	Positive	18.2

Source: Quisumbing *et al.* 1995.

Access to food seems to depend, not just on overall household income, but also on who earns the wage and who controls the family's finances. Research in social environments as varied as Rwanda, Côte d'Ivoire and the Philippines has shown that women spend a much higher proportion of their income on food and child care than men, and in Brazilian cities children have an 18-fold better survival chance when income is controlled by their mother rather than by their father (Table 23.5). Women are also far better geared up for a household food emergency because of their closer involvement with day-to-day coping strategies.

A factor of primary significance seems to be female education. Where girls and women have acquired literacy and the empowerment that goes with the consciousness-raising aspect of a general education, there is evidence that the nutritional and health status of the family rises. Behrman and Wolfe (1984) found that women's schooling was positively and significantly correlated with the provision of a range of nutrients in Nicaragua, while Turner Lomperis (1991) modelled infant mortality in Cali, Colombia, and concluded that mother's education was a stronger correlate even than family income and household size.

Household food security (Chapter 13) is the basis of livelihood systems, but both are being undermined in many parts of the world by the crumbling of community-level buffers against income shocks. The mutuality among villagers that used to be common in parts of Africa is beginning to disappear and structural adjustment is preventing a safety net from being established by the state.

Food aid, at least in the form of a physical transfer of commodities to women, is generally a more effective means of supporting hungry households than the flows of capital and infrastructure investments which dominate the development 'industry' and which are often concentrated in projects that marginalize women.

Intra-household distribution of food

Surveys of food consumption are often pitched at the household level, for the convenience of data collection and because the family is an identifiable decision-making and organizational group. However, this assumes that all members of the household have equal access to assets and resources, which is a fallacy for most societies. For a range of reasons, including cultural norms, patriarchal power and earning capacity, active adult males usually have first call on both quantity and quality of food. The consequences of this for females, the young, the old and the sick, may be a matter of nutritional concern in economically marginal households. Further, evidence from India adduced by Drèze and Sen (1995) suggests that they not only eat less but also have poorer access to medical care. The breastfeeding of girl babies is shorter and their care is subtly but significantly different from that of boys.

We must be careful here not to stray too far into over-generalization because there are many variations by level of income and education, by caste and ethnicity, and particularly by region. Harriss (1990) cites anthropometric dietary survey evidence of gender bias in northern India, Bangladesh and Pakistan and concludes that intra-family disparities of food-intake there are indisputable. Discrimination, particularly against young girls, is a strong undercurrent even in economically more comfortable households.

Indirect evidence of bias is also available in the demographic record. The 1991 Census of India found 407 million girls and women but 439 million boys and men in India (Raju *et al.* 1999). For every 1000 males there were only 927 females, a figure that is called the 'sex ratio'. India has one of the lowest proportions of girls and women in its population in the world. The sex ratio has been falling in India for most of the twentieth century (Table 23.6), becoming more and more unlike most other countries and reaching its lowest ever value in 1991.

Superficially, there are significant regional differences in the proportion of girls and women in the population. The State of Haryana, with 865:1000, has the lowest female/male ratio in the world with the exception

Table 23.6 The Indian sex ratio, 1901–91

1901	972	1951	946
1911	964	1961	941
1921	955	1971	930
1931	950	1981	934
1941	945	1991	927

Source: Census of India.

of a few places with large numbers of male migrant workers, such as the Falkland Islands or countries in the Arabian Gulf. Parts of Gujarat, north western Madhya Pradesh, the border districts of Rajasthan and western Uttar Pradesh (with the exception of the hill districts) also have very low sex ratios.

Conclusion

The central role of women in many aspects of the food chain is only now beginning to receive the attention it deserves. This has meant that the relevant literature on some of the themes we have raised is still in a nascent form. To take one example, Gill Valentine (1999b) has shown that much of the writing on food in western domestic settings has been reductionist, making general assumptions about behaviours that are often not justified. There is an urgent need for more research in this area, not least because food policies are often based upon inadequate knowledge of the role of food in households.

Further reading and references

The books by Charles and Kerr, and DeVault are classic texts on the food provision role of women in rich countries. For low income countries one is thrown back on the journal literature and we await a synthesizing textbook.

Behrman, J.R. and Wolfe, B.L. 1984: More evidence on nutrition demand: income seems overrated and women's schooling underemphasized. *Journal of Development Economics* **14**, 105–128.

Caraher, M., Dixon, P., Lang, T. and Carr-Hill, R. 1999: The state of cooking in England: the relationship of cooking skills to food choice. *British Food Journal* **101**, 590–609.

Charles, N. and Kerr, M. 1988: *Women, food and families*. Manchester: Manchester University Press.

DeVault, M.L. 1991: *Feeding the family: the social organization of caring as gendered work*. Chicago: University of Chicago Press.

Drèze, J. and Sen, A.K. 1995: *India: economic development and social opportunity*. New Delhi: Oxford University Press.

Ekström, M. 1991: Class and gender in the kitchen. In Fürst, E.L., Prättälä, R., Ekström, M., Holm, L. and Kjærnes, U. (eds) *Palatable worlds: sociocultural food studies*. Oslo: Solum Forlag, 145–58.

Food and Agriculture Organization 1998: *Women feed the world*. Rome: FAO.

Fürst, E.L. 1991: Food, gender and identity: a story of ambiguity. In Fürst, E.L., Prättälä, R., Ekström, M., Holm, L. and Kjærnes, U. (eds) *Palatable worlds: sociocultural food studies*. Oslo: Solum Forlag, 111–30.

Harriss, B. 1990: The intrafamily distribution of hunger in South Asia. In Drèze, J. and Sen, A.K. (eds) *The political economy of hunger. Volume 1: entitlement and well-being*. Oxford: Clarendon Press, 351–424.

Jansson, S. 1995: Food practices and division of domestic labour: a comparison between British and Swedish households. *Sociological Review* **43**, 462–77.
Lupton, D. 1996: *Food, the body and the self*. London: Sage.
Mennell, S., Murcott, A. and van Otterloo, A.H. 1992: *The sociology of food: eating, diet and culture*. London: Sage.
Quisumbing, A.R., Brown, L.R., Feldstein, H.S., Haddad, L. and Peña, C. 1995: *Women: the key to food security*. Washington DC: International Food Policy Institute.
Raju, S., Atkins, P.J., Kumar, N. and Townsend, J.G. 1999: *An atlas of women and men in India*. New Delhi: Kali for Women.
Sharman, A. 1991: From generation to generation: resources, experience, and orientation in the dietary patterns of selected urban American households. In Sharman, A., Theophano, J., Curtis, K. and Messer, E. (eds) *Diet and domestic life in society*. Philadelphia: Temple University Press, 173–204.
Turner Lomperis, A.M. 1991: Teaching mothers to read: evidence from Colombia on the key role of maternal education in pre school nutritional health. *Journal of Developing Areas* **26**, 25–52.
Valentine, G. 1999a: A corporeal geography of consumption. *Society and Space* **17**, 329–51.
Valentine, G. 1999b: Eating in: home, consumption and identity. *Sociological Review* **47**, 491–524.
Warde, A. and Hetherington, K. 1994: English households and routine food practices: a research note. *Sociological Review* **42**, 758–78.
World Food Programme, Strategy and Policy Division 1998: Food security, livelihoods and food aid interventions. In *Time for change: food aid and development consultation, 23–24 October 1998, Rome, Italy*. Rome: WFP.

PART VI

24
Conclusion

Margaret Visser (1991) in her book *The Rituals of Dinner* speculates that food is at the basis of much human behaviour. Kinship systems may be interpreted as those groups which share food without question; language originated as a means of planning the acquisition and distribution of food; technology developed in order to improve the efficiency of hunting and food preparation; and morality and politics at first dealt with the fair division of food resources. From our own perspective we would add that much of the earth's surface has been made over into humanized landscapes through the efforts of farmers (Atkins, Simmons and Roberts 1998), and that agriculture was the basis of all economies until relatively recently in human history.

It is difficult, therefore, to deny the importance of food studies to a full understanding of society, although familiarity seems to have bred contempt for the study of mundane items of daily life in some academic circles. This book has been a modest attempt to redress the balance. Although it has been written by two geographers, we have tried not to wear our disciplinary badge too prominently on our sleeves. This is because we believe that interdisciplinary studies are likely to yield the most fruitful results. The ESRC project entitled 'The Nation's Diet' (Chapter 1) was a very significant step along this road. Indeed our book confirms the complex and overlapping analyses of food of different disciplines: it is not possible to classify and categorize particular aspects of food as 'belonging' to a particular discipline, nor has it been possible to assign knowledge about food to particular chapters within this book. The constant cross-referencing of material within our book emphasizes the inter-relationships, or multiple positioning, of knowledge about food in overlapping discourses.

For the future, we would identify four major areas that require scholarly attention. The first is the need for detailed research on individual systems of provision (the structures of production, processing, manufacture, retailing and consumption that characterize each product). This will require an awareness of the particular characteristics of individual food commodities in shaping their marketing/consumption chain. Historical and economic geographers are well positioned for this but integrated studies are needed which bring to bear a range of skills from the social sciences and humanities.

Second, we would argue that studies of food consumption are vital for an in-depth understanding of consumption in general. In addition, while traditional quantitative approaches still have much to offer, the qualitative revolution of the 1990s has given a fresh impetus to work in this area, with a new suite of interpretations, both empirical and theoretical.

Third, the hazards of food production and consumption are a major element in the 'risk society' and seem likely to shape our relationship to food in the 21st century. Academic pursuits on this theme, and the closely related issue of food quality, have only just begun, for instance on GMOs, but they are important because they hold out the prospect of influencing government policies.

A fourth direction concerns the Internet, which has already become an exchange for information about the ingredients required by food processors and which seems likely to influence all aspects of food production, marketing and consumption. Pritchard (1999) has published an article on the web pages of major food corporations and there are many other avenues of research opening up.

The list of web sites at http://www.arnoldpublishers.com/support/food includes some of general interest and a few more specialized ones. A vast amount of information is available on agriculture and food on the World Wide Web, but discrimination is needed to sort the useful from the banal and the downright inaccurate. This information is also dynamic, reflecting the continuing change in food production and consumption, which we have attempted to capture in this book.

References

Atkins, P.J., Simmons, I.G. and Roberts, B.K. 1998: *Land, people and time*. London: Arnold.

Pritchard, W.N. 1999: Local and global in cyberspace: the geographical narratives of US food companies on the internet. *Area* **31**, 9–17.

Visser, M. 1991: *The rituals of dinner: the origins, evolution, eccentricities, and meaning of table manners*. New York: Grove Weidenfeld.

Index